Microfluidics for Cells and Other Organisms

Microfluidics for Cells and Other Organisms

Special Issue Editor

Danny van Noort

MDPI • Basel • Beijing • Wuhan • Barcelona • Belgrade

MDPI

Special Issue Editor
Danny van Noort
Linköping University
Sweden

Editorial Office
MDPI
St. Alban-Anlage 66
4052 Basel, Switzerland

This is a reprint of articles from the Special Issue published online in the open access journal *Micromachines* (ISSN 2072-666X) from 2018 to 2019 (available at: https://www.mdpi.com/journal/micromachines/special_issues/Microfluidics_Cell_Other_Organisms)

For citation purposes, cite each article independently as indicated on the article page online and as indicated below:

LastName, A.A.; LastName, B.B.; LastName, C.C. Article Title. *Journal Name* **Year**, *Article Number, Page Range.*

ISBN 978-3-03921-562-1 (Pbk)
ISBN 978-3-03921-563-8 (PDF)

Contents

About the Special Issue Editor

Danny van Noort is an international researcher with +20 years of experience in advanced sciences at the interface between physics, microtechnology and engineering biology. He holds a MSc in experimental Physics from Leiden University, the Netherlands, and a Ph.D. in applied physics from Linköping University, Sweden. He has since been a researcher at renowned institutes in Germany, the USA, South Korea and Singapore. He has driven frontline research projects on biosensors, microfluidic-based DNA computers and organ-on-chips. In recent years, his focus has been on biomedical applications for creating 3D cell and disease models in vitro. He has a significant publication list, is well cited, and has lectured at conferences and institutes all over the world.

micromachines

MDPI

Editorial

Editorial for the Special Issue on Microfluidics for Cells and Other Organisms

Danny van Noort

Department of Physics, Chemistry and Biology (IFM), Linköping University, 581 83 Linköping, Sweden;
drr.dvn@gmail.com

Received: 29 July 2019; Accepted: 30 July 2019; Published: 5 August 2019

It is my great pleasure to present to you this first volume of 13 papers on the subject of Microfluidics for Cells and other Organisms. By adding "organisms" to the volume title, I was hoping for manuscripts beyond just cells. So, it was good to see that there were submissions of papers on zebrafish [1] and bacteria [2]. This volume highlights a diverse collection of research on single cell manipulation, diagnostics, cell migration, cell flow cytometry, to name a few. I am also happy to see that some papers included automated systems to operate the devices [3,4]. Automation is needed if we want to have a more intensive use of microfluidic based platforms and reproducibility.

Contributions to this volume came from all over the world, from Germany, France, Switzerland, USA, Hong Kong, China, Taiwan, Japan, Singapore and Chile.

This volume shows the importance of using microfluidics as a tool to understand cells and other organisms or even broader, biology better.

As Constantinou et al. [5] showed, hydrodynamic focusing inside a Y-shaped microfluidic device improve the classification of single cells in cytometry. With the aid of image analysis, cells can be identified in cell mixtures.

Analysis of nuclear acids is important for molecular diagnostics as well as automation of the process. Tong et al. [3] introduced a rotating disk device to extract the nuclear acid from cells using magnetic beads.

From my own work on zebra fish embryos, I know how they easily evade the viewing field of the microscope when looking at them in a petri dish. Thus, I developed a chip to keep the embryos in place for real-time observation [6]. Zhu et al. [1] had the same idea but developed a different trapping method to keep the embryos in place.

Cells react to external forces, which is very well studied in the field of mechanobiology. By applying periodic hydrostatic pressure on cells, Horade et al. [7] showed that cells under periodic pressure displayed a faster increase in the size of the cells as compared to atmospheric pressure. Another example in the field of mechanobiology is given by Li et al. [8]. The biomechanical properties of cells can be used for early disease diagnostics. By using a microfluidic device like a Wheatstone Bridge, single cells could be trapped and exposed to precisely controlled pressures.

Performing diagnostics on prenatal fetuses is basically impossible, unless you can do it non-invasively by isolating some cells from the fetus. As it turns out, circulating fetal cells (CFCs) are present in the maternal blood. Ma et al. [4] designed a chip and an automated system to isolate this rare cell from maternal blood.

Cheap and simple in-situ alternatives to standard flow cytometry is making its way to microfluidic devices. I see this as a positive development of more portable devices, which can be deployed remotely to, for example, perform diagnostics. Zhang et al. [2] used conductivity to measure the concentration of bacteria.

Microfluidics can also be used to find binding proteins to specific cancer cells. It is a way to identify the cancer cell. Kaminaga et al. [9] did exactly that, by using micropillar arrays to filter non-target-binding-molecules from specific binding molecules. Liu et al. [10] looked at specific protein interaction on cells, however, at a single cell level. They are specifically looking for oral tumor cells from patients with their microfluidic based cell analyzer.

Single cell cultures do not have the same functionality as co-cultures. Chen et al. [11] fabricated a non-contact co-culture chip with fibroblasts and lung cancer cell lines to study their interaction, with the intention to explore the mechanism of cancer.

Another method to separate cells is to look at their motility, especially when looking at migrating cancer cells. Wang et al. [12] proposed to measure the motility of these cells to access the effect of anti-cancer drugs, by using a paper-based microfluidic device.

Single cell analysis is further highlighted in a review by Luo et al. [13]. It explores various methods for single cell manipulation, analysis as well as the various microfluidic devices available.

Finally, this volume ends with an opinion piece by Grenci et al. [14] highlighting the role of microfluidics or more precise, the role of micro and nanotechnology in biological and biomedical applications. It describes the interdisciplinary processes to develop new biological technologies

Due to the success of this volume of papers, I am now looking forward to the contributions in Volume 2.

References

1. Zhu, Z.; Geng, Y.; Yuan, Z.; Ren, S.; Liu, M.; Meng, Z.; Pan, D. A Bubble-Free Microfluidic Device for Easy-to-Operate Immobilization, Culturing and Monitoring of Zebrafish Embryos. *Micromachines* **2019**, *10*, 168. [CrossRef] [PubMed]
2. Zhang, X.-Y.; Li, Z.-Y.; Zhang, Y.; Zang, X.-Q.; Ueno, K.; Misawa, H.; Sun, K. Bacterial Concentration Detection using a PCB-based Contactless Conductivity Sensor. *Micromachines* **2019**, *10*, 55. [CrossRef] [PubMed]
3. Tong, R.; Zhang, L.; Hu, C.; Chen, X.; Song, Q.; Lou, K.; Tang, X.; Chen, Y.; Gong, X.; Gao, Y.; et al. An Automated and Miniaturized Rotating-Disk Device for Rapid Nucleic Acid Extraction. *Micromachines* **2019**, *10*, 204. [CrossRef] [PubMed]
4. Ma, G.-C.; Lin, W.-H.; Huang, C.-E.; Chang, T.-Y.; Liu, J.-Y.; Yang, Y.-J.; Lee, M.-H.; Wu, W.-J.; Chang, Y.-S.; Chen, M. A Silicon-based Coral-like Nanostructured Microfluidics to Isolate Rare Cells in Human Circulation: Validation by SK-BR-3 Cancer Cell Line and Its Utility in Circulating Fetal Nucleated Red Blood Cells. *Micromachines* **2019**, *10*, 132. [CrossRef] [PubMed]
5. Constantinou, I.; Jendrusch, M.; Aspert, T.; Görlitz, F.; Schulze, A.; Charvin, G.; Knop, M. Self-Learning Microfluidic Platform for Single-Cell Imaging and Classification in Flow. *Micromachines* **2019**, *10*, 311. [CrossRef] [PubMed]
6. Choudhury, D.; van Noort, D.; Iliescu, C.; Zheng, B.X.; Poon, K.-L.; Korzh, S.; Korzh, V.; Yu, H. Fish and Chips: A microfluidic perfusion platform for monitoring the development of early stage zebrafish embryos. *Lab Chip* **2012**, *12*, 892–900. [CrossRef] [PubMed]
7. Horade, M.; Tsai, C.-H.D.; Kaneko, M. On-Chip Cell Incubator for Simultaneous Observation of Culture with and without Periodic Hydrostatic Pressure. *Micromachines* **2019**, *10*, 133. [CrossRef] [PubMed]
8. Li, Y.-J.; Yang, Y.-N.; Zhang, H.-J.; Xue, C.-D.; Zeng, D.-P.; Cao, T.; Qin, K.-R. A Microfluidic Micropipette Aspiration Device to Study Single-Cell Mechanics Inspired by the Principle of Wheatstone Bridge. *Micromachines* **2019**, *10*, 131. [CrossRef] [PubMed]
9. Kaminaga, M.; Ishida, T.; Kadonosono, T.; Kizaka-Kondoh, S.; Omata, T. Microfluidic Device for Screening for Target Cell-Specific Binding Molecules by Using Adherent Cells. *Micromachines* **2019**, *10*, 41. [CrossRef] [PubMed]
10. Liu, L.; Fan, B.; Wang, D.; Li, X.; Song, Y.; Zhang, T.; Chen, D.; Wang, Y.; Wang, J.; Chen, J. Microfluidic Analyzer Enabling Quantitative Measurements of Specific Intracellular Proteins at the Single-Cell Level. *Micromachines* **2018**, *9*, 588. [CrossRef] [PubMed]
11. Chen, H.; Liu, W.; Wang, B.; Zhang, Z. In Situ Analysis of Interactions between Fibroblast and Tumor Cells for Drug Assays with Microfluidic Non-Contact Co-Culture. *Micromachines* **2018**, *9*, 665. [CrossRef] [PubMed]

12. Wang, L.-X.; Zhou, Y.; Fu, J.-J.; Lu, Z.; Yu, L. Separation and Characterization of Prostate Cancer Cell Subtype according to Their Motility Using a Multi-Layer CiGiP Culture. *Micromachines* **2018**, *9*, 660. [CrossRef] [PubMed]

13. Luo, T.; Fan, L.; Zhu, R.; Sun, D. Microfluidic Single-Cell Manipulation and Analysis: Methods and Applications. *Micromachines* **2019**, *10*, 104. [CrossRef] [PubMed]

14. Grenci, G.; Bertocchi, C.; Ravasio, A. Integrating Microfabrication into Biological Investigations: The Benefits of Interdisciplinarity. *Micromachines* **2019**, *10*, 252. [CrossRef] [PubMed]

![micromachines]

Article

Self-Learning Microfluidic Platform for Single-Cell Imaging and Classification in Flow

Iordania Constantinou [1,2,3,†,*], Michael Jendrusch [1,†], Théo Aspert [4,5,6,7], Frederik Görlitz [1], André Schulze [1], Gilles Charvin [4,5,6,7] and Michael Knop [1,8,*]

[1] Zentrum für Molekulare Biologie der Universität Heidelberg (ZMBH), DKFZ-ZMBH Alliance, Universität Heidelberg, 69120 Heidelberg, Germany; jendrusch@stud.uni-heidelberg.de (M.J.); f.goerlitz@gmx.net (F.G.); a.schulze@zmbh.uni-heidelberg.de (A.S.)

[2] Institute of Microtechnology, Technische Universität Braunschweig, 38124 Braunschweig, Germany

[3] Center of Pharmaceutical Engineering (PVZ), Technische Universität Braunschweig, 38106 Braunschweig, Germany

[4] Developmental Biology and Stem Cells Department, Institut de Génétique et de Biologie Moléculaire et Cellulaire, 67400 Illkirch-Graffenstaden, France; aspertt@igbmc.fr (T.A.); charvin@igbmc.fr (G.C.)

[5] Centre National de la Recherche Scientifique, 67400 Illkirch-Graffenstaden, France

[6] Institut National de la Santé et de la Recherche Médicale, 67400 Illkirch-Graffenstaden, France

[7] Université de Strasbourg, 67400 Illkirch, France

[8] Cell Morphogenesis and Signal Transduction, German Cancer Research Center (DKFZ), DKFZ-ZMBH Alliance, 69120 Heidelberg, Germany

[*] Correspondence: i.constantinou@tu-braunschweig.de (I.C.); m.knop@zmbh.uni-heidelberg.de (M.K.); Tel.: +49-0531-391-9769 (I.C.); +49-6221-54-4213 (M.K.)

[†] These authors contributed equally to this work.

Received: 24 April 2019; Accepted: 6 May 2019; Published: 9 May 2019

Abstract: Single-cell analysis commonly requires the confinement of cell suspensions in an analysis chamber or the precise positioning of single cells in small channels. Hydrodynamic flow focusing has been broadly utilized to achieve stream confinement in microchannels for such applications. As imaging flow cytometry gains popularity, the need for imaging-compatible microfluidic devices that allow for precise confinement of single cells in small volumes becomes increasingly important. At the same time, high-throughput single-cell imaging of cell populations produces vast amounts of complex data, which gives rise to the need for versatile algorithms for image analysis. In this work, we present a microfluidics-based platform for single-cell imaging in-flow and subsequent image analysis using variational autoencoders for unsupervised characterization of cellular mixtures. We use simple and robust Y-shaped microfluidic devices and demonstrate precise 3D particle confinement towards the microscope slide for high-resolution imaging. To demonstrate applicability, we use these devices to confine heterogeneous mixtures of yeast species, brightfield-image them in-flow and demonstrate fully unsupervised, as well as few-shot classification of single-cell images with 88% accuracy.

Keywords: microfluidics; 3D flow focusing; 3D particle focusing; particle/cell imaging; bioMEMS; unsupervised learning; neural networks; variational inference

1. Introduction

Phenotypic profiling of cell populations is routinely performed in research and diagnostic laboratories using flow cytometry [1–3]. Flow cytometry provides cellular analysis at an unparalleled throughput and allows for the screening of diverse samples and the isolation of cell subpopulations for further study. Standard applications of flow cytometry employ multi-channel fluorescence detection and sample characterization based on light scattering and fluorescence signal intensity, which provide limited spatial resolution [4]. Imaging flow cytometry combines the speed and sample size of flow cytometry with spatial resolution and allows for the acquisition of images and their use for sample characterization and sorting decisions [5]. While imaging flow cytometry gives researchers the opportunity to conduct multiparametric analysis of cell populations based on single-cell images, acquiring high-resolution images at throughputs common in flow cytometry remains challenging. This is primarily due to the difficulty of precisely positioning cells, and the challenges associated with imaging fast moving objects [6].

Recently, there has been a push towards the development of microfluidic-based flow cytometers with the aim to reduce complexity and sample volume and increase accessibility and portability [7,8]. One challenging aspect in the miniaturization of flow cytometers has been the focusing of fast-moving cells in a small, defined volume. Successful 3D hydrodynamic focusing techniques have been demonstrated over the past decade, but many rely on multi-layer structures, incompatible with polydimethylsiloxane (PDMS)-based soft-lithography [9–11]. Additionally, particles in these devices are focused to the center of tall microfluidic channels, tens, or hundreds of micrometers away from the microscope slide which is suboptimal for imaging flow cytometry [12]. For better signal-to-noise ratio, it is preferable that particles are positioned as close to the microscope slide as possible, minimizing background fluorescence. More importantly, when the specimen is located far from the surface of the cover slip the resolution obtained using high numerical aperture (NA) oil objectives is impaired. The use of such objectives is required for sensitive fluorescence detection, but resolution decreases with distance due to optical aberrations arising from a difference in refractive index between the immersion liquid (oil) and the sample (particles in aqueous solution) [13,14]. Moreover, due to the limited depth of field in conventional microscopy, particles in imaging microfluidic cytometers need to be focused in a small volume along the z-axis for high resolution imaging [15].

Microfluidic devices designed for imaging flow cytometry have been demonstrated with partial success. Devices that utilize inertial forces (i.e., inertial lift and Dean forces that arise from fluid and particle inertia) for cell positioning have been a popular alternative to sheath flow-based particle steering. The most prominent examples include curved and spiral microchannels, but successful positioning in such devices demands high flow rates that often compromise image quality [16,17]. While imaging technologies for flow cytometry have been suggested, imaging particles moving at high velocities while processing captured images in parallel still remains challenging [6]. Another popular device design used for imaging applications utilizes channel heights that are comparable in size to the particles being imaged. Such forced confinement works well for bigger cells, like blood cells, but it's prone to clogging, which makes it unsuitable for smaller cells that often tend to aggregate, such as yeast cells [18]. Recently, the first presentation of a microfluidics-based, imaging flow cytometer capable of cell sorting was demonstrated, and it is expected to lead the way for a new field to emerge in microfabrication [19]. Despite the successful demonstration, this system is complicated to set up and operation still requires trained technicians.

Another big challenge associated with high-throughput imaging flow cytometry is the vast amount of complex data collected. As manual analysis of such data sets would be prohibitively slow and laborious, automated image analysis using advanced algorithms is necessary. In the case of standard flow cytometry, event data is low-dimensional and can be analyzed semiautomatically using gating on fluorescence channel intensities and standard non-parametric clustering of gated events [20–22]. Here dimensions correspond to meaningful visual features such as cell shape and cell focal plane. These approaches break down in the face of high-dimensional data due to what is known as the 'curse of dimensionality' [23]. For complex, high-dimensional data such as single-cell images, distance measures become less useful for clustering and values in single dimensions (i.e., pixel values) become less informative for gating purposes [24]. Using hand-crafted sets of image features, more informative low-dimensional representations of images can be extracted [25,26]. Such features could include texture and image moments, or in more specific cases, cell features like elongation and size. However, these representations are task specific, and may not reflect any kind of biologically relevant properties.

To counteract these restrictions, neural networks are becoming popular for the task of learning biologically interpretable classifiers for imaging data based on which different types of cells can be classified [27–29]. However, in order to apply neural networks to a classification problem, the data need to satisfy two main conditions: classes of data need to be known, and data acquisition and annotation should be efficient. For example, neural networks can be trained and applied for the classification of single-cell imaging data with known categorical factors of variation (e.g., species, or protein localization) and easily acquired and annotated training data sets. Successful applications of neural network classifiers include imaging flow cytometry, image activated cell sorting [19], and offline analysis of single-cell imaging data [27,29]. Other classifiers such us support vector machines (SVM) have also been similarly used for the offline analysis of single-cell imaging data [30,31]. However, such approaches become difficult to use when it comes to analyzing complex populations of *a priori* unknown factors of variation, or performing classification tasks using limited training data.

The characterization of a data set without the use of training examples is known as unsupervised learning. As fully unsupervised classification is a hard problem, a variety of methods focus on simplifying this task by learning meaningful low-dimensional representations of high-dimensional data [32]. For that reason, neural networks are not trained directly for classification, but on related tasks, where it is possible to generate training data artificially [33–35]. A more natural approach to imaging data classification is learning to generate realistic image samples from a data set [36–38]. For example, networks can be trained to predict the relationship between rotations, zooms and crops of a given image, or learn to construct realistic images from a low-dimensional representation. This way, the networks learn low-dimensional features relevant to their training data and by extension to downstream classification tasks, without explicitly being trained on annotated examples. Recent approaches further demand low-dimensional representations to be human-interpretable, such that every dimension corresponds to a single factor of variation of the training dataset. For example, training on single cell images should result in a representation, where one dimension corresponds to cell type, another to cell size and yet another to the position of the cell within the image. Such representations are referred to as disentangled representations. Disentangled representations have been shown to be beneficial for classification using very few training examples (few-shot classification) [39]. A subset of unsupervised learning methods known as variational autoencoder (VAE) provide a foundation for learning disentangled representations that are simple to train and implement [40–45]. In particular, FactorVAE and related methods explicitly modify the VAE training process to promote more interpretable representations.

In this report, we attempt to bridge the gap between technology and biology and present a self-learning microfluidic platform for single-cell imaging and classification in flow. To achieve 3D flow and particle focusing, we use a simple microfluidic device, based on a variation of the commonly used three-inlet, Y-shaped microchannel. We utilize a difference in the height between sheath and sample inlet to confine heterogeneous cells in a small controllable volume directly adjacent to the microscope cover slide, which is ideal for high-resolution imaging of cells in flow. Even though the device design is conceptually similar to previous designs [46–48], controlled 3D hydrodynamic flow focusing has never been fully demonstrated in such devices, nor has particle positioning in focused flow streams been investigated. In this study, we fully characterize different device variations using simulations, and experimentally confirm 3D flow focusing using dye solutions. Additionally, we use a novel, neural network-based regression method to directly measure the distribution of microspheres and highly heterogeneous cells within the focused stream. We confine and image mixtures of different yeast species in flow using bright-field illumination and classify them by species by performing fully unsupervised, as well as few-shot cell classification. To our knowledge, this is the first application of unsupervised learning to classification in imaging flow cytometry.

Micromachines **2019**, *10*, 311

2. Materials and Methods

2.1. Device Design and Fabrication

To achieve sample flow focusing close to the surface of the microscope cover slip we redesigned a simple microfluidic device based on a variation of the commonly used Y-shaped microchannel (Figure 1) [9,46–48]. For the fabrication of the silicon wafer master, we used standard two-layer, SU-8 (MicroChem, Westborough, MA, USA) photolithography [49]. Figure S1a,b show the two layers of photoresist used in the fabrication process. By sequentially combining these two layers, we created a device with three inlets and one outlet, as shown in Figure S1c. The two outer inlets introduce sheath fluids in the device and are taller than the middle inlet, which delivers the sample containing the particles under investigation (i.e., microspheres, cells). The height ratio between the sheath inlets and sample inlet can be controlled by adjusting the height of photoresist and several ratios can be fabricated for testing using the same set of photomasks. Devices for testing were fabricated using single-layer, PDMS-based soft lithography (SYLGARD™ 184 Silicone Elastomer, Dow Chemical, Midland, MI, USA) [50]. Since the difference in inlet height is created at the wafer-level, PDMS devices are fabricated in a single step, with no need of assembling a multi-layer structure.

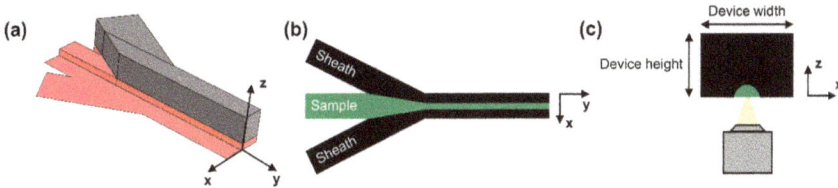

Figure 1. (**a**) Lengthwise 3D device cross-section showing the difference in height between the sheath and sample inlets. Red color is used to show the bottom layer of photoresist, and also the device footprint. The top layer of photoresist is shown in grey (mirror symmetry across the y-axis applies). The difference in height between the sheath inlet and the sample inlet shown is not drawn to scale and only serves as an example; (**b**) Device top view showing the flow focusing mechanism, where the black area is occupied by sheath fluid and the green area is occupied by the sample; (**c**) 2D lengthwise cross-section of the channel (front view) showing sample confinement from both the top and the sides.

2.2. Flow Focusing Principle

A schematic of the full device geometry is shown in Figure 1a. For better visualization of the difference in height between the sheath inlets and the sample inlet, a lengthwise 3D cross section of the device is provided. Any desired height ratio is possible as long as the layer thickness lies within the resolution of the photoresist and the height:width channel aspect ratio remains lower than 1:10 to ensure that channels will not collapse. The flow focusing mechanism behind these devices is shown in Figure 1b (top view; xy-plane) and Figure 1c (cross section; xz-plane). We use a three-inlet, Y-shaped microchannel to introduce the sample along with two sheath flows. The sample enters the device from the middle inlet (shown in green) and is enfolded by two sheath streams (shown in black, Figure 1b). Due to the occurrence of laminar flow in microchannels, the three streams flow parallel to each other without convective mixing [51]. While 2D sample confinement on the xy-plane has been demonstrated in similar devices numerous times [52–54], simultaneous flow confinement along the z-axis (3D flow focusing) has only been briefly investigated [46–48]. One example is a paper by Scott et al., where 2D and 3D flow confinement were achieved in devices of similar geometry, but confinement below the sample inlet height required a geometry modification, a step in the outlet channel after the junction [46]. One other example is a paper by Chung et al. where 3D flow confinement was demonstrated successfully, but control over the degree of confinement also required geometry modifications [48]. Flow focusing along the z-axis is illustrated in Figure 1c. As drawn in the figure, due to the difference in height between the sheath and sample inlet, sheath fluids surround sample flow from all sides (black color), constraining it in a small volume close to the microscope cover slide. Here, the final volume occupied by the sample stream merely depends on the sheath-to-sample flow rate ratio, with higher sheath flow rate resulting in further sample confinement in both the xy- and xz-planes for any given device height.

2.3. Device Simulations

The device geometry was parameterized within COMSOL Multiphysics®, leaving the channel widths, the sample channel height and the sheath fluid channel height as variables. Two-fold symmetry was exploited by simulating only one half of the device split along the x-axis, applying symmetric boundary conditions where appropriate. Steady-state fluid flow through the device was simulated using the computational fluid dynamics (CFD) module, coupled to the transport of diluted species module for all simulations involving fluorescein. We used the particle tracing module for all simulations involving microparticles. Simulations were conducted under the assumption of laminar flow, with no-slip boundary conditions on all walls. Inlets were subjected to laminar inflow constraints, parameterized by the sample flow rate and the sheath flow rate respectively. The outlet pressure was constrained to zero. Fluid parameters were assigned for liquid water at 293 K. For simulations involving fluorescein, the sample inlet was subjected to a concentration constraint, fixing the concentration of fluorescein at the inlet to its experimental value of 1 mM/mL. All other inlets were subjected to zero concentration constraints. Coupled CFD-transport systems were solved using COMSOL's default solver. Maximum sample (fluorescein) height was calibrated using experimental data for a single set of sample and sheath flow rates, yielding a threshold concentration of fluorescein in the model. Fluorescein heights and widths were predicted by thresholding fluorescein concentration at the outlet. Particles used in tracing simulations were assumed to have a diameter of 6 um and density 1.002 kg/L, within the range of the parameters of a *Saccharomyces cerevisiae* cell [55]. Particles were subjected to Stokes' drag, neglecting other force contributions. The simulation was initialized with 500 particles uniformly distributed at the sample inlet and traced for 10 ms. Particle positions were registered at the outlet and used to compute the mean particle position and its standard deviation for comparison with experiment.

2.4. Microscopy

To visualize flow focusing and particle confinement in these devices we used confocal microscopy (Leica TCS SP5 confocal microscope). Threshold was determined for the lowest-flowrate image using a modified version of the iterative intermeans algorithm as implemented in ImageJ (default method) [56]. This threshold was kept fixed for thresholding subsequent images. The concentration of microspheres (Fluoresbrite polychromatic red 6.0 micron, Polysciences Inc., Warrington, PA, USA) dispersed in fluorescein solution was 107 particles/ml. For bright-field imaging we used a Nikon Eclipse Ti-U inverted microscope. Images were acquired using a chromatic aberration free infinity (CFI) Plan Apo Lambda 60× oil objective (NA 1.4). The calculated lateral resolving power of the objective at 380 nm is 139 nm and the focal depth is 515 nm. A Point Gray Grasshopper (GS3-U3-23S6M-C) camera was used for image acquisition. Images of microspheres in flow were captured at 1330 frames/s and 7-μs exposure times. Images of yeast in flow were captured at 1000 frames/s and 5-μs exposure times.

Stationary images for neural network training were captured at the same conditions. Z-stacks were acquired at 0.25 μm offset between slices. Stacks were segmented automatically using a version of the Multiscale Adaptive Nuclei Analysis (MANA) algorithm adapted to bright-field image stacks [57]. Cell-containing frames were automatically detected by keeping images with maximum patch-wise variance at least twice as high as mean image variance. 128 × 128-pixel crops were extracted from cell-containing frames by locating the image-patch of maximum variance and cropping a square of 128 × 128 pixels around the patch center.

2.5. Microsphere Z-Displacement Regression

To determine the offset of each imaged bead from the focal plane, we used neural network-based regression. Z-stacks of static microspheres served as a training set (Supplementary File S1). A neural network was implemented in PyTorch (Figure S2a) [58]. It consisted of three convolutional blocks with leaky rectified linear unit activation [59,60], with batch normalization in every layer but the last. The network was trained on the z-stack data until convergence using the Adam optimizer with mean square error loss and initial learning rate 1×10^{-4} to predict the displacement of bead centers with respect to the focal plane [58,61]. Training images were augmented using random rotations, cropping and addition of Gaussian noise with mean 0 and standard deviation 0.1. Bright-field images of microspheres in flow were evaluated using the network and a z-displacement distribution of microspheres was computed.

2.6. Yeast Cell Z-Distance Regression

To determine the offset between yeast cell images acquired in flow and the focal plane, the strategy used for microspheres (see above) is not easily applicable, since yeast cells exhibit high variability in size and shape. Instead, we used pairs of bright-field cell images captured within the same field of view and at known z-distances from the focal plane. Z-stacks of stationary cells for all yeast species under consideration were acquired with an inter-slice spacing of 0.25 μm. Single-cell stacks were cropped from the acquired fields of view. Image augmentation (rotation, translation, mirroring, addition of noise) was used to yield visually different images at a known z-distance. Using this information, we trained a siamese neural network (Figure S2b) to predict the z-distance between pairs of single-cell images. A siamese neural network yields predictions for inputs relative to a reference by first embedding the reference and inputs using the same neural network module, concatenating the results of the reference-path and target-path, and finally applying a further neural network module for input-reference comparison [62]. This is done to discourage the neural network from learning to compare the input images pixel-by-pixel, instead of on a global scale. The siamese neural network was trained to infer the distance between pairs of z-stack slices for a single cell. Slices were augmented by random rotations, translations, zoom and Gaussian noise with mean 0 and standard deviation 0.1. The network was trained using the Adam optimizer with initial learning rate 1×10^{-4} until convergence [61]. Bright-field images of yeast cells in flow were embedded using the neural network. A well-focused *S. cerevisiae* cell was chosen as a reference for z-distance computation. Z-distances-to-reference were computed for all single-cell images to derive a z-displacement distribution.

2.7. Unsupervised Learning For Cellular Mixtures

To characterize a set of captured single-cell images, we use a probabilistic generative approach. We assume that for single-cell yeast images, x is drawn from a distribution $p_\theta(x) = p_\theta(x|z) \, p(z)$ with parameters θ where z are low-dimensional latent variables, $p(z)$ is the prior distribution of latent variables, and $p_\theta(x|z)$ is the likelihood of an image x given a latent vector z. Here, $p_\theta(x|z)$ is given by a neural network. Following [36], we construct a neural network to give a variational approximation $q_\theta(z|x)$ to the true posterior $p(z|x)$ by reparameterization and optimize the variational lower bound to the marginal log-likelihood $\log p(x)$ with respect to all neural network parameters θ (Figure S3). The neural network q_θ maps single-cell images to samples from the low-dimensional latent distribution, and can thus be understood as an encoder, embedding data points into latent space. Similarly, p_θ can be understood as a decoder, mapping samples from the latent distribution to high-dimensional single-cell images. Optimizing the variational lower bound is then realized as training the encoder and decoder to reconstruct input images well, under the constraint that the latent distribution should be as close as possible to the prior distribution $p(z)$ (Figure S3, purple term). To successfully learn a latent space, where latent dimensions correspond to meaningful visual features (e.g., cell shape, cell focal plane), we follow and implement the FactorVAE term in the variational lower bound promoting the independence of latent dimensions (Figure S3, red term) [40]. This term penalizes the latent distribution's total correlation (TC) as given by the Kullback–Leibler (KL) divergence of the marginal distribution $q(z)$ and its corresponding factored distribution, which is a product of the distributions for each latent dimension [63]. This forces the latent distribution to be close to a product of independent distributions. Therefore, the neural networks are encouraged to learn a more strongly disentangled latent representation.

For our data set, a variational autoencoder with FactorVAE loss was trained on cell-containing crops (size 128 × 128 pixels) from continuous-flow imaging [41]. The encoder consisted of six convolutional kernels of size 3 × 3 with ReLU activation [59], and batch normalization with an increasing number of features (16, 32, 64, 128, 256, 512) [64], followed by reparametrization to yield a sample from a 10-dimensional normal distribution [36,65]. The decoder consisted of transpose convolutions in the reverse order of the encoder, again followed by ReLU activation and batch normalization. The discriminator used for FactorVAE loss computation was a multi-layer perceptron (MLP) with two layers, 64 hidden units and ReLU activation. The networks were trained using the Adam optimizer with initial learning rate 5×10^{-4}, and factor loss balancing parameter $\gamma = 10$ until convergence. K-means clustering as implemented in scikit-learn was applied to the latent space to separate *S. cerevisiae* cells from *S. pombe* cells and compared to ground-truth species labels [66]. Nearest neighbors for sample cells were extracted using euclidean distance in latent space. The latent space was visualized in 2D using t-distributed stochastic neighbor embedding (TSNE) [67].

3. Results and Discussion

3.1. Simulation Results

Our goal is to use these microfluidic devices to deliver and confine yeast cells for in-flow imaging. Since cell diameter for yeast is typically 4–8 μm, we aim to confine sample flow within 10 μm from the microscope cover slide. To optimize our design, we used COMSOL Multiphysics® to simulate the effect of device height and sheath-to-sample flow velocity ratio on the maximum distance between sample and microscope cover slide, referred to here as "sample height". The results of the parametric sweep are shown in Figure 2a. According to the simulation, any sheath-to-sample flow velocity ratio over 20 (Figure 2a, y-axis) should result in sample height below 10 μm (Figure 2a, color scale, darker green color) relative to the coverslip. This appears to be independent of the height of the device used (Figure 2a, x-axis). To test the simulation results, we fabricated two devices with different heights, the cross sections of which are shown in Figure 2b. The first device has a total height of 60 μm (10 μm bottom layer height + 50 μm top layer height), and the second device has a height of 120 μm (10 + 110 μm).

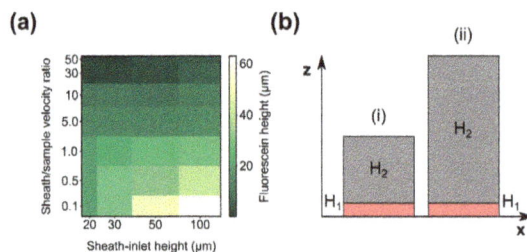

Figure 2. (**a**) Parametric sweep performed in COMSOL Multiphysics®. Flow velocity ratio (y-axis) refers to the ratio between sheath fluid velocity and sample flow velocity. Device height (x-axis) refers to the total height of the device, assuming a constant sample inlet height of 10 μm. Sample height (color scale) refers to the maximum distance between coverslip and sample. Log–log scale has been used to better resolve the areas of interest; (**b**) Cross sections of two-layer device geometries tested, where H_1 is the height of the bottom layer and H_2 is the height of the top layer: (i) H_1 = 10 μm, H_2 = 50 μm, (ii) H_1 = 10 μm, H_2 = 110 μm device.

3.2. Sample Confinement Testing Using Fluorescein

To visualize flow focusing in these devices we used confocal microscopy. Fluorescein dissolved in water was introduced through the sample inlet and was kept at a constant flow rate of 0.25 µL/min (0.87 mm/s). Water was used as sheath fluid and was introduced through the two taller, outer inlets at increasing flow rates. Figure 3a shows a montage of confocal images (channel cross section) for the device with a height of 60 µm. As predicted by simulation, for increasing sheath flow rates the volume occupied by fluorescein in the channel is progressively reduced. We observe fluorescein confinement in all directions, seen in the pictures as a reduction in the width and height of the fluorescein cone, also shown as an image overlay in Figure 3b. The reduction in the height of the fluorescein cone corresponds to confinement towards the microscope cover slip. When sheath flow rate is equal to sample flow rate (0.25 µL/min), fluorescein occupies a large fraction of the channel volume (red). With increasing sheath flow rates, the volume occupied by fluorescein shrinks towards the cover slip (black). The highest sheath flow rate achieved without affecting flow equilibrium was 20 µL/min, corresponding to a sheath-to-sample flow velocity ratio of 16. This is quantified in Figure 3c in a plot showing the measured fluorescein heights for every sheath flow rate tested. As seen in the graph, experimentally measured fluorescein heights correspond well with the equivalent values obtained from simulations, for sheath flow rates up to 20 µL/min. Since our final goal is to use optimized devices for applications in single-cell imaging in flow, sample flow rates were kept low in order to avoid image blur during imaging. For applications in flow cytometry, sample focusing can be optimized at higher flow rates.

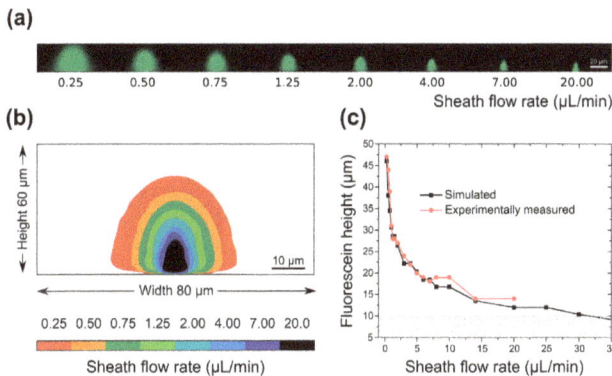

Figure 3. (**a**) Confocal microscopy images at increasing sheath-to-sample flow velocity ratio for 10 + 50 µm device. As sheath flow rate increases, fluorescein is confined both towards the microscope slide (bottom), as well as from the sides. In the z-direction the pixel size is w = 0.41 µm, h = 0.65 µm; (**b**) Z-projection of thresholded confocal microscopy images showing fluorescein confinement for increasing sheath flow rate; (**c**) Simulated (black) and experimentally measured (red) fluorescein heights for constant fluorescein flow rate of 0.25 µL/min and varying sheath flow rates. Shaded area highlights fluorescein height below 10 µm.

Using these devices, we were not able to experimentally reproduce simulated sheath flow rates over 20 µL/min and fluorescein confinement below 10 µm was not achieved (maximum confinement was ~14 µm). Instead, for sheath flow rates over 20 µL/min (34.72 mm/s) we saw fluorescein backflow towards the sample inlet, suggesting a large difference in pressure between the sheath and sample inlets not predicted by the simulation shown in Figure 2a. To predict pressure driven backflow in microchannels using COMSOL Multiphysics®, pressure constraints need to be applied. For all simulations shown in this work, flow velocity constraints that prohibit backflow were applied instead. Indeed, when pressure constraints are applied instead of velocity constrains, we see a negative velocity in the x-direction within the sample channel, which confirms flow towards the sample inlet for sheath flow rates above 20 µL/min (Figure S4). These results demonstrate that devices with a height of 60 µm do not fulfill the necessary requirements for sample flow focusing within 10 µm from the coverslip.

To eliminate the problem of sample backflow towards the inlet we next tested devices with a height of 120 µm. An increase in the area of the channel cross-section is expected to alleviate backflow since channel pressure is expected to drop. The logical way to achieve this would be to increase channel height, as opposed to channel width, as we would otherwise lose horizontal focusing. Similar to the previous device, the fluorescein flow rate was kept at 0.25 µL/min and sheath flow rate was slowly increased from 0.25 µL/min (0.43 mm/s) to 200 µL/min (347.22 mm/s). Again, to evaluate fluorescein confinement we used confocal microscopy and the results are shown in Figure 4a. We found that for a 1:1 sheath-to-sample flow rate ratio, fluorescein occupies a large fraction of the channel volume, also shown in red in the image overlay in Figure 4b. With increasing sheath flow rates, we observe continuous fluorescein confinement, seen in the figures as a reduction in the width and height of the fluorescein cone. Quantification of the fluorescein cone height from the images revealed a confinement within a distance of 10 µm from the cover slip at flow rates over 100 µL/min (173.61 mm/s), shown as the shaded area in the graph in Figure 4c. The distance between the microscope slide and the tip of the fluorescein cone decreased further to a minimum of ~5µm when the highest tested flow rate (200 µL/min) was used. At such high sheath flow rates, sample flow begins to become unstable, which results in a change in the observed fluorescein shape, as seen in Figure 4a. Importantly, due to the larger cross-section of this device, flow equilibrium was maintained, and no fluorescein backflow was present even at very high sheath-to-sample flow rate ratios. As seen in the graph, experimentally measured fluorescein heights correspond well with the equivalent simulated heights, and confinement below ~5 µm was experimentally reproduced. To better visualize confinement, we used COMSOL Multiphysics® to generate animations that show sample flow confinement with increasing sheath flow rates in 2D and 3D, see Supplementary Animation S1 and S2. In summary, devices with a height of 120 µm can robustly confine fluorescein as close as ~5 µm from the microscope slide.

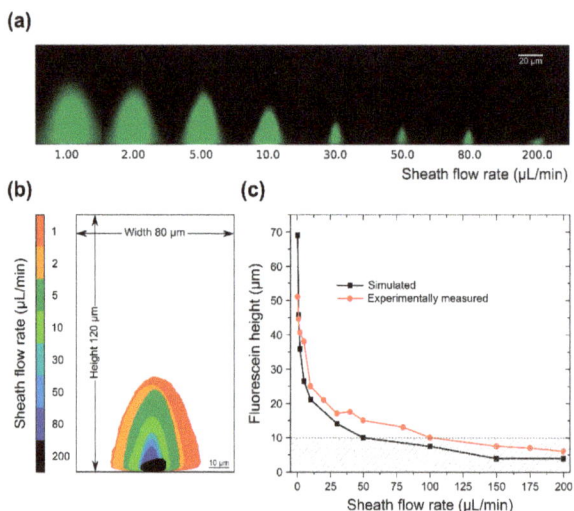

Figure 4. (**a**) Confocal microscopy images at increasing sheath-to-sample flow velocity ratio for the 120 µm (10 + 110 µm) device. In the z-direction, the pixel size is w = 0.43 µm, h = 0.43 µm; (**b**) Z-projection of thresholded confocal microscopy images showing confinement of the fluorescein cone for increasing sheath flow rate; (**c**) Simulated and experimentally measured fluorescein heights for constant fluorescein flow rate of 0.25 µL/min and varying sheath flow rates. Shaded area highlights fluorescein height below 10 µm, which is our target confinement height for live yeast cell in-flow imaging.

Since these devices are optimized for use in single-cell imaging in flow, it is important that flow focusing is maintained for several hundred micrometers after the junction in order to provide enough space for cell detection, cell imaging, and potential integration of cell sorting mechanisms [68]. We therefore investigated four positions within the device with increasing distance from the inlet junction; 100 µm, 500 µm, 1.1 mm, and 2.1 mm. Confocal microscopy images of fluorescein dissolved in water were taken at these positions and are shown in Figure S5i–iv. Flow equilibrium and sample confinement is maintained along the main channel even 2 mm away from the junction ensuring enough space for cell imaging and sorting. Even though the height of the fluorescein cone remains the same along the channel, fluorescein diffuses into the sheath fluid (water), which is to be expected in channels characterized by laminar flows where mass transport by transverse diffusion is the dominant mechanism for mixing [69,70].

3.3. Simulation of Particle Positioning and Validation Using Microspheres

For high-resolution imaging of cells flowing in these devices, it is important to be able to predict how cell position varies along the z-axis according to the flow rates used. To evaluate the equilibrium position of particles flowing in such channels, we first used simulations where we traced 500 microspheres with a diameter of 6 µm uniformly distributed at the sample inlet. We found that at low sheath flow rates, the mean equilibrium position of the microspheres (red squares in Figure 5a) is predicted to be closer to the tip of the fluorescein cone (Figure 5a, black squares) rather than the coverslip. In other words, the center of gravity of the microspheres is predicted to be on average closer to tip of the fluorescein cone than the coverslip. With increasing sheath flow rates (above 100 µL/min) and as the size of the fluorescein cone is decreasing due to flow confinement, the simulations predict that the mean equilibrium position of the microspheres will shift towards the microscope slide, close to the center of the fluorescein cone. This shift in the mean position of microspheres follows a shift in the flow streamlines, which is in turn based on an increase in the shear force exerted by the velocity gradient at higher flow rates (Figures S6 and S7). To validate the results from the simulations, we dispersed red fluorescent microspheres in fluorescein solution and used confocal microscopy to track their location. Due to the limited speed and the line-scanning mode of data acquisition in confocal microscopy, the moving microspheres appear as lines (Figure 5b,c). Even though the confocal microscope image in Figure 5b qualitatively confirms that particles equilibrate on average closer to the tip of the fluorescein cone rather than the base, a quantitative measurement of the position of microspheres is not possible due to limitations of the microscope. Simulations shown in Figure 5c, however, mathematically reproduce the microscopy data shown in Figure 5b. Again, most microspheres are predicted to reach equilibrium closer to the tip of the fluorescein cone (green contour) rather than the cover slip (bottom of the figure). To facilitate comparison between experimental and simulation data we included a juxtaposition of the two as an inset in Figure 5c.

To further quantify the distribution of microsphere positions within the sample phase without line-scanning artefacts, we used bright-field microscopy to image microspheres in flow at a fixed z-position in devices with a height of 120 µm. We set the imaging z-position to match the mean equilibrium z-position of microspheres predicted by simulation. Microspheres were dissolved in water at a concentration of 10^7 particles/ml and were introduced into the device through the sample inlet. To keep the particle velocity within the range we can image without being affected by image blur, we used sample flow rate of 0.1 µL/min and sheath (water) flow rate of 10 µL/min. These flow rates resulted in microspheres being confined within 16 µm from the coverslip, with a mean equilibrium position at approximately 10 µm away from the coverslip. To automatically and accurately quantify the z-positions of all microspheres imaged in flow with respect to the microscope focal plane and the coverslip, we trained a simple neural network on z-stacks of images acquired from static microspheres (Figure 5d and Supplementary File S2). Details about the network can be found in Section 2.4. The distribution of microspheres within the sample stream is depicted as a histogram in Figure 5e. Microsphere z-displacement is shown relative to the true focal plane (z-displacement = 0), as well as relative to the microscope slide (z-displacement = 10 μm). Microspheres located within 2 μm from the focal plane, depicted by the shaded area in Figure 5e, account for 68% of all spheres imaged. As we already demonstrated in Figures 3 and 4, further confinement is possible in these devices by increasing the sheath flow rate. It is expected that this will increase the percentage of microspheres located within 2 μm from the focal plane since there will be less space available for them to move. However, increasing the sheath flow rate also increases the velocity of the microspheres, which in turn results in significant motion blur that no longer allows us to quantify the microsphere distribution.

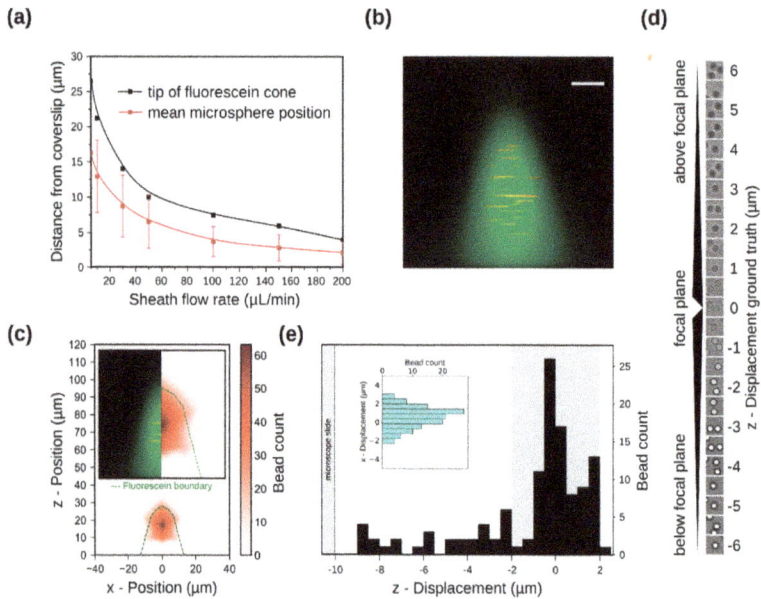

Figure 5. (**a**) Simulated distance between fluorescein cone tip and microscope slide for increasing sheath flow rates and fluorescein flow rate of 0.25 μL/min (black). Mean bead equilibrium position (along with standard deviation) with respect to the microscope slide (red); (**b**) Confocal microscopy image of the cross section of the sample stream (green), sheath fluid (black). Red fluorescent microspheres (diameter ~ 6 μm) dispersed in the sample stream appear as lines due to line scanning in confocal microscopy. This image was taken in a device with a height of 120 μm, at a fluorescein flow rate of 0.25 μL/min (0.43 mm/s) and sheath flow rate of 10 μL/min (17.36 mm/s). The height of the fluorescein cone in this image is 27 μm, and the scale bar is 6 μm; (**c**) Simulation reproducing microscopy data shown in (**b**). Fluorescein contour shown as green dotted line. Microspheres appear to concentrate closer to the fluorescein cone tip rather that the microscope slide. The insert shows a juxtaposition of the confocal image and the equivalent simulated data. (**d**) Microsphere images acquired at known distances from the focal plane, used as part of the training set for bead focal plane regression. (**e**) Histogram of z-displacement of microspheres relative to the focal plane. Bins have a width of 0.5 μm, with the x-axis the displacement relative to the focal plane and the y-axis the bead count for each bin. The shaded region within 2 μm of the focal plane, accounts for 68% of all microspheres. Displacement on the x-y plane is shown in the inset. 87% of sphere centers are located within 2 μm of the stream centerline.

For applications in imaging flow cytometry, it is also important to ensure tight particle focusing on the x-y plane in order to minimize the field-of-view that is imaged and subsequently analyzed. The total width of the sample phase for the flow conditions described above was found to be ~10 μm. The displacement of bead centers from the stream centerline is provided as the inset in Figure 5e (green histogram). Almost all bead centers (87%) were found to lie within 2 μm from the sample stream centerline, which confirms excellent focusing along the x-axis and allows for the capturing of small images for fast read-out and processing. Together, our results demonstrate confinement of microspheres in a narrow stream close to the surface of the coverslip, as predicted by the simulations.

3.4. Single-Cell Imaging in Flow

Next, we replaced microspheres with yeast cells of different species, i.e., *S. cerevisiae*, *Sd. ludwigii* and *S. pombe*. Unlike microspheres, each of these species exhibit characteristic cell shapes in the size-range of 3–12 μm. *S. pombe* cells are rod-shaped, *Sd. ludwigii* are lemon shaped, and *S. cerevisiae* cells are round. Furthermore, cell shapes change through the life cycle since *S. pombe* divides by fission, while *Sd. ludwigii* and *S. cerevisiae* cells divide by forming a bud attached to the so-called mother cell. The irregularity in cell shape and sizes imposes limitations on the tools we can use to determine the z-displacement distribution of yeast cells within the focused sample stream in our device. The simple neural network training strategy used for microspheres (Section 2.4, Figure 5d) could not be used in this case due to the lack of homogeneity. Instead, we used pairs of bright-field cell images captured within the same field of view and at known z-distances from the focal plane. Single-cell stacks were cropped from the acquired fields of view and images were augmented to yield visually different images of cells with a known z-distance. Using this information, a siamese neural network was trained to predict the z-distance between pairs of single-cell images (Figure 6a) [62]. More details about yeast cell z-distance regression have been outlined in Section 2.5. To evaluate the learning success of this strategy, we used the trained network to predict the distance between pairs of images taken from a test data set that has not been used for training purposes. As shown in Figure 6b the neuronal network predicted with high accuracy the z-distance in the test data.

To continue with the analysis of cell focusing, we imaged yeast cells in flow under the same conditions described in Section 3.3 for microsphere imaging. Imaging data for all yeast species are given in Supplementary File S3. Some sample yeast images that highlight cell heterogeneity are shown in Figure 6c. We used bright-field microscopy to image cells in flow at a fixed z-position in devices with a height of 120 μm. The focal plane for yeast cells was found to be approximately 16 um from the coverslip. Using the trained neural network, we determined the position of each imaged cells within the device using well-focused cell images as reference. Similar to what we observed for microspheres in Figure 5e, the z-distribution for all species peaked around the focal plane of the microscope, with a broad tail towards the microscope slide (Figure 6d). When considering cells of all species and sizes, the main peak around the focal plane (z-displacement = 0) contained 51% of imaged cells. The broad tail formed at the bottom of the device, adjacent to the microscope slide can be explained by the broad heterogeneity of the sample, since size and shape distributions are known to influence particle focusing positions in different types of devices [71,72]. Given this information, and in addition to the fact that the tail is much less pronounced when imaging monodisperse microspheres under identical flow conditions in the same devices, we assume that this broad tail is in fact due to cell heterogeneity, examples of which are shown in the inset of Figure 6d. To test this assumption, we limited cell-shape degrees of freedom by considering only *S. cerevisiae* cells in Figure 6e. The distribution now clearly shows two peaks, where the main peak contains 59% of all cells. Closer examination in the composition of the lower peak reveals that 57% of cells are small, single cells and 28% are budding cells. These cases are susceptible to pessimistic neural network predictions. Here, we call a network prediction pessimistic, when the maximum possible z-distance between the imaged cells and the reference cell is predicted. For example, if an input image contains a well-focused mother cell and an out-of-focus daughter cell, the network uses the z-distance between the out-of-focus daughter cell and reference cell, and therefore classifies the entire image as out-of-focus, even though the mother cell is in focus. Examples of these events are shown in the inset of Figure 6e. In total, 85% of lower-peak cells are either differently sized than the main-peak cells, or have pessimistic z-distance predictions. For the purposes of evaluating the cell focusing performance of our device compared to that of microspheres of the same size, we may therefore safely neglect the z-distribution's lower peak. This indicates that the simple microfluidic devices we developed can be successfully used to also focus cells, allowing for imaging and downstream tasks, such as sample characterization and cell sorting.

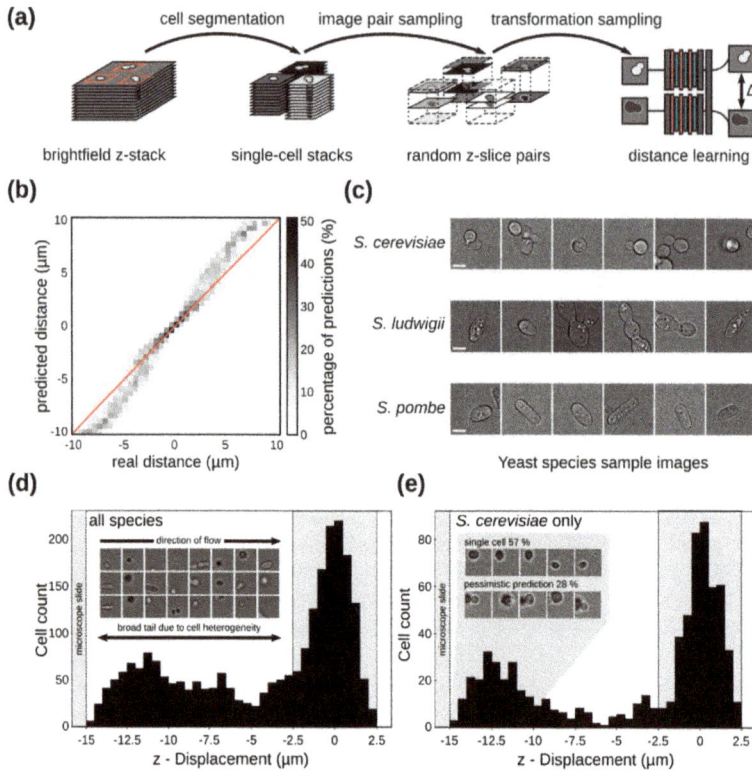

Figure 6. (**a**) Work flow underlying distance learning for automated z-position determination of flowing cells. Stacks of images of large fields of view containing many cells were recorded on the same microscope used for the imaging of flowing cells, but using an ordinary mounting of the cells. Single-cell stacks were cropped out. Random image pairs with known z-distances taken from single-cell stacks were used after image augmentation to train a Siamese neural network. (**b**) Predicted and real z-distance in pairs of cells taken from a test set of single cell stacks. (**c**) Images of the different yeast species used. (**c**) Example images of *S. cerevisiae*, *S. ludwigii* and *S. pombe* cells. The example cells highlight the high within-species and inter-species heterogeneity of the cells used. Scale bar length corresponds to 5 μm. (**d,e**) Histogram of z-displacement of flowing yeast cells in the device relative to an in-focus reference cell. Bins have a width of 0.5 μm, with the x-axis the z-displacement relative to the distribution median. Across all yeast species, example images highlighting their heterogeneity are shown. The shaded region within 2.5 μm of the focal plane contains 51% and 60% of imaged cells for all species (**d**) and *S. cerevisiae* (**e**) respectively. *S. cerevisiae* examples of non-budding cells and pessimistic predictions in the region below −10 μm are shown, accounting for 85% of the distributions lower peak.

3.5. In-Flow Cell Imaging and Cell Classification

Imaging large cell populations using conventional microscopy requires scanning using large mechanical moving stages. Imaging flow cytometry offers the possibility of imaging many cells while avoiding moving mechanical parts. Moreover, each cell is imaged individually, eliminating the need for retrospective cell segmentation and object identification in large fields of view, which is often a daunting task especially when cells are close together. Microfluidic devices like the one described in this work offer the possibility for high-throughput imaging and open up applications that also involve rapid, in-line characterization of the investigated sample. To outline possible applications, we took imaging of different yeast species within our devices another step further and concluded this work by devising a flexible framework for efficient, automated imaging data analysis. To this end, we implemented unsupervised machine learning in the form of a VAE to extract phenotypic information from single-cell images without the need for user intervention.

To demonstrate the capabilities of our framework, we trained a VAE with FactorVAE loss on single-cell images (see Section 2.6) and classified yeast cells by species in a fully unsupervised manner. The model consists of an encoder and a decoder. The encoder maps images to 10-dimensional points in latent space, while the decoder generates images from such points. A schematic of the model and objectives involved is shown in Figure 7a. The model is trained to have generated images match input images and produce a latent space, where each dimension corresponds to a meaningful visual characteristic of cells. For example, the cell species should vary along one dimension, while the cell shape should vary along another (for more details, please refer to Section 2.6). Qualitative evaluation of the resulting latent space by nearest neighbor analysis shows that single-cell images at a similar level of focusing and of similar shape are close together in the learned representation. As such, the four nearest neighbors shown in Figure 7b are budding *S. cerevisiae* cells, if the query cell is a budding *S. cerevisiae* cell, and elongated cells, if the query is an elongated cell. This suggests that our VAE has learned a meaningful latent representation, which could be applied to downstream classification tasks.

As our VAE learns visual similarities and differences among imaged cells, a high within-species variance of cells increases the difficulty of distinguishing different species. One factor resulting in high intraspecies variance is a lack of cell focusing. On the other hand, images of similarly sized, well-focused cells reduce intraspecies variance and as a result greatly simplify and improve classification. Our device achieves such improvements by precisely focusing similarly sized cells close to the microscope slide. In addition to particle focusing along the z-axis, our device tends to orient non-spherical particles in the direction of fluid flow, further reducing intraspecies variance. This tendency can be seen in the cells displayed in Figure 7b. Our model learns this regularity of the data and therefore groups similarly oriented cells.

To allow for the automated analysis of data from our devices, the learned representation should capture phenotypic properties necessary for distinguishing single-cell images of different yeast species. We evaluated the suitability of our learned representation for this task of unsupervised classification by using it to distinguish between images of *S. cerevisiae* and *S. pombe* cells captured in-flow in our microfluidic devices. *S. ludwigii* cells were not used for classification since they tend to form aggregates that disrupt flow and clog the microfluidic channels, resulting in few available examples of *S. ludwigii* cells in flow compared to *S. cerevisiae* and *S. pombe*. A qualitative inspection of the latent space labeled by yeast species (Figure 7c, left) reveals a separation between budding and fission yeast cells, with similar subsets of cells forming smaller clusters. Fully unsupervised classification is performed with an accuracy of 74%. Data-points that were wrongly classified (Figure 7c, center) are mostly located in regions where species labels overlap. Some failure cases are shown on the right in Figure 7c. A qualitative inspection of these failure cases reveals small out-of-focus *S. cerevisiae* cells being misclassified as *S. pombe* cells, while round out-of-focus *S. pombe* cells are misclassified as *S. cerevisiae* cells. Both of these are cases of misclassification, where distinguishing between yeast species becomes hard even for human experts. We have therefore achieved fully unsupervised classification of yeast cell by species with an accuracy of 74%. This complements our device with a strong baseline for cell classification without the need for expert intervention.

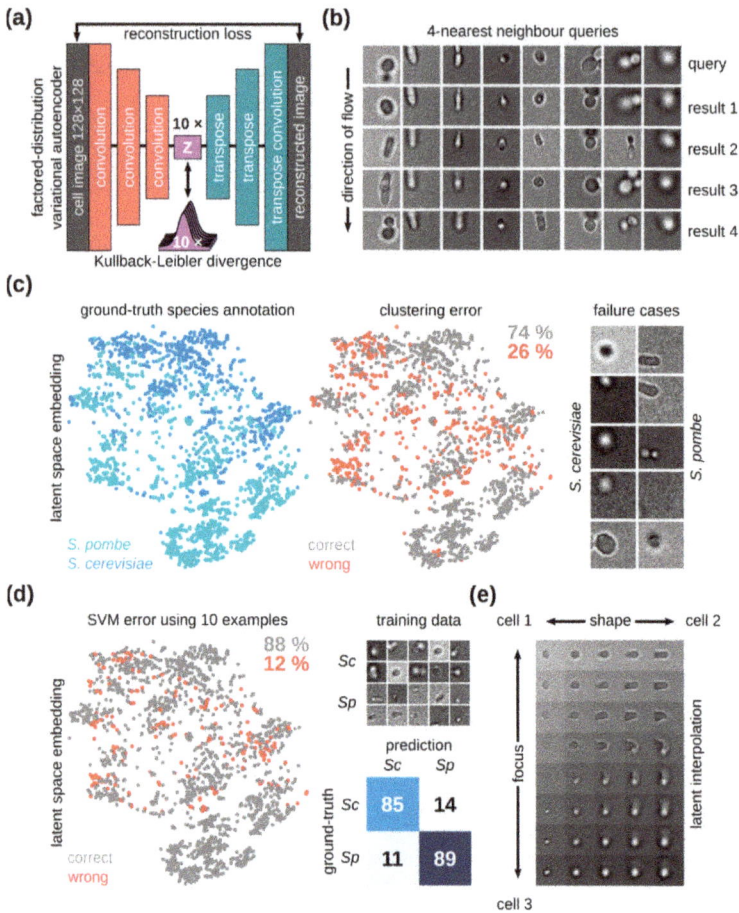

Figure 7. (**a**) Variational autoencoder (VAE) architecture for unsupervised learning. Cell images are convolutionally embedded into 10-dimensional latent space and reconstructed using same-shaped transpose convolutions. The network is trained to perform image reconstruction and is constrained to produce a disentangled latent space by KL divergence relative to a normal distribution and a penalty on total correlation. (**b**) Sample nearest neighbor queries for eight query cells. Query results are displayed in the order of increasing distance in latent-space. (**c**) Assessment of unsupervised classification accuracy. A two-dimensional embedding of data points for *S. cerevisiae* (blue) and *S. pombe* (turquoise) is shown, with ground truth species labels (left), a map of data points wrongly classified (red) by latent space k-means (center), and images of random failure cases for both species (right). Failure cases comprise *S. cerevisiae* cells classified as *S. pombe* cells (left column) and vice versa (right column). k-means on latent space classifies 74% of samples correctly, without the need for supervision. (**d**) Assessment of few-shot classification accuracy. A map of wrongly classified data points (red) using an support vector machines (SVM) classifier on latent space with 10 training examples per species shows an accuracy of 88% (left). The full training set is displayed for both species (top right), together with a confusion matrix showing the percentage of classifications for the classifier (bottom right). *Sc* and *Sp* indicate *S. cerevisiae* and *S. pombe* respectively. (**e**) Latent space interpretability. A latent space interpolation between three cells is shown, indicating latent space vectors encoding for cell focal plane (focus), as well as cell elongation (shape).

While fully unsupervised classification removes the need for large, hand-annotated training data sets, annotating small amounts of training examples is both feasible and beneficial for classification accuracy. We extend our data analysis setup with few-shot cell classification, to improve accuracy and enable expert-guided adaptation to more fine-grained classification tasks. To this end, we trained an SVM classifier on a representative set of 10 annotated single-cell images per species (Figure 7d, top right). The few-shot classifier increases classification performance to an accuracy of 88%, with negligible training time (Figure 7d). While misclassified data-points occur at similar locations in latent space as for fully unsupervised classification, the number of wrongly classified cells is significantly reduced (Figure 7d, left). The confusion matrix for the SVM classifier (Figure 7d, bottom right) shows that both *S. cerevisiae* and *S. pombe* cells are misclassified as the other with a probability of 11–14%, a vast improvement compared to unsupervised classification. Our integrated platform can therefore perform few-shot classification of yeast cell images captured in-flow in our microfluidic device and achieve 88% success in separating yeast cells by species.

In flow cytometry, it is desirable to gate single-cell events by their properties, e.g., fluorescence intensities. In imaging flow cytometry, such properties correspond to spatially resolved features, such as cell morphology and subcellular protein localization. Ideally, our latent space should capture those properties. Qualitative inspection of the latent space reveals interpretable latent dimensions, which can be linked to biologically meaningful morphological features, such as cell focal plane and elongation. These are visualized in a latent interpolation between cells along these dimensions in Figure 7e. The results in this figure also show that our latent space captures valuable semantic information, which could be used to differentiate between different cell phenotypes based on morphology in bright-field images. These indicate that our combined microfluidic and unsupervised learning platform can be used for the biologically-relevant characterization of complex cellular mixtures with minimal human intervention. The learned semantic latent space captures enough phenotypic features to group cells based on their morphology and extract accurate subpopulation classifiers with very few training examples.

4. Discussion

In this work, we have demonstrated 3D flow focusing in simple microfluidic devices for applications in imaging flow cytometry. The devices used utilize a difference in height between the outer sheath inlets and the middle sample inlet and achieve sample flow confinement within a few micrometers from the microscope slide, which makes them suitable for use with high NA oil objectives. In contrast to most previously demonstrated geometries, our devices maintain a simple, single-layer architecture that makes them accessible to non-expert users. Instructed by simulations, we fabricated and tested two devices with different heights, 60 μm and 120 μm, and found that flow equilibrium between sheath and sample is more stable in taller devices, and sample confinement within 5 μm from the microscope slide was achieved. To evaluate device performance for use in single-cell imaging in flow, we introduced 6-μm polymer microspheres dispersed in fluorescein through the sample inlet and monitored their position using confocal microscopy. Further, we used simulation tools to predict bead equilibrium positions in these devices for different sheath flow rate ratios and the results were confirmed using confocal microscopy. Once the mean microsphere equilibrium positions had been calculated and confirmed, we used bright field microscopy to image fast-moving beads traveling through these devices. We applied a novel, neural network-based approach to determine the distance of spheres to the focal plane. We found that 68% of the microspheres were traveling within 2 μm from the focal plane, thus enabling the acquisition of single plane images across their body. These results were reproduced for yeast cells of multiple species travelling in the same devices, further demonstrating the applicability of such devices in imaging flow cytometry. Our optimized microfluidic device can be used for a range of flow rates (0.1 μL/min–200 μL/min have been tested) and a range of particle sizes (3–12 um tested) without the need for geometry modification, which makes it robust and compatible with both imaging and non-imaging flow cytometry. Additionally, such devices could be relevant in other applications where hydrodynamic focusing into a small, well-defined volume is required [73].

Furthermore, we applied state-of-the-art unsupervised learning techniques to classify yeast cells by species. This was done in a fully unsupervised manner and also using SVM in a few-shot setting. We learned a semantically meaningful latent representation of yeast cells, with latent vectors representing visually and biologically meaningful features. Our latent representation allowed for the unsupervised distinction between *S. cerevisiae* and *S. pombe* cells with 74% accuracy, increasing to 88% in the 10-shot setting, which proves it suitable for further downstream classification tasks. We also verified via nearest-neighbor analysis, that biologically and visually similar cells are grouped in latent space. In summary, we have presented the first application of unsupervised learning in imaging flow cytometry. Interpretable latent spaces provide biologically meaningful image parameters that could improve image-activated cell sorting and allow for FACS-like gating on imaging data.

Our demonstration only utilized bright field imaging of microspheres and yeast cells. For practical applications, imaging flow cytometry requires the capability to acquire fluorescent images of cells. This requires acquisition speeds that are fast enough to reveal subcellular structures inside the moving cells, ideally in the range of the diffraction limit of a high NA objective (i.e., ~200 nm lateral and ~800 nm vertical resolution when using a 1.4 NA oil objective). This poses several challenges, some of which can be overcome using different techniques already described in the literature [6].

5. Conclusions

In conclusion, we have demonstrated a learning microfluidic platform capable of imaging live cells in flow and classifying acquired images without the need for human supervision. Cell streaming, confinement and imaging are achieved in a simple and versatile microfluidic device. Such devices can confine flow towards the microscope slide in a controlled manner, which makes them especially suitable for applications in single-cell imaging in flow. We have demonstrated that a large range of flow rates and particle sizes can be used without the need for geometry modification, since both tightness of focusing and line of focusing along the z-direction only depends on the flow rate ratio between sheath and sample. In fact, we were able to successfully stream, focus, and image *S. cerevisiae*, *S. ludwigii* and *S. pombe* yeast. This is the first demonstration of controlled 3D flow confinement well below the inlet height in this simplified version of a single layer Y-shaped microfluidic device and the firsttime particle and cell positioning in the focused stream are studied. We expect that such a robust device will find applications in the quickly growing fields of imaging flow cytometry and flow cytometry on-chip.

Additionally, we achieved image classification using a powerful unsupervised learning paradigm of disentangling variational autoencoders. Our variational autoencoder embeds single-cell images in an interpretable latent space and allows for both similarity-based queries and classification. The biggest advantage of such classification is that it is completely unsupervised, obviating the need for large hand-annotated training data sets prevalent in neural network-based machine learning. To our knowledge, this is the first application of unsupervised representation learning to imaging flow cytometry. In particular, disentangled representation learning has not been applied to single-cell images before and we expect it will play a big role in gating for image activated cell sorting. In conclusion, we presented a simple and affordable platform for continuous-flow single-cell imaging, large-scale data analysis, and image classification.

Supplementary Materials: The following are available online at http://www.mdpi.com/2072-666X/10/5/311/s1, Figure S1: Device fabrication schematic, Figure S2: Schematic of neural network architectures, Figure S3: Schematic of variational autoencoder training, Figure S4: X-direction velocity profile simulation for the 60 μm (10 + 50 μm) device, Figure S5: Flow focusing at four positions within the device with respect to the junction, Figure S6: Qualitative shear force distribution due to xy-velocity gradients in the z direction, Figure S7: Theoretical flow focusing behavior dependent on x-velocity, Supplementary Animation S1, Supplementary Animation S2. Supplementary Files S1–S3 are deposited in the heiDATA Dataverse repository at: https://doi.org/10.11588/data/L5J7WO. Source code and neural network weights for all parts of the project are deposited on GitHub at: https://github.com/mjendrusch/learning-ifc.

Author Contributions: Conceptualization, I.C., M.J. and M.K.; methodology, I.C., M.J.; experiment design and execution I.C., M.J.; simulations M.J.; machine learning M.J.; photolithography, T.A.; microscopy, I.C., M.J., F.G. and A.S.; writing—original draft preparation, I.C.; writing—review and editing, I.C., M.J. and M.K.; supervision, I.C., G.C. and M.K.

Funding: This research was supported by the state of Baden-Württemberg through bwHPC and the German Research Foundation (DFG), grant number INST 35/1134-1 FUGG and DFG N498/12-1. Iordania Constantinou was supported through a fellowship from the Carl Zeiss Stiftung and by the Ministry of Science and Culture (MWK) of Lower Saxony, Germany through the SMART BIOTECS alliance between the Technische Universität Braunschweig and the Leibniz Universität Hannover. Gilles Charvin was supported by a grant from the Foundation pour la Recherche Médicale, and by a grant ANR-10-LABX-0030-INRT, a French State fund managed by the Agence Nationale de la Recherche under the frame program Investissements d'Avenir ANR-10-IDEX-0002-02.

Acknowledgments: Holger Lorenz and the fluorescence microscopy facility at ZMBH.

Conflicts of Interest: The authors declare no conflicts of interest.

References

1. Gross, H.-J.; Verwer, B.; Houck, D.; Recktenwald, D. Detection of rare cells at a frequency of one per million by flow cytometry. *Cytometry* **1993**, *14*, 519–526. [CrossRef] [PubMed]

2. De Rosa, S.C.; Herzenberg, L.A.; Herzenberg, L.A.; Roederer, M. 11-color, 13-parameter flow cytometry: Identification of human naive T cells by phenotype, function, and T-cell receptor diversity. *Nat. Med.* **2001**, *7*, 245–248. [CrossRef]

3. Sandberg, J.; Werne, B.; Dessing, M.; Lundeberg, J. Rapid flow-sorting to simultaneously resolve multiplex massively parallel sequencing products. *Sci. Rep.* **2011**, *1*, 1–7. [CrossRef]

4. Schonbrun, E.; Gorthi, S.S.; Schaak, D. Microfabricated multiple field of view imaging flow cytometry. *Lab Chip* **2012**, *12*, 268–273. [CrossRef]

5. Barteneva, N.S.; Fasler-kan, E.; Vorobjev, I.A. Imaging Flow Cytometry: Coping with Heterogeneity in Biological Systems. *J. Histochem. Cytochem.* **2012**, *60*, 723–733. [CrossRef] [PubMed]

6. Han, Y.; Gu, Y.; Zhang, A.C.; Lo, Y. Review: Imaging technologies for flow cytometry. *Lab Chip* **2016**, *16*, 4639–4647. [CrossRef] [PubMed]

7. Rosenauer, M.; Buchegger, W.; Finoulst, I.; Verhaert, P.; Vellekoop, M. Miniaturized flow cytometer with 3D hydrodynamic particle focusing and integrated optical elements applying silicon photodiodes. *Microfluid. Nanofluid.* **2011**, *10*, 761–771. [CrossRef]

8. Simonnet, C.; Groisman, A. High-throughput and high-resolution flow cytometry in molded microfluidic devices. *Anal. Chem.* **2006**, *78*, 5653–5663. [CrossRef]

9. Sundararajan, N.; Pio, M.S.; Lee, L.P.; Berlin, A.A. Three-Dimensional Hydrodynamic Focusing in Polydimethylsiloxane (PDMS) Microchannels. *J. Microelectromech. Syst.* **2004**, *13*, 559–567. [CrossRef]

10. Chang, C.-C.; Huang, Z.-X.; Yang, R.-J. Three-dimensional hydrodynamic focusing in two-layer polydimethylsiloxane (PDMS) microchannels. *J. Micromech. Microeng.* **2007**, *17*, 1479–1486. [CrossRef]

11. Wu, T.; Chen, Y.; Park, S.; Hong, J.; Teslaa, T.; Zhong, J.F.; Carlo, D.; Teitell, M.A.; Chiou, P. Pulsed laser triggered high speed microfluidic fluorescence activated cell sorter. *Lab Chip* **2012**, *12*, 1378–1383. [CrossRef] [PubMed]

12. Sakuma, S.; Kasai, Y.; Hayakawa, T.; Arai, F. On-chip cell sorting by high-speed local-flow control using dual membrane pumps. *Lab Chip* **2017**, *17*, 2760–2767. [CrossRef]

13. Mao, X.; Lin, S.C.S.; Dong, C.; Huang, T.J. Single-layer planar on-chip flow cytometer using microfluidic drifting based three-dimensional (3D) hydrodynamic focusing. *Lab Chip* **2009**, *9*, 1583–1589. [CrossRef]

14. Eluru, G.; Julius, L.A.N.; Gorthi, S.S. Single-layer microfluidic device to realize hydrodynamic 3D flow focusing. *Lab Chip* **2016**, *16*, 4133–4141. [CrossRef] [PubMed]

15. Gualda, E.J.; Pereira, H.; Martins, G.G.; Gardner, R.; Moreno, N. Three-dimensional imaging flow cytometry through light-sheet fluorescence microscopy. *Cytom. Part A* **2017**, *91*, 144–151. [CrossRef]

16. Nawaz, A.A.; Zhang, X.; Mao, X.; Rufo, J.; Lin, S.C.S.; Guo, F.; Zhao, Y.; Lapsley, M.; Li, P.; McCoy, J.P.; et al. Sub-micrometer-precision, three-dimensional (3D) hydrodynamic focusing via "microfluidic drifting". *Lab Chip Miniaturisation Chem. Biol.* **2014**, *14*, 415–423. [CrossRef]

17. Paiè, P.; Bragheri, F.; Di Carlo, D.; Osellame, R. Particle focusing by 3D inertial microfluidics. *Microsyste. Nanoeng.* **2017**, *3*, 17027. [CrossRef]

18. Rane, A.S.; Rutkauskaite, J.; deMello, A.; Stavrakis, S. High-Throughput Multi-parametric Imaging Flow Cytometry. *Chem* **2017**, *3*, 588–602. [CrossRef]

19. Nitta, N.; Sugimura, T.; Isozaki, A.; Mikami, H.; Hiraki, K.; Sakuma, S.; Iino, T.; Arai, F.; Endo, T.; Fujiwaki, Y.; et al. Intelligent Image-Activated Cell Sorting. *Cell* **2018**, *175*, 266–276.e13. [CrossRef]

20. Normolle, D.P.; Donnenberg, V.S.; Donnenberg, A.D. Statistical classification of multivariate flow cytometry data analyzed by manual gating: Stem, progenitor, and epithelial marker expression in nonsmall cell lung cancer and normal lung. *Cytom. Part A* **2013**, *83A*, 150–160. [CrossRef] [PubMed]

21. Ye, X.; Ho, J.W.K. Ultrafast clustering of single-cell flow cytometry data using FlowGrid. *BMC Syst. Biol.* **2019**, *13* (Suppl. 2), 35. [CrossRef]

22. Pouyan, M.B.; Jindal, V.; Birjandtalab, J.; Nourani, M. Single and multi-subject clustering of flow cytometry data for cell-type identification and anomaly detection. *BMC Med. Genom.* **2016**, *9*, 41. [CrossRef]

23. Kraus, O.Z.; Grys, B.T.; Ba, J.; Chong, Y.T.; Frey, B.J.; Boone, C.; Andrews, B.J.J.; Pärnamaa, T.; Parts, L.; Humeau-Heurtier, A.; et al. The curse of dimensionality. *Mach. Learn.* **2018**, *7*, 18.

24. Köppen, M. The curse of dimensionality. In Proceedings of the 5th Online World Conference on Soft Computing in Industrial Applications, 4–18 September 2000; pp. 4–8.

25. Humeau-Heurtier, A. Texture Feature Extraction Methods: A Survey. *IEEE Access* **2019**, *7*, 8975–9000. [CrossRef]

26. Carpenter, A.E.; Jones, T.R.; Lamprecht, M.R.; Clarke, C.; Kang, I.H.; Friman, O.; Guertin, D.A.; Chang, J.H.; Lindquist, R.A.; Moffat, J.; et al. CellProfiler: Image analysis software for identifying and quantifying cell phenotypes. *Genome Biol.* **2006**, *7*, R100. [CrossRef]

27. Kraus, O.Z.; Grys, B.T.; Ba, J.; Chong, Y.; Frey, B.J.; Boone, C.; Andrews, B.J. Automated analysis of high-content microscopy data with deep learning. *Mol. Syst. Biol.* **2017**, *13*, 924. [CrossRef]

28. Rumetshofer, E.; Hofmarcher, M.; Röhrl, C.; Hochreiter, S.; Klambauer, G. Human-level Protein Localization with Convolutional Neural Networks. In Proceedings of the International Conference on Learning Representations, New Orleans, LA, USA, 6–9 May 2019.

29. Pärnamaa, T.; Parts, L. Accurate Classification of Protein Subcellular Localization from High-Throughput Microscopy Images Using Deep Learning. *G3 (Bethesda)* **2017**, *7*, 1385–1392. [CrossRef]

30. Platt, J. *Sequential Minimal Optimization: A Fast Algorithm for Training Support Vector Machines*; Technical Report MSR-TR-98-14; Microsoft Research: Redmond, WA, USA, 1998.

31. Chong, Y.T.; Koh, J.L.Y.; Friesen, H.; Kaluarachchi Duffy, S.; Cox, M.J.; Moses, A.; Moffat, J.; Boone, C.; Andrews, B.J. Yeast Proteome Dynamics from Single Cell Imaging and Automated Analysis. *Cell* **2015**, *161*, 1413–1424. [CrossRef] [PubMed]

32. Bengio, Y. Deep Learning of Representations: Looking Forward. In *Statistical Language and Speech Processing, Lecture Notes in Computer Science*; Dediu, A.-H., Martín-Vide, C., Mitkov, R., Truthe, B., Eds.; Springer: Berlin/Heidelberg, Germany, 2013; pp. 1–27.

33. Gidaris, S.; Singh, P.; Komodakis, N. Unsupervised Representation Learning by Predicting Image Rotations. *arXiv* **2018**, arXiv:1803.07728.

34. Haeusser, P.; Plapp, J.; Golkov, V.; Aljalbout, E.; Cremers, D. Associative Deep Clustering: Training a Classification Network with no Labels. In Proceedings of the German Conference on Pattern Recognition (GCPR), Stuttgart, Germany, 9–12 October 2018.

35. Caron, M.; Bojanowski, P.; Joulin, A.; Douze, M. Deep Clustering for Unsupervised Learning of Visual Features. *arXiv* **2018**, arXiv:1807.05520.

36. Kingma, D.P.; Welling, M. Auto-Encoding Variational Bayes. *arXiv* **2013**, arXiv:1312.6114.

37. Goodfellow, I.J.; Pouget-Abadie, J.; Mirza, M.; Xu, B.; Warde-Farley, D.; Ozair, S.; Courville, A.; Bengio, Y. Generative Adversarial Networks. *arXiv* **2014**, arXiv:1406.2661.

38. Vincent, P.; Larochelle, H.; Lajoie, I.; Bengio, Y.; Manzagol, P.-A. Stacked Denoising Autoencoders: Learning Useful Representations in a Deep Network with a Local Denoising Criterion. *J. Mach. Learn. Res.* **2010**, *11*, 3371–3408.

39. Higgins, I.; Amos, D.; Pfau, D.; Racaniere, S.; Matthey, L.; Rezende, D.; Lerchner, A. Towards a Definition of Disentangled Representations. *arXiv* **2018**, arXiv:1812.02230.

40. Kim, H.; Mnih, A. Disentangling by Factorising. *arXiv* **2018**, arXiv:1802.05983.

41. Kim, M.; Wang, Y.; Sahu, P.; Pavlovic, V. Relevance Factor VAE: Learning and Identifying Disentangled Factors. *arXiv* **2019**, arXiv:1902.01568.

42. Higgins, I.; Matthey, L.; Pal, A.; Burgess, C.; Glorot, X.; Botvinick, M.; Mohamed, S.; Lerchner, A. beta-VAE: Learning Basic Visual Concepts with a Constrained Variational Framework. In Proceedings of the International Conference on Learning Representations, Toulon, France, 24–26 April 2017.

43. Burgess, C.P.; Higgins, I.; Pal, A.; Matthey, L.; Watters, N.; Desjardins, G.; Lerchner, A. Understanding disentangling in β-VAE. *arXiv* **2018**, arXiv:1804.03599.

44. Chen, R.T.Q.; Li, X.; Grosse, R.; Duvenaud, D. Isolating Sources of Disentanglement in Variational Autoencoders. *arXiv* **2018**, arXiv:1802.04942.

45. Mescheder, L.; Geiger, A.; Nowozin, S. Which Training Methods for GANs do actually Converge? *arXiv* **2018**, arXiv:1801.04406.

46. Scott, R.; Sethu, P.; Harnett, C.K. Three-dimensional hydrodynamic focusing in a microfluidic Coulter counter. *Rev. Sci. Instrum.* **2008**, *79*, 46104. [CrossRef]

47. Hairer, G.; Pärr, G.S.; Svasek, P.; Jachimowicz, A.; Vellekoop, M.J. Investigations of micrometer sample stream profiles in a three-dimensional hydrodynamic focusing device. *Sens. Actuators B Chem.* **2008**, *132*, 518–524. [CrossRef]

48. Chung, S.; Park, S.J.; Kim, J.K.; Chung, C.; Han, D.C.; Chang, J.K. Plastic microchip flow cytometer based on 2- and 3-dimensional hydrodynamic flow focusing. *Microsyst. Technol.* **2003**, *9*, 525–533. [CrossRef]

49. Lake, M.; Narciso, C.; Cowdrick, K.; Storey, T.; Zhang, S.; Zartman, J.; Hoelzle, D. Microfluidic device design, fabrication, and testing protocols. *Protoc. Exch.* **2015**. [CrossRef]

50. Qin, D.; Xia, Y.; Whitesides, G.M. Soft lithography for micro- and nanoscale patterning. *Nat. Protoc.* **2010**, *5*, 491. [CrossRef]

51. Ward, K.; Fan, Z.H. Mixing in microfluidic devices and enhancement methods. *J. Micromech. Microeng.* **2015**, *25*, 094001. [CrossRef]

52. Tan, J.N.; Neild, A. Microfluidic mixing in a Y-junction open channel. *AIP Adv.* **2012**, *2*, 032160. [CrossRef]

53. Ushikubo, F.Y.; Birribilli, F.S.; Oliveira, D.R.B.; Cunha, R.L. Y- and T-junction microfluidic devices: Effect of fluids and interface properties and operating conditions. *Microfluid. Nanofluid.* **2014**, *17*, 711–720. [CrossRef]

54. Watkins, N.; Venkatesan, B.M.; Toner, M.; Rodriguez, W.; Bashir, R. A robust electrical microcytometer with 3-dimensional hydrofocusing. *Lab Chip* **2009**, *9*, 3177–3184. [CrossRef] [PubMed]

55. Bryan, A.K.; Goranov, A.; Amon, A.; Manalis, S.R. Measurement of mass, density, and volume during the cell cycle of yeast. *Proc. Natl. Acad. Sci. USA* **2010**, *107*, 999–1004. [CrossRef]

56. Ridler, T.W.; Calvard, S. Picture Thresholding Using an Iterative Selection Method. *IEEE Trans. Syst. Man Cybern.* **1978**, *8*, 630–632.

57. Salvi, M.; Molinari, F. Multi-tissue and multi-scale approach for nuclei segmentation in H&E stained images. *Biomed. Eng. Online* **2018**, *17*, 89. [PubMed]

58. Paszke, A.; Chanan, G.; Lin, Z.; Gross, S.; Yang, E.; Antiga, L.; Devito, Z. Automatic differentiation in PyTorch. In Proceedings of the 31st Conference Neural Information Processing, Long Beach, CA, USA, 4–9 December 2017; pp. 1–4.

59. Nair, V.; Hinton, G.E. Rectified Linear Units Improve Restricted Boltzmann Machines. In Proceedings of the 27th International Conference on Machine Learning, Haifa, Israel, 21–24 June 2010; p. 8.

60. Xu, B.; Wang, N.; Chen, T.; Li, M. Empirical Evaluation of Rectified Activations in Convolutional Network. *arXiv* **2015**, arXiv:1505.00853.

61. Kingma, D.P.; Ba, J.L. Adam: A Method for Stochastic Optimization. In Proceedings of the 3rd International Conference for Learning Representations, San Diego, CA, USA, 7–9 May 2015; pp. 1–15.

62. Koch, G.; Zemel, R.; Salakhutdinov, R. Siamese Neural Networks for One-shot Image Recognition. In Proceedings of the 32nd International Conference on Machine Learning, Lille, France, 6–11 July 2015; p. 8.

63. Watanabe, S. Information Theoretical Analysis of Multivariate Correlation. *IBM J. Res. Dev.* **1960**, *4*, 66–82. [CrossRef]

64. Ioffe, S.; Szegedy, C. Batch Normalization: Accelerating Deep Network Training by Reducing Internal Covariate Shift. *arXiv* **2015**, arXiv:1502.03167.

65. Rezende, D.J.; Mohamed, S.; Wierstra, D. Stochastic Backpropagation and Approximate Inference in Deep Generative Models. *arXiv* **2014**, arXiv:1401.4082.

66. Pedregosa, F.; Varoquaux, G.; Gramfort, A.; Michel, V.; Thirion, B.; Grisel, O.; Blondel, M.; Prettenhofer, P.; Weiss, R.; Dubourg, V.; et al. Scikit-learn: Machine Learning in Python. *J. Mach. Learn. Res.* **2011**, *12*, 2825–2830.

67. Van der Maaten, L.; Hinton, G. Visualizing Data using t-SNE. *J. Mach. Learn. Res.* **2008**, *9*, 2579–2605.

68. Wyatt Shields, C., IV; Reyes, C.D.; López, G.P. Microfluidic cell sorting: A review of the advances in the separation of cells from debulking to rare cell isolation. *Lab Chip* **2015**, *15*, 1230–1249. [CrossRef]

69. Kuo, J.S.; Chiu, D.T. Controlling Mass Transport in Microfluidic Devices. *Annu. Rev. Anal. Chem.* **2011**, *4*, 275–296. [CrossRef]

70. Salmon, J.B.; Ajdari, A. Transverse transport of solutes between co-flowing pressure-driven streams for microfluidic studies of diffusion/reaction processes. *J. Appl. Phys.* **2007**, *101*, 074902. [CrossRef]

71. Kuntaegowdanahalli, S.S.; Bhagat, A.A.; Kumar, G.; Papautsky, I. Inertial microfluidics for continuous particle separation in spiral microchannels. *Lab Chip* **2009**, *9*, 2973–2980. [CrossRef]

72. Carlo, D. Di Inertial microfluidics. *Lab Chip* **2009**, *9*, 3038–3046. [CrossRef]

73. Yu, C.; Qian, X.; Chen, Y.; Yu, Q.; Ni, K.; Wang, X. Three-Dimensional Electro-Sonic Flow Focusing Ionization Microfluidic Chip for Mass Spectrometry. *Micromachines* **2015**, *6*, 1890–1902. [CrossRef]

micromachines

MDPI

Commentary

Integrating Microfabrication into Biological Investigations: the Benefits of Interdisciplinarity

Gianluca Grenci [1,2,*], Cristina Bertocchi [3] and Andrea Ravasio [4]

[1] Mechanobiology Institute (MBI), National University of Singapore, Singapore 117411, Singapore
[2] Biomedical Engineering Department, National University of Singapore, Singapore 117583, Singapore
[3] Department of Physiology, School of Biological Sciences, Pontificia Universidad Católica de Chile, Santiago 8330025, Chile; cbertocchi@bio.puc.cl
[4] Institute for Biological and Medical Engineering, Schools of Engineering, Medicine and Biological Sciences, Pontificia Universidad Católica de Chile, Santiago 7820436, Chile; andrea.ravasio@uc.cl
* Correspondence: mbigg@nus.edu.sg; Tel.: +65-6601-1903

Received: 25 March 2019; Accepted: 13 April 2019; Published: 16 April 2019

Abstract: The advent of micro and nanotechnologies, such as microfabrication, have impacted scientific research and contributed to meaningful real-world applications, to a degree seen during historic technological revolutions. Some key areas benefitting from the invention and advancement of microfabrication platforms are those of biological and biomedical sciences. Modern therapeutic approaches, involving point-of-care, precision or personalized medicine, are transitioning from the experimental phase to becoming the standard of care. At the same time, biological research benefits from the contribution of microfluidics at every level from single cell to tissue engineering and organoids studies. The aim of this commentary is to describe, through proven examples, the interdisciplinary process used to develop novel biological technologies and to emphasize the role of technical knowledge in empowering researchers who are specialized in a niche area to look beyond and innovate.

Keywords: microscopy; microfluidics; microfabrication; biomedical engineering

---

1. Introduction

Our understanding of physiological functions and diseases states, and therefore our ability to develop effective therapeutic strategies, are limited by the overwhelming system-level complexity of biological networks and the spatiotemporal integration of their multiscale and multiparametric components [1]. To deal with such complexity, biologists have developed a large toolbox of complementary approaches. Historically, biologists preferentially adopted reductionist experimental strategies where each experiment aims to test one parameter [2]. Such an approach stringently requires that the on-test biological parameter, being a gene, a protein or a whole biological function, is effectively isolated from its network. Typically, this is implemented by genetic manipulation and by controlling the experimental conditions. Thus, careful design and supervision over the experimental settings are necessary for successfully conduct the experiment and the use of pairwise controls to validate and generalize the conclusions. However, the very same redundancies that provide robustness to life are the major drawback of this strategy as rarely a node in a biological network can be pinpointed by a single manipulation to an unambiguous conclusion [2]. In recent years, more holistic and quantitative approaches to biological research have emerged [3–7]. For instance, synthetic biology and mechanobiology have opened the way for engineering the complexity of biological systems by providing biomimetic models of bioreactions and biological functions [4]. Various microfluidic and on-chip technologies (lab-on-chip, organ-on-chip, body-on-chip) have allowed highly sophisticated, and yet simple, bottom-up strategies to recapitulate intricate biological circuits in a miniaturized and customized fashion [8–11]. Finally, a family of technologies collectively termed "omics" (i.e., genomics, proteomics, lipidomics, metabolomics and functional omics), combined with multiparametric data analysis, provide an entirely new perspective and technological platform to simultaneously investigate dependent and independent parameters of biological systems [7]. For instance, in genomics, one gene microarray experiment can provide an accurate snapshot of the expression level of thousands of genes, simultaneously. Thanks to the combined technological and scientific efforts mentioned above, clinically important innovations such as personalized medicine and point-of-care diagnostic are becoming part of the everyday experience of doctors and patients [12–14]. Despite the existence of diverse approaches, innovation in the biological field is driven by the common need for parallelization, miniaturization, and precise customization of the experiments. To this end, microfabrication and lithographic tools have arguably played an essential role.

In this commentary, we will describe the technology context that drove the current tech revolution in biological research and discuss the factors that enabled such breakthroughs. Thereafter, we will illustrate some of our own key developments that emerged from close interaction and collaboration between scientists of different expertise working under the same roof.

2. Development of Microfabrication for Biological Research: A Brief Historical Perspective

For the better part of last 70 years, the main driving force for the development of micro/nano technologies has been the need to scale down the size of micro-components. This has resulted in the development of microchips with increased computational capacity, and usually reduced power consumption. Our generation has witnessed the incredible technological revolution that this race has led to. More interestingly, the development of new technologies and implementation of new approaches for the patterning of small features on a large-scale flat substrate has kept pace with Moore's law [15]. This was achieved thanks to the use of ultra-violet light (UV) of decreasing wavelength in order to improve features resolution: a 1:1 contact printing scheme was employed first, followed by image projection and step-and-scan exposure schemes that allowed reduction in the size of the image through an optical system (see e.g., [16]). Each time a shift in wavelength or in exposure was introduced, a huge initial investment cost was sustained by manufacturers in order to equip their plants with the latest new technology. Nevertheless, the global market size and the pervasive nature of electronic components in almost every aspect of modern society made the investments worthwhile. The huge push towards developing new patterning strategies that allow for faster and cheaper production of integrated circuits of ever-increasing resolution has not always been successful. Take for instance X-ray lithography (XRL), a good example of a promising technology with less than satisfactory insertion in production lines to date [17]. X-ray photons appear as a natural extension of UV lithography with their extremely short wavelengths (≤ 1 nm), but the technical difficulties in realizing exposure systems for XRL that are suitable for industrial scale application practically stopped any further exploration in that direction [18]. For every example of successful innovation scaled from the lab to the industrial application, many more examples of "failed" technologies can be found. However, those efforts were never entirely wasted, as they contributed to expand the micro/nano-patterning toolbox that we have in our hands.

Today many different, perhaps unintended applications or research fields are benefiting from such developments. One such application is soft lithography as introduced by the seminal work of Whitesides for biological and biomedical purposes [19,20]. When soft lithography was first described [21,22], the critical step for rendering these technologies for life-science applications was incidentally already included, as molding involves the use of biocompatible and optically clear material such as poly dimethyl siloxane (PDMS). Thus, soft lithography became popular and since then has paved the way for fabrication of micro-devices for biological and bio-medical applications. As a result of these initial groundbreaking innovations, a large toolbox of enabling technologies has been developed [23]. For instance, microfluidic culture technology associated with tissue engineering has been at the forefront for innovation in cell biology research, as it allows a deeper understanding of physiology and disease [24]. Stem cells and cancer cells are examples of cells whose functions are extremely dependent on their surrounding microenvironment [25,26]. The important factors defining the microenvironment, and thus cell function are: cell-cell and cell-matrix interaction and physicochemical factors such as temperature, pH and mechanical loads. Microfluidics, as compared to conventional culture methods allow precise control of the microenvironment, and thus enable mimicry of the in-vivo milieu. In experiments aimed at differentiation of metastable cells like cancer and stem cells, 3D microfluidic culture models have proven to be a powerful tool to improve the physiological relevance of in vitro models [27,28]. Furthermore, investigation of stem cells and their interactions with their environment has a high potential for translational regenerative medicine and stem cell therapies [29]. Another application of 3D microfluidic culture model is the chip-based model for cancer invasiveness. Understanding the intra- and extravasation of cancer cells through biological tissues is important to design effective cancer therapies. Thus, to study the biophysical barriers to the metastatic process, where cancer cells cross cell tissues such as the endothelium or blood-brain barrier is of prime importance [30]. For instance, chip-based models allowed to engineer the formation of an endothelial barrier in a 3D biomimetic environment [31–33], which recapitulates the physiological conditions of the process in a controlled manner. Furthermore, hollow structures in a microfluidic platform can be used to mimic the biophysical properties of mammary ducts and blood vessels [34]. Application of microfabrication to study tissue- and organ-level processes (tissues-on-a-chip and organs-on-a-chip) is a growing field of research and could serve as a platform for carrying out tightly controlled, high-throughput drugs toxicity screening studies [8,35,36]. Besides controlling the cellular microenvironment, microfabrication and soft lithography techniques have found excellent uses in imaging biological specimens [37–39], cell counting and sorting [40–47], and engineering of micro bioreactors [48–50], amongst other applications. Furthermore, soft lithography have allowed to propose and develop revolutionary approaches for biomedical applications; one such example is its impact on diagnostics and global health care enabled by miniaturization, costs reduction and integration of multiple functionalities in portable and reliable platforms [51,52]. Production of nano-particles for drug-delivery is another example [53,54], but the list continues to grow and a comprehensive discussion of all the impactful application of soft-lithography a micro-fabrication methods to the biomedical field is out of scope for this commentary.

3. Working in an Interdisciplinary Environment: Bridging Biologists and Engineers

As a branch of the life sciences, biological research adopts the scientific method, thereby hypotheses are formulated to address scientific questions and experiments are designed to draw factual conclusions that verify the original hypotheses (Figure 1). In this process, a key bottleneck is the design of adequate experiments that can provide feasible strategies and satisfy rigorous scientific standards. Two courses of action are feasible in this regard: researchers resort to commercially available instrument, devices and packaged kits when available (Figure 1, empty-arrows path). Alternatively, in the absence of commercial solutions, biologists adopt an interdisciplinary approach and devise the necessary enabling tools by collaborating with engineers (Figure 1, full-arrows path). Besides commercial availability, the choice for one or the other solution may also depend on a large variety of parameters such as convenience, time- and cost-effectiveness, feasibility, and the need to customize the experiment, standardization, to name a few. In our experience, the preference for one solution over the other is also influenced by the researcher's inclination for innovation that leads to development of novel technologies.

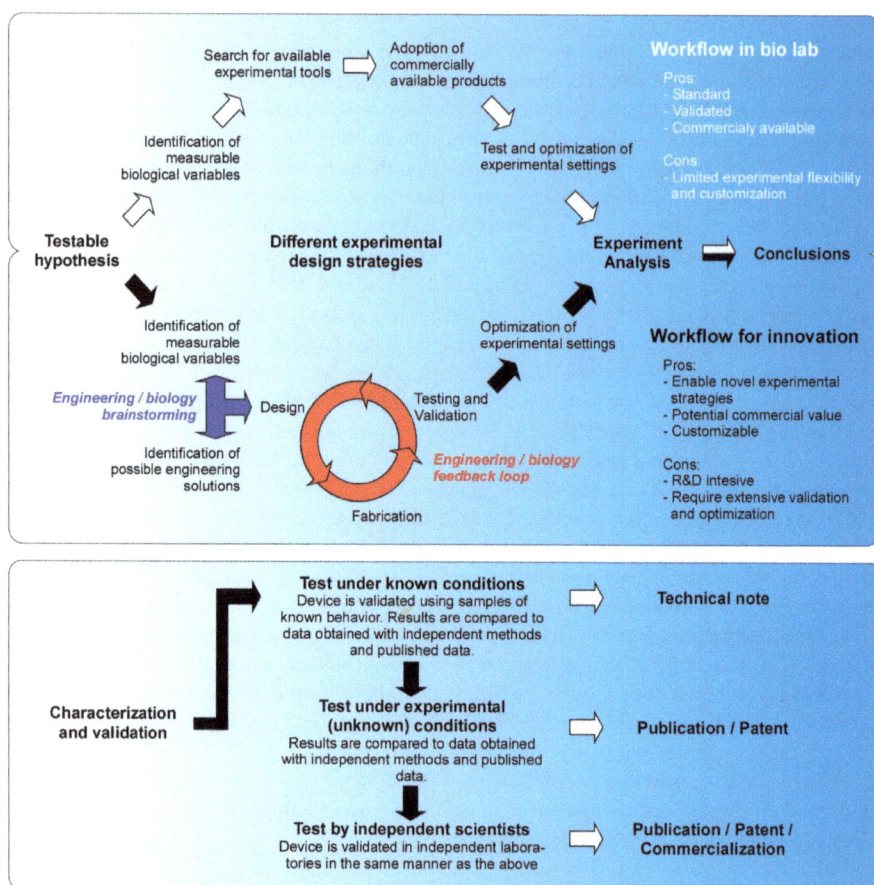

Figure 1. Top panel: Schematic comparison of two different workflows in biological laboratories. Workflow in biomed lab—the path following the empty arrows describes a common experimental approach in biological laboratories relying on commercially available tools. Workflow for innovations—innate inquisitiveness and lack of adequate experimental devices may lead to development of innovative solutions (full arrows). Interdisciplinary team work is often necessary and advisable. Bottom panel: Levels of device characterization and validation (full arrows) and possible outcome as result of successful testing (empty arrows).

Nowadays, researchers have access to an extensive toolbox of instruments, devices and experimental kits that can be purchased from specialized biotech companies. Thus, once they identify the biological parameter that needs to be quantified, scientists can search for commercially available products (Figure 1—top panel, empty arrows path) and adopt them into their research workflow. This path of least resistance offers multiple advantages, such as being time- and cost-effective. Furthermore, mass-produced commercial products are standardized in accordance to international standards for good manufacturing practice and are supplemented with detailed operating protocols. Additionally, commercial solutions go through extensive validation processes carried out by many users over a long period of time, which increases the reliability of the results that they generate and the understanding of possible pitfalls. This allows immediate adoption of the technology by investigators with minimal to no knowledge of the device´s engineering aspects, an ideal situation for biomed specialists who can devote their time to perform experiments and produce quality results. Obviously commercial products also come with their inherent limitations. For example, commercial products typically perform best within a narrow window of specifications. Thus, their applicability is limited to the designated purpose and their customization could be technically challenging, incomplete and ultimately unsatisfactory.

Since biomedicine still remains a frontier science, adoption of lab-ready technologies may not be an option as they may not be available yet. In such cases, an interdisciplinary collaborative approach —combining expertise of scientists in biomedicine and engineering fields—is required to develop the necessary experimental solution (Figure 1—top panel, full arrows path). This typically consists of a two-steps process: firstly, extensive discussion between biologists and engineers must be initiated to identify the biological parameters that need to be quantified and the engineering strategies that could be implemented to develop suitable experimental tools (Figure 1—top panel, blue three-headed arrow). In our own experience, this initial brainstorming step is by far the most challenging part of the process as it requires the translation of biological concepts into feasible engineering strategies and processes. All of this is further complicated by the need to effectively communicate with experts across disciplines, address expectation discrepancies and understand different work style and standards of the various disciplines involved. Therefore, a great deal of interpersonal tolerance, intellectual effort and investment of time for discussions and clarification are key to the success of such collaborative approaches. Next, an iterative process of implementation and optimization is applied during the development and validation of the novel tool (Figure 1—top panel, red three-steps circular workflow). The designing and prototyping steps are critical in finding suitable engineering solutions for the intended biological functionality. In some cases, not often though, already established microfabrication techniques can be directly translated without major variations. Incompatibility of materials, complex sequence of functionalization steps and noncompliance with biological research procedures are all examples of critical designing problems. After successful prototyping, the product is then handed over to the experimentalists to characterize the device performance and validate the results (Figure 1—bottom panel). To choose how to validate the device is a very critical and unintuitive step and it strongly depends on the type of device, its working principle, the intended use and the type of sample. In general, three levels of testing are required for full validation: initially the performance of the device is tested by using sample of well characterized behavior. For instance, polystyrene beads and soluble markers can be used to test the flow in microfluidics [55]. If the performance is not satisfactory, it might be necessary to go back to the drawing board and repeat the previous steps multiple times until optimal performance is achieved. Thereafter, the product's designated function is probed in real-case biological scenarios of unknown outcome, e.g., by using pairs of treated and untreated samples that allows paired analysis [56]. Those results are then validated using independent experimental conditions and compared to published data for consistency. After this extensive testing phase, the novel experimental device is considered ready for use in its intended scope and it can be disclosed to the public though a publication and the invention protected by a patent. At this point, the device is independently validated by multiple users who serve as beta-tester of the product. If the product performs consistently throughout all the validation process, the innovation may become commercially available and enrich the toolbox available to biologists (Figure 1—top panel, empty arrows path). Obviously, commercialization potential does not necessarily reflect the research value of the innovation as successful commercialization depends on factors like fabrication scalability, size of the potential market, and the intrinsic and perceived monetary value of the product. Once the innovation is made available to the scientific community through publication and/or commercialization, it can be used in any biological lab and hence it enters the workflow described by empty arrows in Figure 1.

4. Microfabrication for Biological Investigations: A Few Examples from Our Toolbox

There is a wealth of examples of micro-fabricated devices or methodologies derived from micro-fabrication which have been adapted to bio-medical applications. On one side, we have a long list of researches that take advantage of simple micro-fabrication, where in this context simple is meant as in "established and universally known micro-fabrication procedures". In Table 1, we propose a short list of such cases. A more extensive discussion comprising of microfluidic examples is out of the scope of this contribution and we kindly refer the readers to reviews on the subject (see for example [57–59]).

Table 1. Examples of microfabrication technologies applied to biological research.

Application	Method(s)	Required Micro-Fabricated System	References
Protein micro-patterning	Micro-contact printing	PDMS stamps	[60–63]
	Stenciling	Micro-stencils	[64–67]
	UV-patterning	UV mask	[68–70]
Traction force microscopy	Micro-pillars deflection tracking	Soft micro-pillars	[71,72]
Cells response to topology	Cells culturing on substrates with different topologies	Micro/nano patterned substrates	[73–77]
Collective cells migration	Engineering of environmental cues for collective migration	Micro-stencils Micro-patterning	[78–81]
Cells response to micro-structured environment	Cells culturing in micro-patterned niches	Substrates with micro-wells	[82,83]

On the other side, there are those cases where the requirements for the experiment pose a challenge to the micro-fabrication experts and new methodologies or original combinations of established ones need to be developed for the fabrication of the suitable micro-structures. In order to better clarify this claim, in the following sections we will showcase three exemplary cases where newly conceived micro-fabricated devices provided an original platform to address unanswered biological questions. These examples are:

- soSPIM: a light sheet microscopy system that needs only one single objective with high numerical aperture owing to the use of microfabricated disposable chips;
- IR-live: a microfluidic platform for label-free chemical imaging of live cells by infra-red absorption spectroscopy;
- Microfabricated microwells for expansion of circulating tumor cells (CTC).

4.1. soSPIM: Advanced Microscopy on A Chip

Selective plane illumination microscopy (SPIM) is a type of fluorescence microscopy which uses a light-sheet to illuminate the sample from the side. The light sheet is normally generated by a low numerical aperture (NA) objective, which is mounted orthogonally to a high NA objective that collects the fluorescent light [84].

In this configuration, planar excitation and the absence of out-of-focus light ensure high contrast, and reduced photo-bleaching and photo-toxicity compared to other fluorescent imaging techniques. However, the requirement for at least two orthogonally aligned objectives makes it difficult to integrate a high NA immersion objective, hindering SPIM use at the single cell level. However such a set-up has been implemented in structured illumination microscopes [85]. Single-molecule detection has also been demonstrated using dedicated and complicated two-objective-based experimental set-ups, e.g., with a movable mirrored cantilever that is placed close to mammalian cells in culture [86] or using lattice light-sheets [87].

Despite its evident benefits, SPIM suffers from the lack of easy solutions for handling samples within the confined space between the two objectives. Observing specimens such as embryos or organoids, with sizes of hundreds of micrometers, is a practical challenge in a standard commercially available SPIM. With this technical problem in mind, a micro-engineering approach was proposed by Galland et al. [88], where a single high NA objective was used for both the generation of the light sheet and the collection of the fluorescent light, giving the instrument its name single-objective SPIM (soSPIM) (Figure 2A). This imaging approach can be implemented on a standard inverted microscope by using disposable cell culture dishes with arrayed micro-mirrors (Figure 2B). The disposable culture dishes are a key component of the soSPIM set-up, which can be exclusively produced using micro-fabrication techniques.

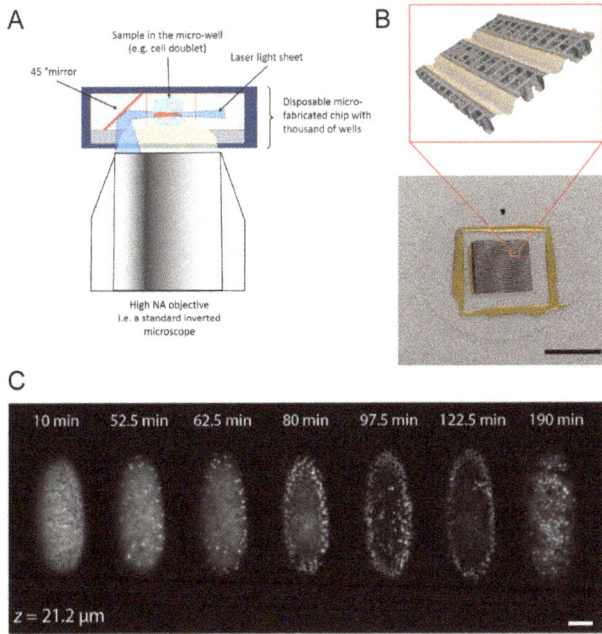

Figure 2. (**A**) Working principle of the soSPIM technique. The excitation laser beam is passed through the same high numerical objective that collect the fluorescent light emitted by the excited sample (shown as a cell doublet in the cartoon). The incoming beam produces a horizontal light sheet by scanning over the surface of the flanking micro-mirror, which is at exactly 45° inclination. By scanning the beam at different height on the mirrors, Z sectioning can be achieved. (**B**) Picture of an actual coverslip (bottom, scale bar is 1 cm) and zoom-in 3D reconstruction image acquired with a Keyence VHX 6000 microscope of the soSPIM device with 40 µm × 40 µm wide micro-wells, 50 µm deep. (**C**) Time-lapse imaging of the early stage of development of a *Drosophila* embryo. Time sequence of soSPIM optical sections 21.2 µm deep within a drosophila embryo expressing the nuclear protein Histone-mCherry imaged with a 20X magnification, 0.5 NA objective and a 4.3 µm thick light-sheet. A 35 µm Z-stack with a 1.35 µm z-step was acquired every 150 seconds for 220 minutes.

The working principle of the soSPIM (Figure 2A) is simple: micro-mirrors at exactly 45° of inclination are placed next to arrays of micro-wells. The micro-wells (Figure 2B), which can accommodate samples of different size, are the location for cells seeding and observation. The flanking micro-mirrors reflect a laser beam projected from the objective, and a light sheet is generated by horizontally scanning the laser on the mirror surface. Fluorophores in the sample are then excited by this light sheet and images are acquired with the same high NA objective. Sectioning in the Z direction is achieved by producing the light sheet on the mirrors at different planes and by de-focusing to different distances.

The details of the micro-fabrication steps can be found in the original publication [88], but the key aspect is that the disposable micro-mirrored coverslips are produced by a molding procedure starting with a primary mold made of silicon. This primary mold is fabricated through a combination of lithographic and etching processes, taking advantage of the crystal properties of silicon to produce optically smooth mirror surfaces. These surfaces are of the required inclination and can be aligned few micrometers apart from to the multi-well arrays. Moreover, the molding approach enables scaling up the production from handmade laboratory-based to semi-automated industrial, which makes the technique feasible for commercialization. In Figure 2C, we show an example of the potential of soSPIM (reprinted with permission from [88]). The use of larger mirrors adjusted according to the size of the samples, and a horizontal rotation stage for multiview imaging, enabled us to expand the capabilities of our system to perform time-lapse imaging of a *Drosophila* embryo with enhanced long-term stability (thanks to the use of a single objective) and without the requirement of perfect mechanical alignment of the two objectives as in traditional SPIM.

The soSPIM project highlights few of the key elements in the "path for innovations" described in Figure 1. Based on an original realization of an already disclosed microscopy approach, the implementation of this original idea required the design, fabrication and testing of a newly conceived micro-optical device. Once identified the lay-out of the device and its functional requirements (e.g., optical smoothness of the mirroring surfaces, size and arrangement of the micro-wells, materials compatibility), the "Engineering/biological validation" red loop was run across few times before reaching a final protocol capable of producing viable devices in high enough numbers for practical utilization in biological experiments. At this point, internal validation (a still on-going process) was conducted *via* experimental collaboration within the team that developed the soSPIM and colleagues working at the same institution [88,89]. The project is presently being evaluated for possible commercial exploitation with the help of several external groups acting as beta-testers.

4.2. IR-live: Infra-red Spectro-microscopy On Live Cells

Infrared (IR) spectroscopy, also known as Fourier transform IR spectroscopy (FTIR), uses the absorption of IR photons as a way to characterize the chemical content of a sample with little to none preparation. Illumination with infrared light promotes energy exchange between the inherent vibrational modes of molecular bonds and incident photons. The exchange results in distinct, fingerprint-like spectral bands that appear in absorption measured as a function of wavelength of incident light, while the energy exchange in the form of heat is negligible. The precise position, line shape, and intensity of infrared absorption bands depend on the molecular structure and conformation as well as intra- and inter- molecular interactions [90].

Despite some major improvements witnessed in recent years, such as the development of bright light sources like Synchrotrons [91] and quantum cascade lasers [92] and the availability of arrays of IR detectors in a configuration similar to that in a CCD camera (IR Focal Plane Arrays (FPA)), biologists have yet to use FTIR for live-cells imaging on a regular basis. And for at least one good reason: the strong water absorption in the mid-IR range. Even layer as shallow as 10 μm thick can completely obscure the features of live-cells due to the characteristic water absorption spectrum in the same IR range where cell's components are to be found. To tackle this problem, several groups have proposed microfluidic devices [93–95] or confined liquid compartments [96] which enable investigations into cellular processes, such as cell death [97], cell cycle [98], stem cell differentiation [99] or protein misfolding [100] at a single cell level and a subcellular spatial resolution [101]. However, the strategies adopted by these groups are often hampered by slow and expensive fabrication processes leading to limited experimental flexibility [102].

To facilitate the application of FTIR for live-cells imaging there is a need for easy-to-use and standardized microfluidic devices. With this in mind, Birarda et al. [55] proposed and demonstrated a soft-lithographic approach, wherein plastic devices with embedded transparent view ports (CaF$_2$ disks with 10 mm diameter and 1 mm thickness) are fabricated. The full fabrication strategy is detailed in [55,103]. It is interesting to highlight that this strategy requires accessing a clean-room micro-fabrication facility only for the generation of a silicon primary mold, which is used as the base for the lay-out of the final device. The actual production of the device can be performed in any laboratory equipped with standard non-lithographic tools (e.g., a plasma system, a hot plate and a UV light).

Conveniently, the fabrication of silicon molds can be outsourced to commercial facilities, therefore, in principle, such microfluidic devices can be fabricated with knowledge of only soft lithographic procedures and without an extensive understanding of lithographic processes. Figure 3 shows the resulting device (Figure 3A) with a plastic adaptor for mounting under the microscope (Figure 3B). The standardization provided by the proposed approach, allowed to further explore the design and production of prototypes for the mounting jigs through 3D printing techniques, thanks to the precise positioning of inlet and outlet ports and precise control of the overall dimensions of the devices. This is a step further in the direction of evolving FTIR as a suitable imaging technique for biological applications, as it simplifies the setting up of experiments and does not require advanced skills in handling microfluidic devices.

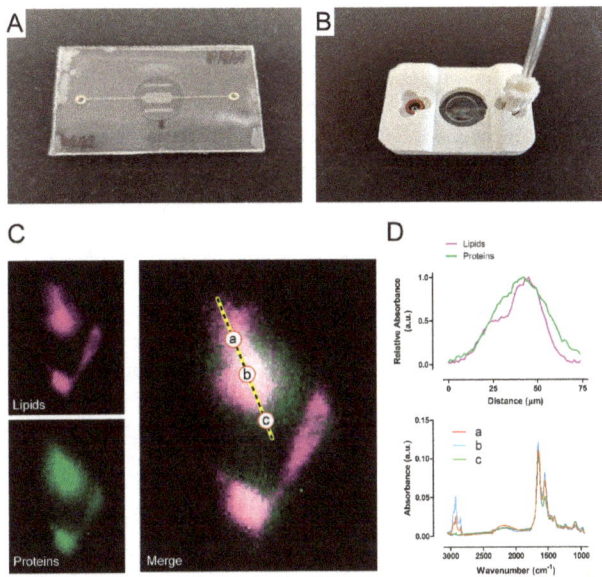

Figure 3. (**A**) A picture of the device for IR spectroscopy. (**B**) The device is shown mounted within the 3D printed plastic jig that allows for easy connection to external fluid management system and that it is compatible with a standard FTIR microscope set-up. (**C**) High-spatial resolution chemical maps of protein (magenta) and lipids (green) as measured in live REF52 cells (re-printed with permission from [55]. (**D**) re-printed with permission from [55]: top panel shows the line profile intensity of proteins (green) and lipids (magenta) as measure along the dashed yellow line in the merged chemical map shown in C. In the bottom panel, 3 punctual absorption spectra are shown for the pixel marked as a, b and c along the same line.

Using this new kind of devices, live cells IR spectromicroscopy on rat embryo fibroblast cell line REF-52 with high spatial resolution was proven. The chemical maps, line profile and punctual spectra shown in Figure 3C,D are re-printed with permission from [55]. The spatial distributions of two major components of the cells, proteins and lipids, are reconstructed with a pixel size of 1.1 μm × 1.1 μm from spectroscopic data acquired with an FPA at the BSISB program facility (Advanced Light Source beamline 5.4, Berkeley, CA, USA). Details of the data acquisition and analysis procedures are described in the same publication. In short, proteins and lipids content are quantified by calculating the integral below the normalized absorption spectra (as shown in Figure 3D, lower panel) in their respective peak area inside spectral regions: Amide II 1480–1600 cm^{-1} for proteins and 2800–3000 cm^{-1} for lipids.

Looking back to the initial attempts at producing FTIR compatible devices for live cells chemical mapping [104,105], the IR-live project is a good example, in our opinion, of the potential for innovation provided by micro-fabrication and in particular soft-lithography. The utilization of a micro-device in a biological research context is always setting for challenging requirements, such as biocompatibility and suitability for the experimental procedures. Here, a supplement of difficulty is provided by the previously unexplored properties of IR-compatible materials (calcium or barium fluoride, as used in the discussed examples), together with a required attention to a production scheme that would makes it possible for the final user to be independent from an advanced micro-fabrication facility. We argue that moving from the initial standard lithographic to the soft-lithographic approach enables a wider application of FTIR methods to biological research and its requirements, in that it makes easier to design and fabricate microfluidic devices with more advanced functions.

4.3. Microfabricated Microwells for Expansion of Circulating Tumor Cells (CTC)

Metastasis is a multi-step process characterized by proliferation of tumor cells, their intravasation into the blood or lymph, followed by their extravasation into the surrounding tissue and subsequent outgrowth in the new microenvironment [106]. Once released from the primary tumor, tumor cells enter the circulatory system; here they can be detected in the bloodstream as circulating tumor cells (CTCs) or in the bone marrow as disseminated tumor cells (DTCs). Since bone marrow sampling is a fairly invasive procedure, it is not generally favored for the clinical management of cancers [107,108]. Thus, in recent years, the focus has been on detecting CTCs in peripheral blood, as there is a clear association of CTCs with metastasis, clinical stage, prognosis of cancers and the response of patients to treatment [109,110]. The detection of CTCs could therefore be a promising minimally invasive diagnostic test for screening patients with metastatic cancers. The main challenge in detecting CTCs is their low levels in blood [111], which makes their recovery from patient samples extremely difficult. Thus, the idea is to design tools that can enrich viable CTCs in the sample for metastatic cancer diagnosis, treatment monitoring, personalized drug screening, and subsequent research studies.

Although several methods for culturing CTCs in vitro have been established in the past [111–115], a few critical issues, such as long culturing time (up to a month), need for pre-enrichment procedures, and low efficiency in CTC isolation (< 20%), still remain to be solved before diagnostic tools based on CTC counts can be routinely used.

Some of these major concerns were addressed in [116], in which the use of culture dishes with tapered micro-wells and an optimized protocol for enrichment were proposed as an improved method for effective CTC isolation. The tapered shape of the micro-wells was identified as one of the key factors contributing to the efficiency of this technique. Indeed, the morphology of the spheroids and CTC clusters created using this approach was found to conform to the micro-well boundaries. Unfortunately, the wells were produced by laser ablation, which resulted in micro-wells with inconsistent morphologies, de facto limiting the accuracy of clusters comparison. Hence, a photolithographic micro-fabrication approach was explored to control the shape of the wells with better precision.

This was achieved using a methodology detailed in [117–119] that is based on an unusual photolithographic approach, referred to as diffuser back-side lithography [120]. This method allows the fabrication of photo-resist micro-structures (pillars) with rounded profile (Figure 4A) which were considered ideal for this application. By molding a replica of these pillars in a PDMS layer, an array of micro-wells with the designed geometry and distribution, as shown in Figure 4B was generated; a complete device (as presented in the published results) can be produced as an assembly of 3 independently fabricated PDMS functional layers: a gradient generator to produce different concentrations of chemicals of interest, a micro-channels layer that confines the different concentrations and directs them to the layer containing the ellipsoidal micro-wells. As a result, each channel with micro-wells will contain cells clusters subjected to a different chemical mix within the same platform.

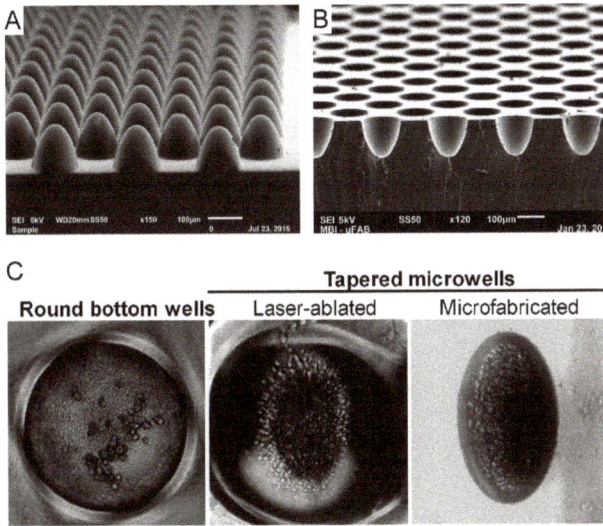

Figure 4. (**A**) SEM images of a PDMS working mold presenting ellipsoidal domes. (**B**) SEM of the PDMS device presenting ellipsoidal wells produced by replica of the mold shown in (**A**). (**C**) Reprinted with permission from [118]. Optical images of resulting CTC cluster in 3 different types of micro-wells. Clinical samples do not form clusters in conventional flat bottom, cylindrical wells (left) but are able to develop clusters consistently in the tapered ones (center and right-side pictures). The ellipsoidal wells produced by micro-fabrication techniques were proven to give the highest efficiency in the production of homogeneous clusters.

It is worth mentioning that the overall fabrication strategy for this 3-layered device is aimed at enabling flexibility in design according to the final application. The micro-wells array can be arranged in as many separated channels and with as many wells as required, while the gradient generator layer can be incorporated with modified designs and functional elements (e.g., pre-mixing, reservoirs, valves layers, and so on) exploiting thus the rich arsenal developed by microfluidic technologies.

Figure 4C is re-printed with permission from [118] and it shows optical microscopy pictures of exemplary CTC clusters as produced in the micro-wells of three different types. In the left-side picture the use of cylindrical micro-wells, i.e., with vertical walls and flat bottom, proved to be inefficient in the production of clusters, while the laser ablated wells (middle picture) are capable of promoting the clustering of the cells but with the limitations already introduced. Micro-fabricated ellipsoidal wells as shown in the right picture give instead consistent clusters with high efficiency.

The key element of this CTC liquid biopsy platform is arguably the design of the micro-wells array. The previous experiments with cylindrical and irregular wells suggested to the researchers the need to explore a system with better defined and homogeneously shaped wells for the growth of the CTC clusters. Under the hypothesis that this would provide for a more efficient control of the biological system, the discussion with the microfabrication team led to a viable strategy for the fabrication of the device, which suitability for clinical purposes is undergoing validation.

5. Concluding Remarks

The three examples we have discussed are a small selection of the many instances of fruitful interdisciplinary collaborations that we experienced at the Mechanobiology Institute (MBI). At MBI, researchers with backgrounds as diverse as in engineering and biology work in an open lab environment and share common laboratory facilities. The continuous dialogue that ensues gives rise to a common "mechanobiology" language and encourages sharing of expertise.

Historically, successful scientists used to be 'jack-of-all-trades': knowledgeable about science and, at the same time, experienced craftsmen who could build their own experimental tools. For instance, astronomers and microscopists of the past were often skillful craftsmen who showed astonishing degrees of perfection in glass working, producing perfectly shaped mirrors and lenses for their magnifying tubes. Besides legends like Galileo Galilei and Sir Isaac Newton that we instantly think of, several prominent scientists-cum-craftsmen existed throughout the 19th century. Nowadays, this is seldom the case and professional, including scientists and engineers, usually specialize in their own fields, contributing to high throughput, better-quality results. Unfortunately, specialization can sometimes lead to isolation into one's own area of study. This process has been aggravated by the typical organization of academic institutions into schools and department that are separated by physical structures and independent administrative bodies and by the current status of job markets that are more commonly seeking applicants with specialized skills than those with broader experience.

While the quality and relevance of research conducted by biologists is unquestionable, overspecialization and isolation clearly do not facilitate technical innovation and scientific breakthroughs. As a matter of fact, there are many biological laboratories that routinely use a very limited number of experimental methods and seldom venture into uncharted territories, limiting opportunities for innovation. Therefore, in such units that lack people with diverse skillsets, there is a natural tendency to formulate scientific hypotheses that could be addressed by 'technology' preferred by the investigator. This will have a central influence not only in defining the type of experiment that will be conducted, but also the type of questions asked, and hypotheses formulated. Abraham Maslow gets this point across in his law of the hammer: "I suppose it is tempting, if the only tool you have is a hammer, to treat everything as if it were a nail" [121]. Similar to how learning a new language gives the brain new tools for expression, learning and developing new technologies can greatly expand the range of questions and hypotheses put forth by the researcher while investigating a topic of interest.

6. Patents

Applications for patent have been filed for the soSPIM project (US 2016/0214107 A1, July 28, 2016) and the CTC liquid biopsy microfluidic assay (WO 2017/188890 A1, November 02 2017).

Author Contributions: G.G., C.B. and A.R. conceived and wrote the paper.

Acknowledgments: The authors acknowledge financial support for the research activities here disclosed by Mechanobiology Institute (MBI) and Pontificia Universidad Católica de Chile through internal seed grants (Fondo Semilla 2019). The publication of this review as an Open Access contribution is also supported by MBI seed grant. We thank Andrew Wong and Ravimuthula Sruthi Jagannathan of MBI' Science Communication Core for critical reading and for professional editing of the manuscript.

Conflicts of Interest: The authors declare no conflict of interest.

References

1. Novikoff, A.B. The concept of integrative levels and biology. *Science* **1945**, *101*, 209–215. [CrossRef] [PubMed]
2. Regenmortel, M.H.V.V. Reductionism and complexity in molecular biology. *EMBO Rep.* **2004**, *5*, 1016. [CrossRef]
3. Ahn, A.C.; Tewari, M.; Poon, C.-S.; Phillips, R.S. The Limits of Reductionism in Medicine: Could Systems Biology Offer an Alternative? *PLOS Med.* **2006**, *3*, e208. [CrossRef] [PubMed]
4. Hughes, J.H.; Kumar, S. Synthetic mechanobiology: Engineering cellular force generation and signaling. *Curr. Opin. Biotechnol.* **2016**, *40*, 82–89. [CrossRef] [PubMed]
5. Black, J.B.; Perez-Pinera, P.; Gersbach, C.A. Mammalian Synthetic Biology: Engineering Biological Systems. *Annu. Rev. Biomed. Eng.* **2017**, *19*, 249–277. [CrossRef]
6. Iskratsch, T.; Wolfenson, H.; Sheetz, M.P. Appreciating force and shape-the rise of mechanotransduction in cell biology. *Nat. Rev. Mol. Cell Biol.* **2014**, *15*, 825–833. [CrossRef]
7. Karczewski, K.J.; Snyder, M.P. Integrative omics for health and disease. *Nat. Rev. Genet.* **2018**, *19*, 299–310. [CrossRef]

8.   Sosa-Hernandez, J.E.; Villalba-Rodriguez, A.M.; Romero-Castillo, K.D.; Aguilar-Aguila-Isaias, M.A.; Garcia-Reyes, I.E.; Hernandez-Antonio, A.; Ahmed, I.; Sharma, A.; Parra-Saldivar, R.; Iqbal, H.M.N. Organs-on-a-Chip Module: A Review from the Development and Applications Perspective. *Micromachines* **2018**, *9*, 536. [CrossRef] [PubMed]

9.   Kimura, H.; Sakai, Y.; Fujii, T. Organ/body-on-a-chip based on microfluidic technology for drug discovery. *Drug Metab. Pharmacokinet.* **2018**, *33*, 43–48. [CrossRef] [PubMed]

10.   Zhang, C.; Zhao, Z.; Abdul Rahim, N.A.; van Noort, D.; Yu, H. Towards a human-on-chip: Culturing multiple cell types on a chip with compartmentalized microenvironments. *Lab Chip* **2009**, *9*, 3185–3192. [CrossRef]

11.   Yum, K.; Hong, S.G.; Healy, K.E.; Lee, L.P. Physiologically relevant organs on chips. *Biotechnol. J.* **2014**, *9*, 16–27. [CrossRef]

12.   Sia, D.; Llovet, J.M. Translating '–omics' results into precision medicine for hepatocellular carcinoma. *Nat. Rev. Gastroenterol. Hepatol.* **2017**, *14*, 571. [CrossRef] [PubMed]

13.   Reddy, B.; Hassan, U.; Seymour, C.; Angus, D.C.; Isbell, T.S.; White, K.; Weir, W.; Yeh, L.; Vincent, A.; Bashir, R. Point-of-care sensors for the management of sepsis. *Nat. Biomed. Eng.* **2018**, *2*, 640–648. [CrossRef]

14.   Food Drug Administration Center for Drugs Evaluation Research. *Self-MonitoringBlood Glucose Test Systems for Over-the-Counter Use*; Food Drug Administration Center for Drugs Evaluation Research: Silver Spring, MD, USA, 2016.

15.   Moore, G.E. Cramming more components onto integrated circuits, Reprinted from Electronics, volume 38, number 8, April 19, 1965, pp.114 ff. *IEEE Solid-State Circuits Soc. Newsl.* **2006**, *11*, 33–35. [CrossRef]

16.   Ronse, K. Optical lithography—A historical perspective. *C. R. Phys.* **2006**, *7*, 844–857. [CrossRef]

17.   Cerrina, F. X-ray imaging: Applications to patterning and lithography. *J. Phys. D Appl. Phys.* **2000**, *33*, R103–R116. [CrossRef]

18.   Tormen, M.; Grenci, G.; Marmiroli, B.; Romanato, F. X-ray Lithography. In *Nano-Lithography*; Landis, S., Ed.; Wiley Online Books: Hoboken, NJ, USA, 2013; Volume 1, pp. 1–81.

19.   Xia, Y.; Whitesides, G.M. Soft Lithography. *Angew. Chem.* **1998**, *37*, 550–575. [CrossRef]

20.   Duffy, D.C.; McDonald, J.C.; Schueller, O.J.; Whitesides, G.M. Rapid Prototyping of Microfluidic Systems in Poly(dimethylsiloxane). *Anal. Chem.* **1998**, *70*, 4974–4984. [CrossRef]

21.   Kane, R.S.; Takayama, S.; Ostuni, E.; Ingber, D.E.; Whitesides, G.M. Patterning proteins and cells using soft lithography. *Biomaterials* **1999**, *20*, 2363–2376. [CrossRef]

22.   Xia, Y.; Rogers, J.A.; Paul, K.E.; Whitesides, G.M. Unconventional Methods for Fabricating and Patterning Nanostructures. *Chem. Rev.* **1999**, *99*, 1823–1848. [CrossRef] [PubMed]

23.   Whitesides, G.M.; Ostuni, E.; Takayama, S.; Jiang, X.; Ingber, D.E. Soft lithography in biology and biochemistry. *Annu. Rev. Biomed. Eng.* **2001**, *3*, 335–373. [CrossRef]

24.   Verpoorte, E.; Rooij, N.F.D. Microfluidics meets MEMS. *Proc. IEEE* **2003**, *91*, 930–953. [CrossRef]

25.   Zemel, A.; Rehfeldt, F.; Brown, A.E.; Discher, D.E.; Safran, S.A. Optimal matrix rigidity for stress fiber polarization in stem cells. *Nature physics* **2010**, *6*, 468–473. [CrossRef] [PubMed]

26.   Quail, D.F.; Joyce, J.A. Microenvironmental regulation of tumor progression and metastasis. *Nat. Med.* **2013**, *19*, 1423–1437. [CrossRef]

27.   Lv, D.; Hu, Z.; Lu, L.; Lu, H.; Xu, X. Three-dimensional cell culture: A powerful tool in tumor research and drug discovery. *Oncol. Lett.* **2017**, *14*, 6999–7010. [CrossRef]

28.   Bao, M.; Xie, J.; Piruska, A.; Huck, W.T.S. 3D microniches reveal the importance of cell size and shape. *Nat. Commun.* **2017**, *8*, 1962. [CrossRef]

29.   Liu, F.D.; Tam, K.; Pishesha, N.; Poon, Z.; Van Vliet, K.J. Improving hematopoietic recovery through modeling and modulation of the mesenchymal stromal cell secretome. *Stem Cell Res. Ther.* **2018**, *9*, 268. [CrossRef]

30.   Adriani, G.; Pavesi, A.; Tan, A.T.; Bertoletti, A.; Thiery, J.P.; Kamm, R.D. Microfluidic models for adoptive cell-mediated cancer immunotherapies. *Drug Discov. Today* **2016**, *21*, 1472–1478. [CrossRef]

31.   Wang, X.; Sun, Q.; Pei, J. Microfluidic-Based 3D Engineered Microvascular Networks and Their Applications in Vascularized Microtumor Models. *Micromachines* **2018**, *9*, 493. [CrossRef] [PubMed]

32.   Jeon, J.S.; Bersini, S.; Whisler, J.A.; Chen, M.B.; Dubini, G.; Charest, J.L.; Moretti, M.; Kamm, R.D. Generation of 3D functional microvascular networks with human mesenchymal stem cells in microfluidic systems. *Integr. Biol. Quant. Biosci. Nano Macro* **2014**, *6*, 555–563. [CrossRef]

33. Bersini, S.; Miermont, A.; Pavesi, A.; Kamm, R.D.; Thiery, J.P.; Moretti, M.; Adriani, G. A combined microfluidic-transcriptomic approach to characterize the extravasation potential of cancer cells. *Oncotarget* **2018**, *9*, 36110–36125. [CrossRef]

34. Cho, Y.M.; Moon, W.K.; Kim, H.S.; Na, K.; Yang, J.H.; Huh, Y.H.; Kim, J.A.; Chung, S.; Lee, S.H. Construction of a 3D mammary duct based on spatial localization of the extracellular matrix. *NPG Asia Mater.* **2018**, *10*. [CrossRef]

35. Baker, M. A living system on a chip. *Nature* **2011**, *471*, 661. [CrossRef]

36. Willyard, C. Channeling chip power: Tissue chips are being put to the test by industry. *Nat. Med.* **2017**, *23*, 138. [CrossRef]

37. Zaretsky, I.; Polonsky, M.; Shifrut, E.; Reich-Zeliger, S.; Antebi, Y.; Aidelberg, G.; Waysbort, N.; Friedman, N. Monitoring the dynamics of primary T cell activation and differentiation using long term live cell imaging in microwell arrays. *Lab Chip* **2012**, *12*, 5007–5015. [CrossRef]

38. Jung, J.H.; Han, C.; Lee, S.A.; Kim, J.; Yang, C. Microfluidic-integrated laser-controlled microactuators with on-chip microscopy imaging functionality. *Lab Chip* **2014**, *14*, 3781–3789. [CrossRef]

39. Levario, T.J.; Zhan, M.; Lim, B.; Shvartsman, S.Y.; Lu, H. Microfluidic trap array for massively parallel imaging of Drosophila embryos. *Nat. Protoc.* **2013**, *8*, 721. [CrossRef]

40. Holzner, G.; Du, Y.; Cao, X.; Choo, J.; deMello, A.J.; Stavrakis, S. An optofluidic system with integrated microlens arrays for parallel imaging flow cytometry. *Lab Chip* **2018**, *18*, 3631–3637. [CrossRef]

41. Vaidyanathan, R.; Yeo, T.; Lim, C.T. Microfluidics for cell sorting and single cell analysis from whole blood. *Methods Cell Biol.* **2018**, *147*, 151–173. [CrossRef]

42. Wasserberg, D.; Zhang, X.; Breukers, C.; Connell, B.J.; Baeten, E.; van den Blink, D.; Èlia, S.; Bloem, A.C.; Nijhuis, M.; Wensing, A.M.J.; et al. All-printed cell counting chambers with on-chip sample preparation for point-of-care CD4 counting. *Biosens. bioelectron.* **2018**, *117*, 659–668. [CrossRef]

43. Shields, C.W., IV; Reyes, C.D.; López, G.P. Microfluidic cell sorting: A review of the advances in the separation of cells from debulking to rare cell isolation. *Lab Chip* **2015**, *15*, 1230–1249. [CrossRef] [PubMed]

44. Kung, Y.C.; Huang, K.W.; Chong, W.; Chiou, P.Y. Tunnel Dielectrophoresis for Tunable, Single-Stream Cell Focusing in Physiological Buffers in High-Speed Microfluidic Flows. *Small* **2016**, *12*, 4343–4348. [CrossRef] [PubMed]

45. Kung, Y.-C.; Huang, K.-W.; Fan, Y.-J.; Chiou, P.-Y. Fabrication of 3D high aspect ratio PDMS microfluidic networks with a hybrid stamp. *Lab Chip* **2015**, *15*, 1861–1868. [CrossRef] [PubMed]

46. Fan, Y.J.; Wu, Y.C.; Chen, Y.; Kung, Y.C.; Wu, T.H.; Huang, K.W.; Sheen, H.J.; Chiou, P.Y. Three dimensional microfluidics with embedded microball lenses for parallel and high throughput multicolor fluorescence detection. *Biomicrofluidics* **2013**, *7*, 044121. [CrossRef] [PubMed]

47. Chen, Y.; Wu, T.-H.; Kung, Y.-C.; Teitell, M.A.; Chiou, P.-Y. 3D pulsed laser-triggered high-speed microfluidic fluorescence-activated cell sorter. *Analyst* **2013**, *138*, 7308–7315. [CrossRef]

48. Chang, Y.H.; Wu, M.H. Microbioreactors for Cartilage Tissue Engineering. *Methods Mol. Biol.* **2015**, *1340*, 235–244. [CrossRef]

49. Huang, S.B.; Wang, S.S.; Hsieh, C.H.; Lin, Y.C.; Lai, C.S.; Wu, M.H. An integrated microfluidic cell culture system for high-throughput perfusion three-dimensional cell culture-based assays: Effect of cell culture model on the results of chemosensitivity assays. *Lab Chip* **2013**, *13*, 1133–1143. [CrossRef] [PubMed]

50. Abaci, H.E.; Devendra, R.; Smith, Q.; Gerecht, S.; Drazer, G. Design and development of microbioreactors for long-term cell culture in controlled oxygen microenvironments. *Biomed. Microdev.* **2012**, *14*, 145–152. [CrossRef]

51. Yager, P.; Edwards, T.; Fu, E.; Helton, K.; Nelson, K.; Tam, M.R.; Weigl, B.H. Microfluidic diagnostic technologies for global public health. *Nature* **2006**, *442*, 412–418. [CrossRef]

52. Fan, R.; Vermesh, O.; Srivastava, A.; Yen, B.K.; Qin, L.; Ahmad, H.; Kwong, G.A.; Liu, C.C.; Gould, J.; Hood, L.; et al. Integrated barcode chips for rapid, multiplexed analysis of proteins in microliter quantities of blood. *Nat. Biotechnol.* **2008**, *26*, 1373–1378. [CrossRef]

53. McHugh, K.J.; Nguyen, T.D.; Linehan, A.R.; Yang, D.; Behrens, A.M.; Rose, S.; Tochka, Z.L.; Tzeng, S.Y.; Norman, J.J.; Anselmo, A.C.; et al. Fabrication of fillable microparticles and other complex 3D microstructures. *Science* **2017**, *357*, 1138–1142. [CrossRef]

54. Rolland, J.P.; Maynor, B.W.; Euliss, L.E.; Exner, A.E.; Denison, G.M.; DeSimone, J.M. Direct Fabrication and Harvesting of Monodisperse, Shape-Specific Nanobiomaterials. *J. Am. Chem. Soc.* **2005**, *127*, 10096–10100. [CrossRef] [PubMed]

55. Birarda, G.; Ravasio, A.; Suryana, M.; Maniam, S.; Holman, H.Y.N.; Grenci, G. IR-Live: Fabrication of a low-cost plastic microfluidic device for infrared spectromicroscopy of living cells. *Lab Chip* **2016**, *16*, 1644–1651. [CrossRef] [PubMed]

56. Islam, M.; Brink, H.; Blanche, S.; DiPrete, C.; Bongiorno, T.; Stone, N.; Liu, A.; Philip, A.; Wang, G.; Lam, W.; et al. Microfluidic Sorting of Cells by Viability Based on Differences in Cell Stiffness. *Sci. Rep.* **2017**, *7*, 1997. [CrossRef] [PubMed]

57. Weibel, D.B.; DiLuzio, W.R.; Whitesides, G.M. Microfabrication meets microbiology. *Nat. Rev. Microbiol.* **2007**, *5*, 209. [CrossRef] [PubMed]

58. Rothbauer, M.; Zirath, H.; Ertl, P. Recent advances in microfluidic technologies for cell-to-cell interaction studies. *Lab Chip* **2018**, *18*, 249–270. [CrossRef] [PubMed]

59. Luo, T.; Fan, L.; Zhu, R.; Sun, D. Microfluidic Single-Cell Manipulation and Analysis: Methods and Applications. *Micromachines* **2019**, *10*, 104. [CrossRef] [PubMed]

60. Dirscherl, C.; Springer, S. Protein micropatterns printed on glass: Novel tools for protein-ligand binding assays in live cells. *Eng. Life Sci.* **2018**, *18*, 124–131. [CrossRef]

61. Von Philipsborn, A.C.; Lang, S.; Bernard, A.; Loeschinger, J.; David, C.; Lehnert, D.; Bastmeyer, M.; Bonhoeffer, F. Microcontact printing of axon guidance molecules for generation of graded patterns. *Nat. Protoc.* **2006**, *1*, 1322. [CrossRef]

62. Chen, C.S.; Mrksich, M.; Huang, S.; Whitesides, G.M.; Ingber, D.E. Geometric Control of Cell Life and Death. *Science* **1997**, *276*, 1425–1428. [CrossRef]

63. Jain, N.; Iyer, K.V.; Kumar, A.; Shivashankar, G.V. Cell geometric constraints induce modular gene-expression patterns via redistribution of HDAC3 regulated by actomyosin contractility. *Proc. Natl. Acad. Sci. USA* **2013**, *110*, 11349. [CrossRef]

64. Sahni, G.; Yuan, J.; Toh, Y.-C. Stencil Micropatterning of Human Pluripotent Stem Cells for Probing Spatial Organization of Differentiation Fates. *J. Vis. Exp. JoVE* **2016**. [CrossRef] [PubMed]

65. Gao, Y.; Tian, J.; Wu, J.; Cao, W.; Zhou, B.; Shen, R.; Wen, W. Digital microfluidic programmable stencil (dMPS) for protein and cell patterning. *RSC Adv.* **2016**, *6*, 101760–101769. [CrossRef]

66. Wright, D.; Rajalingam, B.; Selvarasah, S.; Dokmeci, M.R.; Khademhosseini, A. Generation of static and dynamic patterned co-cultures using microfabricated parylene-C stencils. *Lab Chip* **2007**, *7*, 1272–1279. [CrossRef] [PubMed]

67. Masters, T.; Engl, W.; Weng, Z.L.; Arasi, B.; Gauthier, N.; Viasnoff, V. Easy Fabrication of Thin Membranes with Through Holes. Application to Protein Patterning. *PLoS ONE* **2012**, *7*, e44261. [CrossRef] [PubMed]

68. Azioune, A.; Carpi, N.; Tseng, Q.; Théry, M.; Piel, M. Protein Micropatterns: A Direct Printing Protocol Using Deep UVs. In *Methods in Cell Biology*; Cassimeris, L., Tran, P., Eds.; Academic Press: Waltham, MA, USA, 2010; Chapter 8; Volume 97, pp. 133–146.

69. Azioune, A.; Storch, M.; Bornens, M.; Théry, M.; Piel, M. Simple and rapid process for single cell micro-patterning. *Lab Chip* **2009**, *9*, 1640–1642. [CrossRef] [PubMed]

70. Lin, L.; Chu, Y.-S.; Thiery, J.P.; Lim, C.T.; Rodriguez, I. Microfluidic cell trap array for controlled positioning of single cells on adhesive micropatterns. *Lab Chip* **2013**, *13*, 714–721. [CrossRef] [PubMed]

71. Gupta, M.; Kocgozlu, L.; Sarangi, B.R.; Margadant, F.; Ashraf, M.; Ladoux, B. Micropillar substrates: A tool for studying cell mechanobiology. *Methods Cell Biol.* **2015**, *125*, 289–308. [CrossRef]

72. Ravasio, A.; Vaishnavi, S.; Ladoux, B.; Viasnoff, V. High-resolution imaging of cellular processes across textured surfaces using an indexed-matched elastomer. *Acta Biomater.* **2015**, *14*, 53–60. [CrossRef]

73. Moe, A.A.; Suryana, M.; Marcy, G.; Lim, S.K.; Ankam, S.; Goh, J.Z.; Jin, J.; Teo, B.K.; Law, J.B.; Low, H.Y.; et al. Microarray with micro- and nano-topographies enables identification of the optimal topography for directing the differentiation of primary murine neural progenitor cells. *Small* **2012**, *8*, 3050–3061. [CrossRef]

74. Yim, E.K.; Darling, E.M.; Kulangara, K.; Guilak, F.; Leong, K.W. Nanotopography-induced changes in focal adhesions, cytoskeletal organization, and mechanical properties of human mesenchymal stem cells. *Biomaterials* **2010**, *31*, 1299–1306. [CrossRef]

75. Migliorini, E.; Grenci, G.; Ban, J.; Pozzato, A.; Tormen, M.; Lazzarino, M.; Torre, V.; Ruaro, M.E. Acceleration of neuronal precursors differentiation induced by substrate nanotopography. *Biotechnol. Bioeng.* **2011**, *108*, 2736–2746. [CrossRef]

76. Sperling, L.E.; Reis, K.P.; Pozzobon, L.G.; Girardi, C.S.; Pranke, P. Influence of random and oriented electrospun fibrous poly(lactic-co-glycolic acid) scaffolds on neural differentiation of mouse embryonic stem cells. *J. Biomed. Mater. Res. Part A* **2017**, *105*, 1333–1345. [CrossRef]

77. Abagnale, G.; Steger, M.; Nguyen, V.H.; Hersch, N.; Sechi, A.; Joussen, S.; Denecke, B.; Merkel, R.; Hoffmann, B.; Dreser, A.; et al. Surface topography enhances differentiation of mesenchymal stem cells towards osteogenic and adipogenic lineages. *Biomaterials* **2015**, *61*, 316–326. [CrossRef] [PubMed]

78. Ravasio, A.; Cheddadi, I.; Chen, T.; Pereira, T.; Ong, H.T.; Bertocchi, C.; Brugues, A.; Jacinto, A.; Kabla, A.J.; Toyama, Y.; et al. Gap geometry dictates epithelial closure efficiency. *Nat. Commun.* **2015**, *6*, 7683. [CrossRef] [PubMed]

79. Chen, T.; Callan-Jones, A.; Fedorov, E.; Ravasio, A.; Brugués, A.; Ong, H.T.; Toyama, Y.; Low, B.C.; Trepat, X.; Shemesh, T.; et al. Large-scale curvature sensing by directional actin flow drives cellular migration mode switching. *Nat. Phys.* **2019**. [CrossRef] [PubMed]

80. Xi, W.; Sonam, S.; Beng Saw, T.; Ladoux, B.; Teck Lim, C. Emergent patterns of collective cell migration under tubular confinement. *Nat. Commun.* **2017**, *8*, 1517. [CrossRef] [PubMed]

81. Vedula, S.R.K.; Ravasio, A.; Anon, E.; Chen, T.; Peyret, G.; Ashraf, M.; Ladoux, B. Microfabricated Environments to Study Collective Cell Behaviors. In *Methods in Cell Biolology*; Piel, M., Théry, M., Eds.; Academic Press: Waltham, MA, USA, 2014; Chapter 16; Volume 120, pp. 235–252.

82. Gao, X.; Stoecklin, C.; Zhang, Y.; Weng, Z.; De Mets, R.; Grenci, G.; Viasnoff, V. Artificial Microniche Array with Spatially Structured Biochemical Cues. *Methods Mol. Biol.* **2018**, *1771*, 55–66. [CrossRef]

83. Folch, A.; Jo, B.-H.; Hurtado, O.; Beebe, D.J.; Toner, M. Microfabricated elastomeric stencils for micropatterning cell cultures. *J. Biomed. Mater. Res.* **2000**, *52*, 346–353. [CrossRef]

84. Huisken, J.; Swoger, J.; Del Bene, F.; Wittbrodt, J.; Stelzer, E.H.K. Optical Sectioning Deep Inside Live Embryos by Selective Plane Illumination Microscopy. *Science* **2004**, *305*, 1007. [CrossRef]

85. Planchon, T.A.; Gao, L.; Milkie, D.E.; Davidson, M.W.; Galbraith, J.A.; Galbraith, C.G.; Betzig, E. Rapid three-dimensional isotropic imaging of living cells using Bessel beam plane illumination. *Nat. Methods* **2011**, *8*, 417–423. [CrossRef] [PubMed]

86. Gebhardt, J.C.; Suter, D.M.; Roy, R.; Zhao, Z.W.; Chapman, A.R.; Basu, S.; Maniatis, T.; Xie, X.S. Single-molecule imaging of transcription factor binding to DNA in live mammalian cells. *Nat. Methods* **2013**, *10*, 421–426. [CrossRef] [PubMed]

87. Chen, B.C.; Legant, W.R.; Wang, K.; Shao, L.; Milkie, D.E.; Davidson, M.W.; Janetopoulos, C.; Wu, X.S.; Hammer, J.A., 3rd; Liu, Z.; et al. Lattice light-sheet microscopy: Imaging molecules to embryos at high spatiotemporal resolution. *Science* **2014**, *346*, 1257998. [CrossRef] [PubMed]

88. Galland, R.; Grenci, G.; Aravind, A.; Viasnoff, V.; Studer, V.; Sibarita, J.B. 3D high- and super-resolution imaging using single-objective SPIM. *Nat. Methods* **2015**, *12*, 641–644. [CrossRef] [PubMed]

89. Singh, A.P.; Galland, R.; Finch-Edmondson, M.L.; Grenci, G.; Sibarita, J.-B.; Studer, V.; Viasnoff, V.; Saunders, T.E. 3D Protein Dynamics in the Cell Nucleus. *Biophys. J.* **2017**, *112*, 133–142. [CrossRef]

90. Mantsch, H.H.; Chapman, A.R. *Infrared Spectroscopy of Biomolecules*; Wiley-Liss: New York, NY, USA, 1996; 359p.

91. Nasse, M.J.; Walsh, M.J.; Mattson, E.C.; Reininger, R.; Kajdacsy-Balla, A.; Macias, V.; Bhargava, R.; Hirschmugl, C.J. High-resolution Fourier-transform infrared chemical imaging with multiple synchrotron beams. *Nat. Methods* **2011**, *8*, 413. [CrossRef]

92. Kröger-Lui, N.; Gretz, N.; Haase, K.; Kränzlin, B.; Neudecker, S.; Pucci, A.; Regenscheit, A.; Schönhals, A.; Petrich, W. Rapid identification of goblet cells in unstained colon thin sections by means of quantum cascade laser-based infrared microspectroscopy. *Analyst* **2015**, *140*, 2086–2092. [CrossRef]

93. Mitri, E.; Birarda, G.; Vaccari, L.; Kenig, S.; Tormen, M.; Grenci, G. SU-8 bonding protocol for the fabrication of microfluidic devices dedicated to FTIR microspectroscopy of live cells. *Lab Chip* **2014**, *14*, 210–218. [CrossRef]

94. Marcsisin, E.J.; Uttero, C.M.; Miljković, M.; Diem, M. Infrared microspectroscopy of live cells in aqueous media. *Analyst* **2010**, *135*, 3227–3232. [CrossRef]

95. Holman, H.-Y.N.; Bechtel, H.A.; Hao, Z.; Martin, M.C. Synchrotron IR Spectromicroscopy: Chemistry of Living Cells. *Anal. Chem.* **2010**, *82*, 8757–8765. [CrossRef]

96. Nasse, M.J.; Ratti, S.; Giordano, M.; Hirschmugl, C.J. Demountable Liquid/Flow Cell for in vivo Infrared Microspectroscopy of Biological Specimens. *Appl. Spectrosc.* **2009**, *63*, 1181–1186. [CrossRef] [PubMed]

97. Birarda, G.; Bedolla, D.E.; Mitri, E.; Pacor, S.; Grenci, G.; Vaccari, L. Apoptotic pathways of U937 leukemic monocytes investigated by infrared microspectroscopy and flow cytometry. *Analyst* **2014**, *139*, 3097–3106. [CrossRef] [PubMed]

98. Bedolla, D.E.; Kenig, S.; Mitri, E.; Ferraris, P.; Marcello, A.; Grenci, G.; Vaccari, L. Determination of cell cycle phases in live B16 melanoma cells using IRMS. *Analyst* **2013**, *138*, 4015–4021. [CrossRef]

99. Sandt, C.; Frederick, J.; Dumas, P. Profiling pluripotent stem cells and organelles using synchrotron radiation infrared microspectroscopy. *J. Biophotonics* **2013**, *6*, 60–72. [CrossRef]

100. Hoffner, G.; André, W.; Sandt, C.; Djian, P. Synchrotron-based infrared spectroscopy brings to light the structure of protein aggregates in neurodegenerative diseases. *Rev. Anal. Chem.* **2014**, *33*, 231. [CrossRef]

101. Miller, L.M.; Bourassa, M.W.; Smith, R.J. FTIR spectroscopic imaging of protein aggregation in living cells. *Biochim. Biophys. Acta (BBA) Biomembr.* **2013**, *1828*, 2339–2346. [CrossRef] [PubMed]

102. Tobin, M.J.; Puskar, L.; Barber, R.L.; Harvey, E.C.; Heraud, P.; Wood, B.R.; Bambery, K.R.; Dillon, C.T.; Munro, K.L. FTIR spectroscopy of single live cells in aqueous media by synchrotron IR microscopy using microfabricated sample holders. *Vib. Spectrosc.* **2010**, *53*, 34–38. [CrossRef]

103. Suryana, M.; Shanmugarajah, J.V.; Maniam, S.M.; Grenci, G. Soft Lithographic Procedure for Producing Plastic Microfluidic Devices with View-ports Transparent to Visible and Infrared Light. *JoVE* **2017**. [CrossRef] [PubMed]

104. Vaccari, L.; Birarda, G.; Businaro, L.; Pacor, S.; Grenci, G. Infrared Microspectroscopy of Live Cells in Microfluidic Devices (MD-IRMS): Toward a Powerful Label-Free Cell-Based Assay. *Anal. Chem.* **2012**, *84*, 4768–4775. [CrossRef]

105. Birarda, G.; Grenci, G.; Businaro, L.; Marmiroli, B.; Pacor, S.; Vaccari, L. Fabrication of a microfluidic platform for investigating dynamic biochemical processes in living samples by FTIR microspectroscopy. *Microelectron. Eng.* **2010**, *87*, 806–809. [CrossRef]

106. Valastyan, S.; Weinberg, R.A. Tumor metastasis: Molecular insights and evolving paradigms. *Cell* **2011**, *147*, 275–292. [CrossRef] [PubMed]

107. Fehm, T.; Braun, S.; Muller, V.; Janni, W.; Gebauer, G.; Marth, C.; Schindlbeck, C.; Wallwiener, D.; Borgen, E.; Naume, B.; et al. A concept for the standardized detection of disseminated tumor cells in bone marrow from patients with primary breast cancer and its clinical implementation. *Cancer* **2006**, *107*, 885–892. [CrossRef]

108. Nieva, J.J.; Kuhn, P. Fluid biopsy for solid tumors: A patient's companion for lifelong characterization of their disease. *Future Oncol.* **2012**, *8*, 989–998. [CrossRef] [PubMed]

109. Goldkorn, A.; Ely, B.; Quinn, D.I.; Tangen, C.M.; Fink, L.M.; Xu, T.; Twardowski, P.; Van Veldhuizen, P.J.; Agarwal, N.; Carducci, M.A.; et al. Circulating tumor cell counts are prognostic of overall survival in SWOG S0421: A phase III trial of docetaxel with or without atrasentan for metastatic castration-resistant prostate cancer. *J. Clin. Oncol. Off. J. Am. Soc. Clin. Oncol.* **2014**, *32*, 1136–1142. [CrossRef] [PubMed]

110. Cui, L.; Kwong, J.; Wang, C.C. Prognostic value of circulating tumor cells and disseminated tumor cells in patients with ovarian cancer: A systematic review and meta-analysis. *J. Ovarian Res.* **2015**, *8*, 38. [CrossRef] [PubMed]

111. Van der Toom, E.E.; Verdone, J.E.; Gorin, M.A.; Pienta, K.J. Technical challenges in the isolation and analysis of circulating tumor cells. *Oncotarget* **2016**, *7*, 62754–62766. [CrossRef]

112. Zhang, L.; Ridgway, L.D.; Wetzel, M.D.; Ngo, J.; Yin, W.; Kumar, D.; Goodman, J.C.; Groves, M.D.; Marchetti, D. The Identification and Characterization of Breast Cancer CTCs Competent for Brain Metastasis. *Sci. Transl. Med.* **2013**, *5*, 180ra148. [CrossRef]

113. Yu, M.; Bardia, A.; Aceto, N.; Bersani, F.; Madden, M.W.; Donaldson, M.C.; Desai, R.; Zhu, H.; Comaills, V.; Zheng, Z.; et al. Ex vivo culture of circulating breast tumor cells for individualized testing of drug susceptibility. *Science* **2014**, *345*, 216. [CrossRef] [PubMed]

114. Gao, D.; Vela, I.; Sboner, A.; Iaquinta, P.J.; Karthaus, W.R.; Gopalan, A.; Dowling, C.; Wanjala, J.N.; Undvall, E.A.; Arora, V.K.; et al. Organoid cultures derived from patients with advanced prostate cancer. *Cell* **2014**, *159*, 176–187. [CrossRef]

115. Maheswaran, S.; Haber, D.A. Ex Vivo Culture of CTCs: An Emerging Resource to Guide Cancer Therapy. *Cancer Res.* **2015**. [CrossRef] [PubMed]

116. Khoo, B.L.; Lee, S.C.; Kumar, P.; Tan, T.Z.; Warkiani, M.E.; Ow, S.G.; Nandi, S.; Lim, C.T.; Thiery, J.P. Short-term expansion of breast circulating cancer cells predicts response to anti-cancer therapy. *Oncotarget* **2015**, *6*, 15578–15593. [CrossRef] [PubMed]

117. Khoo, B.L.; Grenci, G.; Jing, T.; Lim, Y.B.; Lee, S.C.; Thiery, J.P.; Han, J.; Lim, C.T. Liquid biopsy and therapeutic response: Circulating tumor cell cultures for evaluation of anticancer treatment. *Sci. Adv.* **2016**, *2*, e1600274. [CrossRef] [PubMed]

118. Khoo, B.L.; Grenci, G.; Lim, Y.B.; Lee, S.C.; Han, J.; Lim, C.T. Expansion of patient-derived circulating tumor cells from liquid biopsies using a CTC microfluidic culture device. *Nat. Protoc.* **2018**, *13*, 34–58. [CrossRef] [PubMed]

119. Khoo, B.L.; Grenci, G.; Lim, J.S.Y.; Lim, Y.P.; Fong, J.; Yeap, W.H.; Bin Lim, S.; Chua, S.L.; Wong, S.C.; Yap, Y.-S.; et al. Low-dose anti-inflammatory combinatorial therapy reduced cancer stem cell formation in patient-derived preclinical models for tumour relapse prevention. *Br. J. Cancer* **2019**, *120*, 407–423. [CrossRef] [PubMed]

120. Lee, J.-H.; Choi, W.-S.; Lee, K.-H.; Yoon, J.-B. A simple and effective fabrication method for various 3D microstructures: Backside 3D diffuser lithography. *J. Micromech. Microeng.* **2008**, *18*, 125015. [CrossRef]

121. Maslow, A.H. *The Psychology of Science*; Harper & Row: Manhattan, NY, USA, 1966.

micromachines

MDPI

Article

An Automated and Miniaturized Rotating-Disk Device for Rapid Nucleic Acid Extraction

Rui Tong [1], Lijuan Zhang [2], Chuandeng Hu [3], Xuee Chen [1], Qi Song [4], Kai Lou [2], Xin Tang [4], Yongsheng Chen [5], Xiuqing Gong [6], Yibo Gao [2,*] and Weijia Wen [3,*]

[1] The Nano Science and Technology (NSNT) Program, The Hong Kong University of Science and Technology, Clear Water Bay, Kowloon, Hong Kong; rtongaa@connect.ust.hk (R.T.); xchendi@connect.ust.hk (X.C.)

[2] Shenzhen Shineway Hi-Tech Co., Ltd., Shenzhen 518112, China; zlj@swtech.me (L.Z.); lk@swtech.me (K.L.)

[3] Department of Physics, The Hong Kong University of Science and Technology, Clear Water Bay, Kowloon, Hong Kong; chuae@connect.ust.hk

[4] Guangzhou HKUST Fok Ying Tung Research Institute, Nansha, Guangzhou 511458, China; songqi2712@gmail.com (Q.S.); xintang@ust.hk (X.T.)

[5] Department of Ocean Science, The Hong Kong University of Science and Technology, Clear Water Bay, Kowloon, Hong Kong; yschen@connect.ust.hk

[6] Materials Genome Institute, University of Shanghai, Shanghai 200444, China; gongxiuqing@shu.edu.cn

* Correspondences: gyb@swtech.me (Y.G.); phwen@ust.hk (W.W.);
   Tel.: +86-755-8958-7395 (Y.G.); +852-2358-7979 (W.W.)

Received: 7 February 2019; Accepted: 20 March 2019; Published: 22 March 2019

Abstract: The result of molecular diagnostic and detection greatly dependent on the quality and integrity of the isolated nucleic acid. In this work, we developed an automated miniaturized nucleic acid extraction device based on magnetic beads method, consisting of four components including a sample processing disc and its associated rotary power output mechanism, a pipetting module, a magnet module and an external central controller to enable a customizable and automated robust nucleic acid sample preparation. The extracted nucleic acid using 293T cells were verified using real-time polymerase chain reaction (PCR) and the data implies a comparable efficiency to a manual process, with the advantages of performing a flexible, time-saving (~10 min), and simple nucleic acid sample preparation.

Keywords: sample preparation; nucleic acid; DNA; RNA

---

1. Introduction

With the rapid development of molecular biology technology in recent years, molecular diagnostic and detection technologies represented by nucleic acid hybridization, nucleic acid amplification and nucleic acid sequence analysis have become increasingly significant in many fields. However, the fundamental challenge facing all modern molecular biology detection techniques, such as polymerase chain reaction (PCR), high-throughput sequencing, etc. is how to promptly and efficiently separate and extract the required genomic nucleic acid from complex and diverse biological samples, since the quality and integrity (state of degradation) of the isolated nucleic acid directly affects the subsequent experimental results [1]. At present, researchers all over the world have made many breakthroughs in the technology of nucleic acid separation and extraction.

Nucleic acids are broadly classified into deoxyribonucleic acid (DNA) and ribonucleic acid (RNA). Ever since it was first discovered in 1869, many researchers have made unremitting efforts in the extraction of nucleic acids, and have improved various materials and reagents for nucleic acid extraction. Milestone research findings includes: phenol extraction technique [2], sodium dodecyl sulfate (SDS) method [3], and acid guanidinium phenolchoroform (AGPC) extraction technique [4,5].

Numerous well-known biological reagent companies have developed various nucleic acid extraction kits based on these conventional nucleic acid extraction methods for the separation and extraction of DNA and RNA from a wide variety of tissue samples. The conventional methods of nucleic acid extraction often include precipitation and centrifugation, which require an extensive number of steps, and thus are complicated, time-consuming (requiring up to 3 h, and much longer if incubated overnight) [6], and difficult to achieve miniaturized automation. Most of the methods require operators to be in direct contact with toxic chemical reagents. Therefore, with the rapid development of molecular biology and polymer materials science, the conventional method of separating and extracting nucleic acids from liquid phase systems has been gradually replaced by new methods based on solid phase adsorbate carriers [7,8]. Such emerging nucleic acid separation and extraction methods mainly include: Glass particles method [9], silica matrices method [10,11], anion exchange method [12], and magnetic beads-based extraction method [13]. Regardless of which method is used to separate and extract nucleic acids, in general, the operation steps of such methods can be mainly divided into four parts [7,8,14]. The first part is to use the lysis to promote cell disruption and release the nucleic acids. The second part is to specifically adsorb the released nucleic acids on a specific carrier, with this specific carrier has strong affinity and adsorption only for nucleic acids, but has no affinity for other biochemical components such as proteins, polysaccharides, and lipids. The third part is to wash with a specific washing buffer to remove non-nucleic acid impurities, and the last part is to elute the nucleic acid adsorbed on the specified carrier to obtain purified nucleic acid [15].

The extraction of nucleic acids by the spin column-based method has been widely used, and most of the plasmid DNA extraction kits on the market have been developed based on the spin column-based method. The method adopts a special silicon matrix adsorption material, which is characterized as follows: In the presence of a high hydrochloric acid buffer, the DNA can be specifically adsorbed, the impurities can be removed with a series of washing steps, and the low-salt alkaline buffer can elute the DNA bound to the adsorption column [14]. However, the disadvantage of this method is that the sample required is large, thus consuming a lot of samples. Furthermore, the application of this method on some rare samples is greatly limited. At the same time, the spin column method requires repeated centrifugation during the process, which is not suitable for a high-throughput, automated operation. Especially in the field of genetic diagnosis, monitoring and control of sudden outbreaks, the use of the spin column-based method to extract nucleic acids requires a large number of operators and equipment to meet the demand. Since the 1990s, due to various deficiencies in the spin column method, in order to adapt to the high-throughput, high-sensitivity, and automated operation requirements of modern molecular biology testing experiments, the method of extracting nucleic acids using magnetic beads emerged [16]. This method is the perfect combination of nanotechnology and biotechnology since magnetic beads are high-affinity composite magnetic microspheres (typically 1 to 100 nm) formed by combining an inorganic magnetic particle with a polymeric material. This method advantages that other nucleic acid extraction methods cannot match, which are mainly reflected in: (1) It can realize high-throughput operation and automation; (2) The operation is simple and time-saving, the entire extraction process consists only four steps, and the whole process can be completed within 40 min; (3) It is safe and non-toxic; (4) The specific binding of magnetic beads and nucleic acid makes the extracted nucleic acid high in purity and concentration; (5) Low in cost and can be applied in a wide number of applications. Since magnetic bead synthesis uses low-cost inorganic and organic raw materials, no special equipment is required, which makes the final synthesis, research, and development costs very low [17,18]. A major improvement in nucleic acid extraction by magnetic beads is the use of a high-affinity composite magnetic microsphere, formed by the combination of inorganic magnetic particles and polymer materials. Because of its many properties of polymer microspheres and magnetic particles, it is uniformly and stably dispersed in the solution without an external magnetic field, and can be easily and quickly separated once the external magnetic field is added.

Microfluidic technology originated from the concept of a micro-electro-mechanical system based on MEMS technology, as proposed by Manz et al. in the early 1990s [19]. The purpose of this system is to transfer functions of the lab onto portable devices, and even chips, through miniaturization and integration of chemical analysis systems. The core of this technology is microfluidic chips, in which a series of microchannels are fabricated on a small chip. Through the manipulation and control of the microfluids in the microchannel, the entire chemical and biological laboratory functions are realized. Several microfluidic chip nucleic acid extraction techniques have been reported to isolate target cells from blood samples by using micromachined 'weir-type' filters [20], pore filters [21], or pathogen-specific immunomagnetic beads [22]. The captured target cells are introduced into a PCR reaction chamber on a chip, and the DNA is released by cell thermal lysis. However, since the mixture present in the cell debris may inhibit the PCR process, most of the microfluidic chips reported for DNA extraction require preliminary off-chip sample processing steps. For DNA extraction on the chip, the most common method currently used is to extract DNA from cell lysates using magnetic beads coated with silica or functional groups (carboxy [22], amine [23], biotin [24], nucleotide probes [25]). In addition, there have been successful studies on dielectrophoretic trapping [26] and isotachophoresis [27] on DNA purification microfluidic chips.

Although most developments in the field of molecular diagnostics have focused on improving methods for detecting and identifying disease-related target analytes, less attention has been devoted to developing systems for purifying samples. Our goal is to develop a customizable, automated, and miniaturized system for nucleic acid sample preparation based on the magnetic beads method. In this work, we developed an automated miniaturized nucleic acid extraction device consisting of four components including a sample processing disc and its associated rotary power output mechanism, a pipetting module, a magnet module, and an external central controller. The device was designed discreetly to ensure its portability ($220 \times 165 \times 210$ mm dimension and 3 kg weight), efficiency, and easiness to operate. Our platform integrates the functions of lysis, binding, washing and elution, and can be customized according to different extraction protocols to meet various user demands. We analyzed the performance of the device using 293T cells, the extracted nucleic acids were detected using real-time PCR and verified by electrophoresis. Manual process of nucleic acid extraction usually requires 40 to 50 min, and the previously reported microchip-based DNA extraction costs 12 min [28] to 15 min [22], our device further curtailed the process duration within 10 min. The commercialized automated nucleic acid extraction instruments (e.g., QIAcube from QIAGEN, Hilden, Germany) and liquid handling robots or platforms (e.g., QIAgility from QIAGEN) have similar functions as our device but with a much bulkier size and heavier weight (QIAcube is 71.5 kg and QIAgility is 41 kg). The small dimension, light weight, and rapid processing time makes our device suitable for future use in point-of-care testing.

2. Materials and Methods

2.1. Design Concept

Compared with the conventional nucleic acid extraction method, magnetic beads extraction and purification of nucleic acid has incomparable advantages. The experiment process is simple, easy to operate, and can save time since the time-consuming centrifugation and precipitation steps required in the conventional methods are not required when using magnetic particles. Moreover, this system can be used on both manual and automated processes. The process is also safe and nontoxic, since it does not use the toxic reagents such as benzene and chloroform required in the conventional methods. The specific binding of magnetic particles and nucleic acid makes the extracted nucleic acid high in purity and concentration. Magnetic bead nucleic acid extraction can generally be divided into four steps: Lysis, binding, washing, and elution. Based on the process flow of the magnetic bead nucleic acid extraction as shown in Figure 1, we designed and fabricated an automated miniaturized device in realizing of the automation of all the steps and functions, thereby reducing manpower and improving purification efficiency.

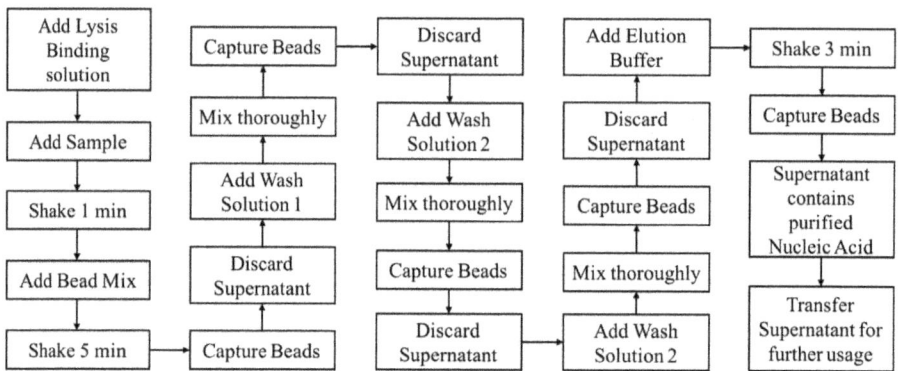

Figure 1. Protocol for nucleic acid extraction using the magnetic bead method.

2.2. Configuration and Mechanism of the Device

The device consists of a sample processing disc and its associated rotary power output mechanism, a pipetting module, a magnet module, and an external central controller to control the execute commands of all modules in tandem or in parallel. The schematic diagram of the device is shown in Figure 1. The sample processing disc, as shown is Figure 2A, has numerous round holes of different sizes on the outermost circumference. In particular, holes with the smallest size are for pipette tip placement, holes with the largest size are for the placement of 1.5 mL microcentrifuge tubes, and there is a medium sized hole with a circumference slightly greater than the maximum circumference of the pipette tip, which is employed for the disposal of used pipette tips and reagent waste. The rectangular hole is for placing the microfluidic chip, which is used for further PCR. The rotary power output mechanism is located underneath the sample processing disc and is connected to its center to regulate the rotating angle of the disc, as shown in Figure 2B. The pipetting module comprises of a shaft for holding pipette tips, a tip ejector arm for removal of tips, and a pump used to draw up or dispense the liquid from the disposable pipette tip. In addition, two stepping motors each connect to a slip belt (Figure 2C) are coupled to the external central controller separately to drive the shaft and the pump to move vertically relative to the horizontal plane. The magnet module includes a cone magnet (Figure 2A,B) colored in purple) to supply an external magnetic field, and a servo motor connects to the external central controller to dominate the presence and absence of the magnetic field. Furthermore, the sample processing disc, the rotary power output mechanism, the pipetting module, and the magnet module are integrated into a portable box with a dimension of 220 × 165 × 210 mm, as shown in Figure 2D, and the whole device has a weight of approximately 3 kg. Inside the portable box, a UV light controlled by the external controller is located on top to expose and decontaminate the device, therefore eliminating contamination between runs.

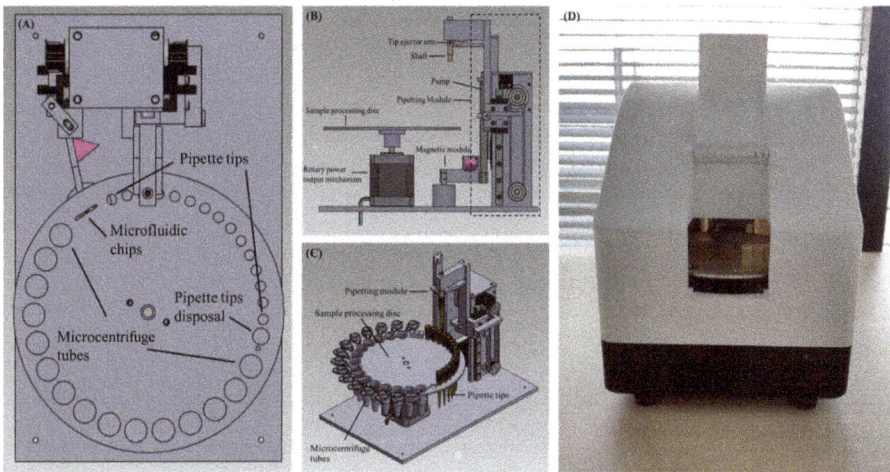

Figure 2. Schematic diagram of the automated miniaturized rotary sample preparation device. (**A–C**) Top, front, side and isometric view of the device. (**D**) Actual image of the device with all modules integrated in a portable box.

2.3. Experiment Process

As described in the previous section, the sample processing disc has various sizes of round holes located on the outermost circumference. For experimental purposes, we assigned labels for each hole according to their sizes, as shown in Figure 3A. The smallest holes for placing pipette tips were labeled S, the medium sized hole for waste disposal was labeled M, and the largest holes for placing microcentrifuge tubes were labeled L. Prior to starting the experiment, we were required to add each reagent with a specified volume to the microcentrifuge tubes and then place them on the predetermined labeled holes, as shown in the initial setup of Figure 3B. The cell sample we used was 293T cells cultured in Dulbecco's modified Eagle's medium (DMEM; Gibco, Carlsbad, CA, USA) with 10% fetal bovine serum (FBS; Gibco, USA) and 1% Penicillin/Streptomycin at 37 °C in 5% CO_2 with an initial amount of 2×10^6 cells. The software program in the external central controller, which controls each module independently had been pre-set in accordance with the protocol of this experiment, and the program could be conveniently customized if using different protocols or kit. After placing all the required pipette tips and microcentrifuge tubes on the corresponding holes, the sample preparation process could be started by simply pressing the start button on the external central controller. The entire process could then be automatically conducted by the pre-set program. The detail processes ran by the device are described below, and the process flow is shown in Figure 3B. First, the sample processing disc rotated so that S1 was right underneath the shaft, the shaft moved vertically down to take in a pipette tip and moved upward after. To transfer reagent from L2 to L1, the sample processing disc rotated so that L2 was underneath the shaft, the shaft moved downward to a position where the pipette tip was merged in the reagent and the pump moved upward to aspirate the reagent into the pipette tip. The sample processing disc then rotated to L1, the shaft and pump moved downward to dispense the reagent, the pump then moved up and down repeatedly to aspirate and dispense the liquid to mix the reagents thoroughly. The used pipette tip must be changed before next step, thus the sample processing disc rotated so that M1 was underneath the shaft, and the shaft moved downward to trigger the tip ejector arm to dispose the used pipette tip and a waste container was located beneath to store used tips and reagent waste. After a new tip was introduced from S2, liquid in L4 was transferred to L3 and mixed thoroughly. Following this, the used tip was ejected through M1, a new was taken tip from S3 and liquid was transferred from L1 to L9, and then from L3 to L9 and mixed thoroughly. The magnet was then shifted to L9 to apply the magnetic field in order to magnetically capture the binding beads, with the magnetic field on, the shaft moved downward to aspirate the supernatant then discarded the used tip along with the supernatant into the waste container through M1. A new tip was taken from S4, and liquid was transferred from L5 to L9 and mixed thoroughly. Following this, the magnetic field was applied to L9 and the supernatant was aspirated and disposed. A new tip was taken from S5, and liquid was transferred from L6 to L9 and the previous steps were repeated. After this, a new tip was taken from S6 and liquid was transferred from L7 to L9 with the process described above repeated again. Until this point in the process, all the reagent additions, mixing, and washing steps required in this protocol had been completed by the system. Finally, a new tip was taken from S7 and the elution buffer located in L8 to L9 was transferred and mixed thoroughly. The used tip was then ejected through M1 and a new tip was taken from S8, a magnetic field was applied to L9, and the supernatant was aspirated, the purified RNA was also in the supernatant for further downstream applications. During this experiment, we dispensed the supernatant, which contained purified RNA in an empty microcentrifuge tube placed on L10 for additional RNA reverse transcription and PCR verification experiments. Furthermore, the sample acquired could be directly dispensed into the microfluidic chip placed on the sample processing disc for downstream experiments in the future.

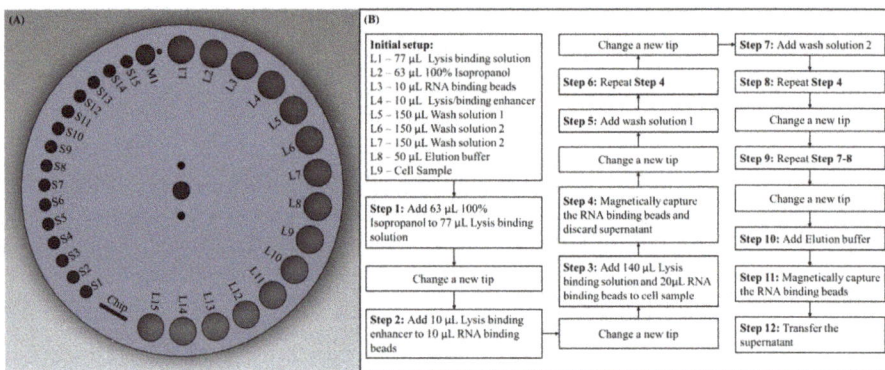

Figure 3. Experiment setup and procedure. (**A**) Detailed labels of each placement holes on the sample processing disc. (**B**) Process flow for automated sample preparation.

However, the process flow shown in Figure 3B is designed in accordance with the reagents and protocol we chose in our experiment. It can be easily seen that we had only occupied L1 to L9 and S1 to S8, there were still many placement holes left for use if the protocol required more reagents. Since each step of the experiment consisted of many micro steps preprogrammed in the external controller, the whole experiment process could be easily customized to meet different demands. For example, in step 3, in order to add 140 μL lysis binding solution in L1 to cell sample L9, the sample processing disc was programmed to rotate so that L1 was under the shaft, the shaft then moved vertically down until the pipette tip was merged in the reagent. Following this, the pump moved upward to aspirate the liquid into the pipette, and the shaft moved back up so that the sample processing disc could be rotated until L9 was under the shaft. Finally, the shaft moved downward and so did the pump to dispense the reagent in L9. The rotation degree of the sample processing disc, moving distance of the shaft, and the pump were all controlled by the external central controller by entering a preset value. Therefore, if the user would like to transfer reagent from L10 to L11, the user could simply program it in the external central controller. If the user only needed two washing steps, he or she can easily delete the program relating to step 9 (in Figure 3B) in the programmable controller.

2.4. Microchip Fabrication

The microchip was fabricated by standard lithography technology, as shown in Figure 4. Following this, 4-inch double sided polished silicon wafers were cleaned in piranha solution and hydrofluoric acid solution. The photolithographically patterned substrate first subjected to descum process in O_2 plasma. Then, induced coupled plasma-deep reaction ion etching (ICP-DRIE) was applied to etch silicon to form wells and channels for carrying out real-time PCR (RT-PCR) reactions. After photoresist was stripped by plasma, the wafer grown thermal oxide with a thickness of 1000 Å as a passivation layer used to avoid non-specific adsorption of PCR components. The oxide wafer was bonded with a 4-inch glass wafer by anodic bonding to form a closed PCR chamber and finally the wafer was diced into individual microchips. It is worth noting that we designed the chip inlet and outlet on the side of the chip rather than its top and the inlet and outlet port could be opened just after wafer dicing. This design could avoid drilling or punching processes of the inlet and outlet port, as well as making it easier for our device to add extracted nucleic acid into the microchip. The microchip designed for this experiment has three reaction chambers each had a reaction volume of 2 μL.

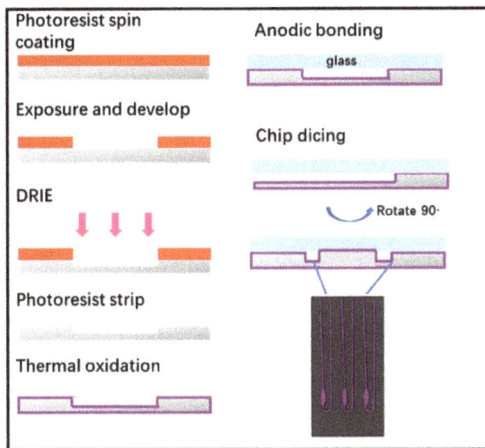

Figure 4. Fabrication process flow of the microchip.

2.5. RNA Reverse Transcription

Total RNA was reversely transcribed into cDNA with the following protocol: RNA mixture contained 2 μL of template (extracted using the automated miniaturized device as described above), 1 μL primer (Thermal Fisher Scientific, Carlsbad, CA, USA), 1 μL dNTPs (Thermal Fisher Scientific, USA) and 6 μL ddH$_2$O to bring the volume up to 10 μL. The mixture was then heated to 65 °C for 5 min and the sample was quickly chilled on ice for 2 min. A reaction mixture was prepared in a new and separate tube including 4 μL 5 × buffer (Thermal Fisher Scientific, USA), 0.5 μL RT enzyme (Thermal Fisher Scientific, USA), 0.5 μL inhibitor (Thermal Fisher Scientific, USA), and 5 μL ddH$_2$O to form a 10 μL solution. Following this, 10 μL reaction mixture was added to 10 μL RNA mixture then the sample was incubated at 42 °C for 60 min followed by 25 °C for 5 min.

2.6. Detection of Gene Expression by Real-Time PCR

The cDNA was amplified by real-time PCR, and the amplification target genes Actin and GAPDH were verified by 2% agarose gel electrophoresis. The GAPDH primers consist of a forward primer CAT GAG AAG TAT GAC AAC GCC T and a reverse primer AGT CCT TCC ACG ATA CCA AAG T, which produce a PCR product with a size of 113 bp. The Actin primers were forward primer GAG CAC AGA GCC TCG CCT TT and reverse primer TCA TCA TCC ATG GTG AGC TGG C resulted in a PCR product with a size of 70 bp. Each RT-PCR reaction was performed in 10 μL 2 × Bio-Rad super mix, 1 μL of RT reaction as cDNA, 0.5 μL primers (10 μM) and 8.5 μL ddH$_2$O. The cycling parameters involved denaturation at 95 °C for 3 min, followed by 40 cycles of 95 °C for 30 s and 60 °C for 1 min.

Each target gene expression was amplified by two repeated experiments. The RT-PCR was performed using LineGene 9600 Plus (Bioer) and its cycle threshold (Ct) value was determined.

2.7. Materials

293T cells were cultured in DMEM (DMEM; Gibco, USA) with 10% FBS (FBS; Gibco, USA) and 1% Penicillin/Streptomycin at 37 °C in 5% CO_2, MagMAX™-96 Total RNA Isolation Kit (Thermal Fisher Scientific, Carlsbad, CA, USA), 100% isopropanol, 100% ethanol, Maxima H Minus First Strand cDNA Synthesis Kit (Thermal Fisher Scientific, USA), iQ™ SYBR Green Supermix (BIO-RAD, Foster City, CA, USA), NanoDrop™ One (Thermal Fisher Scientific, USA).

3. Results and Discussion

In order to test the performance of the automated miniaturized device for nucleic acid sample preparation, we conducted reverse transcript and real-time PCR using the sample obtained from the device. The results are shown in Figure 5. Prior to testing, the total RNA was extracted from 293T cells by a complete automated process using our automated miniaturized nucleic acid sample preparation device. The concentration of total RNA was quantified by NanoDrop™ One and was found to be 500 ng/μL. Total RNA was then reversely transcribed into cDNA and we performed real-time PCR using the protocol as described in Materials and Methods. Since 2 μL total RNA was used in transcription, the concentration of cDNA is 50 ng/μL, thus add 1 μL cDNA into each PCR reaction is equivalent to adding 50 ng of sample into each PCR reaction. We then used Actin and GAPDH gene expression as standard in PCR amplification. Figure 5A shows the real-time PCR amplification results on commercial real-time PCR instrument (BIOER) of Actin and GAPDH gene expression. The upper black and the light blue curve represent two repeated RT-PCR amplification result of Actin, and the lower blue and red curve represent two repeated RT-PCR amplification result of Glyceraldehyde 3-phosphate dehydrogenase (GAPDH). The Ct value of Actin and GAPDH was 14 and 16, respectively, demonstrating that RNA was successfully extracted from the sample cell using completely automated process without human intervention. The Ct value also shows a rather high concentration of the extracted product.

To verify the extracted RNA from 293T cells, we conducted a downstream analysis using conventional electrophoresis gel separation. Figure 5B shows the electrophoresis results of Actin and GAPDH gene expression as an amplification result verification. The marker we used, as shown in the far right line in Figure 5B, was Thermo Fisher Scientific 100 bp DNA Ladder. The results show that the extracted RNA via post-PCR from automated sample preparation device could be successfully separated and clearly detected using the conventional gel electrophoresis, indicating the feasibility of our device for efficient nucleic acid sample preparation.

In order to further verify the performance of our developed device, we compared the real-time PCR amplification results from large commercialized real-time PCR instruments with portable microchip real-time PCR device, as shown in Figure 5C. The amplification curve shown in Figure 5C was attained using the same total RNA extracted from our automated miniaturized device followed by reverse transcription into cDNA, as previously described. However, the DNA was amplified using another of our devices, the portable microchip real-time PCR device, as shown in Figure 5C, rather than a commercialized PCR instrument. Each PCR reaction was carried out using 2 μL PCR mixture, as previously described. The Ct value of Actin and GAPDH was 14 and 16, respectively, which was identical to the experiment results obtained from large commercial PCR instruments. These consistent results illustrate that the total RNA extracted had high purity, as even 2 μL volume of PCR mixture could demonstrate the same result, thus proving the efficiency of our automated miniaturized device. Moreover, our portable microchip real-time PCR instrument generated the same Ct value as commercialized PCR instruments.

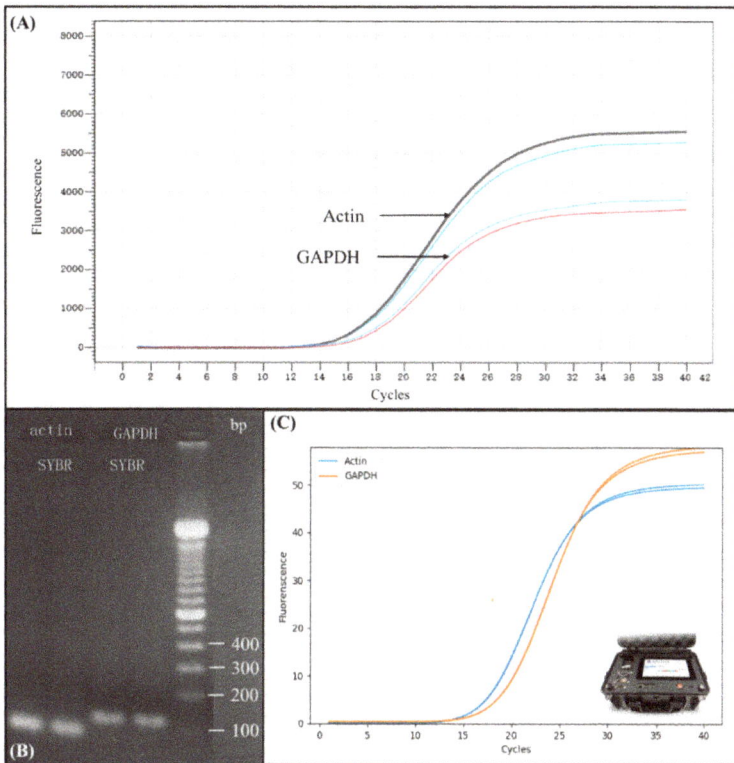

Figure 5. Verification of real-time polymerase chain reaction (PCR) results using total RNA extracted from 293T cells using automated miniaturized nucleic acid sample preparation device. (**A**) Real-time PCR amplification results on commercial real-time PCR instrument (BIOER) of Actin and GAPDH gene expression. (**B**) Electrophoresis gel separation and detection of Actin and GAPDH gene expression as a successful amplification result verification. (**C**) Comparable real-time PCR amplification results on portable microchip real-time PCR instrument using the same sample and protocol.

4. Conclusions

In conclusion, we have developed an automated miniaturized device to achieve automation of nucleic acid purification from actual samples. This device can adapt to different extraction protocols by configuring the software in the external central controller to meet personalized needs, and the extracted nucleic acid sample could be directly introduced into the microchip for further downstream applications. This method enables an easy and time-saving nucleic acid extraction, regardless of the experience of the operator. The data indicated that the automated total RNA extraction was equivalent to the performance using a manual process, but the automated extraction is more time-saving with the whole process curtailed to 10 min. The extracted total RNA from 293T cells can be verified by either PCR with post gel electrophoresis or qPCR. The process does not require external instrument for centrifugation or precipitation ensures its portability. In the future, we can integrate the automated miniaturized device with our portable microchip real-time PCR instrument to achieve a fully automated experimental process for practical biomedical applications.

Author Contributions: Conceptualization, R.T. and Y.G.; software, R.T.; validation, L.Z., C.H. and X.C.; resources, X.T. and Y.C.; data curation, Q.S. and K.L.; writing—original draft preparation, R.T.; writing—review and editing, R.T.; supervision, W.W.; funding acquisition, W.W. and X.G.

Funding: This research was funded by "Hong Kong RGC, Grant number 16219216", "Shanghai Pujiang Program (17PJ1402800)" and "the National Natural Science Foundation of China (21775101)".

Acknowledgments: The authors would like to acknowledge the support of the technicians and researchers from Design and Manufacturing Services Facility (DMSF), Nanosystem Fabrication Facility (NFF) at the Hong Kong University of Science and Technology, and Shenzhen Shineway Hi-tech Co., Ltd.

Conflicts of Interest: The authors declare no conflict of interest.

References

1. Eads, B.; Cash, A.; Bogart, K.; Costello, J.; Andrews, J. [3] Troubleshooting Microarray Hybridizations. *Methods Enzymol.* **2006**, *411*, 34–49. [PubMed]
2. Kirby, K. A new method for the isolation of deoxyribonucleic acids: Evidence on the nature of bonds between deoxyribonucleic acid and protein. *Biochem. J.* **1957**, *66*, 495. [CrossRef] [PubMed]
3. Marmur, J. A procedure for the isolation of deoxyribonucleic acid from micro-organisms. *J. Mol. Biol.* **1961**, *3*, 208–218. [CrossRef]
4. Chirgwin, J.M.; Przybyla, A.E.; MacDonald, R.J.; Rutter, W.J. Isolation of biologically active ribonucleic acid from sources enriched in ribonuclease. *Biochemistry* **1979**, *18*, 5294–5299. [CrossRef] [PubMed]
5. Chomczynski, P.; Sacchi, N. Single-step method of RNA isolation by acid guanidinium thiocyanate-phenol-chloroform extraction. *Anal. Biochem.* **1987**, *162*, 156–159. [CrossRef]
6. Yahya, A.; Firmansyah, M.; Arlisyah, A.; Risandiansyah, R. Comparison of DNA Extraction Methods Between Conventional, Kit, Alkali and Buffer-Only for PCR Amplification on Raw and Boiled Bovine and Porcine Meat. *J. Exp. Life Sci.* **2017**, *7*, 110–114. [CrossRef]
7. Tan, S.C.; Yiap, B.C. DNA, RNA, and protein extraction: The past and the present. *BioMed Res. Int.* **2009**, *2009*. [CrossRef]
8. Price, C.W.; Leslie, D.C.; Landers, J.P. Nucleic acid extraction techniques and application to the microchip. *Lab Chip* **2009**, *9*, 2484–2494. [CrossRef]
9. Padhye, V.V.; York, C.; Burkiewicz, A. Nucleic Acid Purification on Silica Gel and Glass Mixtures. U.S. Patent 5,658,548, 19 August 1997.
10. Vogelstein, B.; Gillespie, D. Preparative and analytical purification of DNA from agarose. *Proc. Natl. Acad. Sci. USA* **1979**, *76*, 615–619. [CrossRef]
11. Esser, K.-H.; Marx, W.H.; Lisowsky, T. maxXbond: First regeneration system for DNA binding silica matrices. *Nat. Methods* **2006**, *3*, 68. [CrossRef]
12. Seligson, D.B.; Shrawder, E.J. Method of Isolating and Purifying Nucleic Acids from Biological Samples. U.S. Patent 4,935,342, 19 June 1990.
13. Archer, M.J.; Lin, B.; Wang, Z.; Stenger, D.A. Magnetic bead-based solid phase for selective extraction of genomic DNA. *Anal. Biochem.* **2006**, *355*, 285–297. [CrossRef] [PubMed]
14. Chacon-Cortes, D.; Griffiths, L.R. Methods for extracting genomic DNA from whole blood samples: Current perspectives. *J. Biorepos. Sci. Appl. Med.* **2014**, *2014*, 1–9.
15. Goldberg, S. Mechanical/physical methods of cell distribution and tissue homogenization. In *Proteomic Profiling*; Springer: Berlin, Germany, 2015; pp. 1–20.
16. Hawkins, T. DNA Purification and Isolation Using Magnetic Particles. U.S. Patent 5,705,628, 6 January 1998.
17. Berensmeier, S. Magnetic particles for the separation and purification of nucleic acids. *Appl. Microbiol. Biotechnol.* **2006**, *73*, 495–504. [CrossRef] [PubMed]
18. Franzreb, M.; Siemann-Herzberg, M.; Hobley, T.J.; Thomas, O.R. Protein purification using magnetic adsorbent particles. *Appl. Microbiol. Biotechnol.* **2006**, *70*, 505–516. [CrossRef]
19. Manz, A.; Graber, N.; Widmer, H. áM Miniaturized total chemical analysis systems: A novel concept for chemical sensing. *Sens. Actuators B Chem.* **1990**, *1*, 244–248. [CrossRef]
20. Yuen, P.K.; Kricka, L.J.; Fortina, P.; Panaro, N.J.; Sakazume, T.; Wilding, P. Microchip module for blood sample preparation and nucleic acid amplification reactions. *Genome Res.* **2001**, *11*, 405–412. [CrossRef]
21. Hui, W.C.; Yobas, L.; Samper, V.D.; Heng, C.-K.; Liw, S.; Ji, H.; Chen, Y.; Cong, L.; Li, J.; Lim, T.M. Microfluidic systems for extracting nucleic acids for DNA and RNA analysis. *Sens. Actuators Phys.* **2007**, *133*, 335–339. [CrossRef]

22. Lee, J.-G.; Cheong, K.H.; Huh, N.; Kim, S.; Choi, J.-W.; Ko, C. Microchip-based one step DNA extraction and real-time PCR in one chamber for rapid pathogen identification. *Lab Chip* **2006**, *6*, 886–895. [CrossRef]
23. Nakagawa, T.; Hashimoto, R.; Maruyama, K.; Tanaka, T.; Takeyama, H.; Matsunaga, T. Capture and release of DNA using aminosilane-modified bacterial magnetic particles for automated detection system of single nucleotide polymorphisms. *Biotechnol. Bioeng.* **2006**, *94*, 862–868. [CrossRef] [PubMed]
24. Yeung, S.W.; Hsing, I.-M. Manipulation and extraction of genomic DNA from cell lysate by functionalized magnetic particles for lab on a chip applications. *Biosens. Bioelectron.* **2006**, *21*, 989–997. [CrossRef] [PubMed]
25. Wang, C.-H.; Lien, K.-Y.; Wu, J.-J.; Lee, G.-B. A magnetic bead-based assay for the rapid detection of methicillin-resistant Staphylococcus aureus by using a microfluidic system with integrated loop-mediated isothermal amplification. *Lab Chip* **2011**, *11*, 1521–1531. [CrossRef] [PubMed]
26. Prinz, C.; Tegenfeldt, J.O.; Austin, R.H.; Cox, E.C.; Sturm, J.C. Bacterial chromosome extraction and isolation. *Lab Chip* **2002**, *2*, 207–212. [CrossRef] [PubMed]
27. Wainright, A.; Nguyen, U.T.; Bjornson, T.; Boone, T.D. Preconcentration and separation of double-stranded DNA fragments by electrophoresis in plastic microfluidic devices. *Electrophoresis* **2003**, *24*, 3784–3792. [CrossRef] [PubMed]
28. Cho, Y.-K.; Lee, J.-G.; Park, J.-M.; Lee, B.-S.; Lee, Y.; Ko, C. One-step pathogen specific DNA extraction from whole blood on a centrifugal microfluidic device. *Lab Chip* **2007**, *7*, 565–573. [CrossRef] [PubMed]

micromachines

MDPI

Article

A Bubble-Free Microfluidic Device for Easy-to-Operate Immobilization, Culturing and Monitoring of Zebrafish Embryos

Zhen Zhu [1,*], **Yangye Geng** [1], **Zhangyi Yuan** [1], **Siqi Ren** [1], **Meijing Liu** [2], **Zhaozheng Meng** [3] and **Dejing Pan** [4,*]

1 Key Laboratory of MEMS of Ministry of Education, School of Electronic Science and Engineering, Southeast University, Nanjing 210096, China; gengyangye@seu.edu.cn (Y.G.); zy-yuan16@mails.tsinghua.edu.cn (Z.Y.); ren.siqi.43s@st.kyoto-u.ac.jp (S.R.)
2 Model Animal Research Center, Nanjing University, Nanjing 210061, China; liumj@nicemice.cn
3 Department of Biosystems Science and Engineering, Bio Engineering Laboratory, ETH Zurich, CH-4058 Basel, Switzerland; zhaozheng.meng@bsse.ethz.ch
4 Cambridge-Suda Genomic Resource Center and Jiangsu Key Laboratory of Neuropsychiatric Diseases, Soochow University, Suzhou 215213, China
* Correspondence: zhuzhen@seu.edu.cn (Z.Z.); pandejing@suda.edu.cn (D.P.); Tel.: +86-25-837-92632 (ext. 8830) (Z.Z.); +86-512-658-83562 (D.P.)

Received: 28 January 2019; Accepted: 25 February 2019; Published: 28 February 2019

Abstract: The development of miniaturized devices for studying zebrafish embryos has been limited due to complicated fabrication and operation processes. Here, we reported on a microfluidic device that enabled the capture and culture of zebrafish embryos and real-time monitoring of dynamic embryonic development. The device was simply fabricated by bonding two layers of polydimethylsiloxane (PDMS) structures replicated from three-dimensional (3D) printed reusable molds onto a flat glass substrate. Embryos were easily loaded into the device with a pipette, docked in traps by gravity, and then retained in traps with hydrodynamic forces for long-term culturing. A degassing chamber bonded on top was used to remove air bubbles from the embryo-culturing channel and traps so that any embryo movement caused by air bubbles was eliminated during live imaging. Computational fluid dynamics simulations suggested this embryo-trapping and -retention regime to exert low shear stress on the immobilized embryos. Monitoring of the zebrafish embryogenesis over 20 h during the early stages successfully verified the performance of the microfluidic device for culturing the immobilized zebrafish embryos. Therefore, this rapid-prototyping, low-cost and easy-to-operate microfluidic device offers a promising platform for the long-term culturing of immobilized zebrafish embryos under continuous medium perfusion and the high-quality screening of the developmental dynamics.

Keywords: microfluidics; 3D printing; zebrafish embryo; embryogenesis

---

1. Introduction

The zebrafish, *Danio rerio*, has become a prominent vertebrate model for disease modeling and drug discovery [1]. Approximately 82% of disease-related human genes have at least one zebrafish orthologue [2]. Zebrafish embryos feature a small size, an optically-transparent body, rapid development, and cost-efficient husbandry, and thus have been increasingly used as ideal model organisms in research areas such as embryogenesis, developmental biology and chemical genetics [3–5]. To date, most routine experiments with zebrafish embryos are still performed by means of cell-culturing protocols in conventional microplates, which have several drawbacks [6]. For instance, the static culturing environment of embryos in sample wells may cause accumulative surface absorption of chemical compounds and cross-contamination of embryo metabolites. The water flow for embryos developed in a natural fluid environment cannot be imitated by the static culture. Embryos moving in wells may impose restrictions on the resolution and quality of live imaging.

In the past decade, microfluidics has developed rapidly due to its unique merits, such as small geometric features in micrometer-scale and high surface-to-volume ratio, ability to handle small volume of fluids (microliter to picoliter) in laminar flow regime, and requiring low reagents with a fast response. In addition, the advantages of microfluidics in portability, automation, high throughput, and the ability to integrate multiple functions on a single chip make it an exceptional platform in a variety of fields such as chemical analysis, cell biology, and medicine. Microfluidics can even enhance the propagation, development and potency of some biological organisms, especially in assisted reproductive technology (ART) [7,8].

Furthermore, with the development of advanced manufacturing technologies such as microfabrication, laser micromachining and three-dimensional (3D) printing, dedicated microfluidic devices for the immobilization, flow-perfusion culture, dosing and time-lapse imaging of zebrafish embryos have emerged in the past decade [9–11]. A set of microfluidic devices were presented for zebrafish embryos' culturing and monitoring based on specific modifications in 3D structure design, glass/silicon etching and bonding processes, and medium perfusion [12–14]. Although these devices have controllable perfusion systems and enable the culturing of single zebrafish embryos in each chamber, complicated fabrication and operation processes that require dedicated clean-room equipment and instruments limit availability in most biological laboratories, and any movement of embryos in chambers may affect the imaging quality.

To resolve the problem of embryo movement during optical screening, Akagi et al. developed a microfluidic array with horizontal traps for the immobilization and perfusion of zebrafish embryos [15]. The design and operation mechanism of this array was similar to that of the worm-encapsulated droplet trap array [16], but required a high flow rate of up to 2 mL/min to drag and dock millimeter-scale embryos in traps. Zhu et al. developed a 3D high-throughput microfluidic platform, which enables the stable immobilization of single embryos by combining continuous medium perfusion at a flow rate of 400 μL/min and aspiration via horizontal tunnels embedded between the traps and suction channel [17]. Moreover, in order to avoid the high-flow-rate perfusion that may exert high shear stress on embryos and potentially affect the embryogenesis and embryonic development, laser micromachining was introduced to fabricate a multilayer 3D array of embryo traps with vertical tunnels embedded beneath to immobilize single zebrafish embryos by combined gravitational sedimentation and low-pressure suction [18,19]. However, the surface roughness of vertical traps processed by laser micromachining limited the image resolution and quality of immobilized embryos. Therefore, an easily fabricated and straightforward-to-handle microfluidic system that enables the gentle immobilization and perfusion, stable culturing, and high-resolution imaging of zebrafish embryos is desirable.

In this work, we presented a proof-of-concept microfluidic device for the immobilization, culturing and imaging of zebrafish embryos. The device was comprised of a flat glass substrate and two layers of polydimethylsiloxane (PDMS) structures replicated from 3D printed masters. A bottom PDMS layer was constructed with an embryo-culturing channel to load embryos and five traps to capture embryos for long-term culturing and real-time imaging during their development. Taking advantage of the gas permeability of PDMS materials, a degassing chamber patterned in the top PDMS layer was used to remove air bubbles from the fluidic channel and traps in the bottom PDMS layer by applying a vacuum in the degassing chamber. Computational fluid dynamics (CFD) simulations were performed to estimate the shear stress on the immobilized embryos that were perfused under the continuous medium in the device. Culturing and monitoring of zebrafish embryonic development was carried out over 20 h to verify the performance of the microfluidic device.

2. Materials and Methods

2.1. Microfluidic Device

The microfluidic device consisted of a glass substrate and two PDMS layers patterned with microstructures (Figure 1). The bottom layer, i.e., the embryo-culturing channel, was designed to trap and culture zebrafish embryos. It was composed of an inlet for embryo loading, an inlet and an outlet for medium perfusion, and five horizontal-funnel-like traps for embryo immobilization. Considering that the diameter of zebrafish embryos with chorion is about 1.2 mm, the embryo inlet was 2 mm in diameter, the embryo-culturing channel was 2.5 mm wide and 2 mm high, and the trap featured a wide opening of 1.8 mm and a narrow opening of 0.6 mm to dock and retain an embryo without being flushed away. After the loading of embryos in the device, the embryo inlet was inserted with a 2 mm diameter polytetrafluoroethylene (PTFE) plug to avoid liquid leakage during culturing-medium perfusion. The top PDMS layer, which was irreversibly bonded onto the upper surface of the bottom PDMS layer, was a degassing chamber with several posts to support it. The degassing chamber had a height of 1 mm and fully covered the embryo-culturing channel and traps.

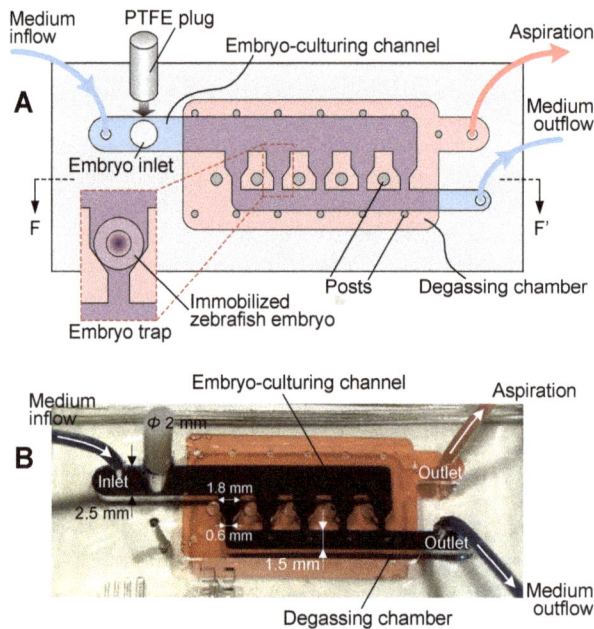

Figure 1. Overview of the microfluidic device for zebrafish embryo immobilization, culturing and monitoring. (**A**) Schematic of the microfluidic device with peripheral interfaces for embryo loading and trapping, culturing-medium perfusion, and degassing. (**B**) Device photo illustrating the embryo-culturing channel (dark blue), the degassing chamber (red), and the polytetrafluoroethylene (PTFE) plug (white) inserted in the embryo inlet.

The operation process of the microfluidic device is schematically shown in Figure 2. To load embryos into the device, a 1 mL pipette tip was cut to form a big opening (>1.2 mm) so that embryos could be sucked up and released easily without being damaged. Then, embryos could be transferred into the embryo-culturing channel through the embryo inlet using a pipette (Figure 2A). Afterwards, a 2 mm diameter PTFE plug was inserted into the embryo inlet so that no liquid leakage occurred during medium perfusion (Figure 2B). The medium inflow was then started by controlling a syringe pump. To immobilize embryos, the device was tilted slightly to roll the embryos in the channel and gravitationally dock them in the traps (Figure 2C). The space of each trap could house only one embryo to ensure single-embryo immobilization. After docking, embryos were perfused continuously with the culturing medium to stably maintain the immobilization status by hydrodynamic forces (Figure 2D,E). During the whole process of embryo loading, immobilization and culturing, a vacuum was always applied to the degassing chamber via aspiration to remove any air bubbles from the fluidic channel.

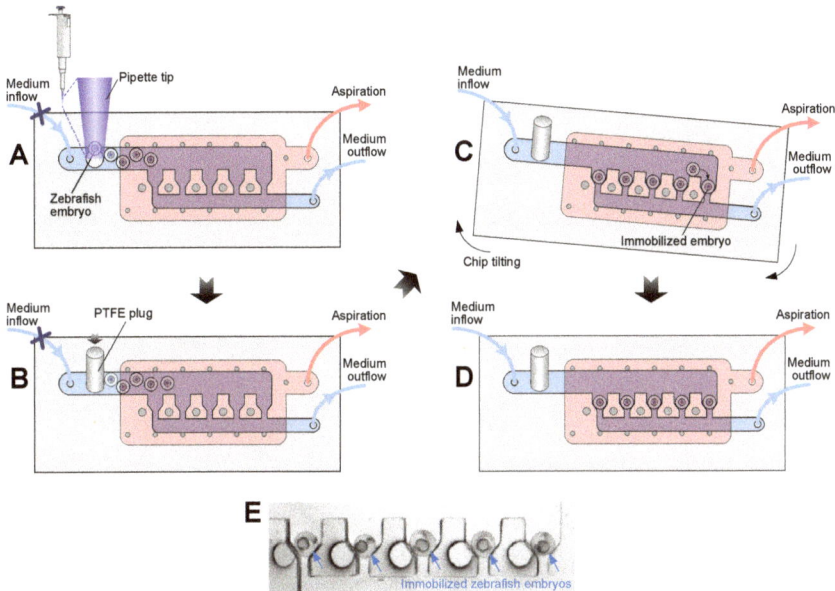

Figure 2. Operation process of the microfluidic device for embryo loading, trapping, and medium perfusion. (**A**) Loading embryos through the embryo inlet using a pipette. (**B**) Blocking the embryo inlet with the PTFE plug. (**C**) Tilting the device to roll embryos in the embryo-culturing channel and dock one embryo in each trap. (**D**) Retaining embryos in traps for culturing and real-time monitoring. (**E**) A micrograph showing five embryos immobilized in the device.

2.2. Device Fabrication

The microfluidic device was simply fabricated using a glass-PDMS multi-layer process. Since zebrafish embryos have diameters of around 1.2 mm and require a millimeter- and submillimeter-scale feature of microstructures for entrapment, the standard SU-8-based soft-lithography process was not feasible. Herein, we used an affordable 3D printer (MiiCraft+, MiiCraft, Hsinchu, Taiwan) with its proprietary photopolymer resin (BV007, MiiCraft) to fabricate molds for PDMS replication [20–23]. Masters for both the embryo-culturing channel and degassing chamber were designed using SolidWorks software (Dassault Systèmes, Waltham, MA, USA). The printer features an *x-y* resolution of about 56 μm and a z resolution of 50 μm, which was defined by the minimum upward step of the stage. For each layer of resin with a thickness of 50 μm, the ultraviolet (UV) exposure time was set to 5 s. After printing, the master was soaked in ethanol for 5 min then rinsed with fresh ethanol for another 1 min, followed by UV post-curing for 20 min. After post-curing, the master was soaked in ethanol again for 2 h then baked on a hotplate at 60 °C for 12 h to finalize the master fabrication (Figure 3A,B-1). Both masters were then silanized with trichloro(1H,1H,2H,2H-perfluorooctyl)silane (Sigma-Aldrich, Saint Louis, MO, USA) in vapor phase to prevent PDMS (Sylgard 184, Dow Corning, Midland, MI, USA) from adhering to resin molds (Figure 3B-2). Afterwards, masters were transferred to PDMS with a mixture of 10:1 w/w base to curing agent (Figure 3B-3). After peeling the PDMS replicas from masters, the degassing chamber was first punched with a hole as the inlet for aspiration and subsequently bonded onto the upper surface of the PDMS embryo-culturing channel (Figure 3B-4) through oxygen plasma surface modification (PDC-002-HP, Harrick Plasma, Ithaca, NY, USA). Then, the bonded PDMS stamp was punched with holes as inlets and outlets for embryo loading and medium perfusion. Lastly, the PDMS stamp was irreversibly bonded to a bare glass slide (Figure 3B-5) to finalize the device fabrication.

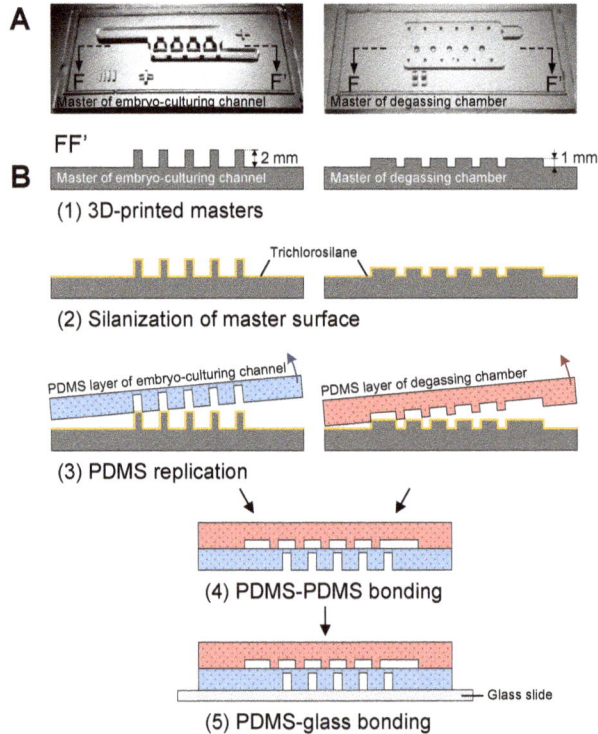

Figure 3. Fabrication process of the microfluidic device. (**A**) Photos of 3D printed masters for embryo-culturing channel and degassing chamber. (**B**-1) Schematic side view along FF' of Figures 1A and 3A. (**B**-2) Surface silanization of masters with the trichlorosilane. (**B**-3) PDMS replication from masters. (**B**-4) Bonding the degassing chamber onto the embryo-culturing channel. (**B**-5) Bonding the PDMS stamp onto the glass slide.

2.3. Computational Fluid Dynamics (CFD) Modelling and Simulations

In order to investigate the fluid dynamics for trapping and maintaining embryos and shear stress exerted on the immobilized embryos in the microfluidic device, 3D CFD simulations were performed using COMSOL Multiphysics software (COMSOL Inc., Burlington, MA, USA) using 'Laminar Flow' physics from the CFD Module. Geometric structures and parameters in the model were derived from the microfluidic device in Figure 1. According to the geometric parameters and applied flow rate of 50 μL/min, the Reynolds number in the embryo-culturing channel was calculated to be 0.37, which confirmed that the microfluidic device followed the laminar flow regime. Embryos immobilized by the traps were simply modelled as rigid and non-deformable spheres in the fluidic domain. Subdomains were assigned with a density of 1000 kg/m^3 and a dynamic viscosity of 0.001 Pa·s (for water at 20 °C).

For the incompressible fluid under laminar flow regime that follows the Navier–Stokes equation, the pressure and flow velocity were conducted by the governing equations as follows:

$$\rho(\mathbf{u}\cdot\nabla)\mathbf{u} = \nabla\cdot\left[-p\mathbf{I} + \mu\left(\nabla\mathbf{u} + (\nabla\mathbf{u})^{T}\right)\right] + \mathbf{F} \tag{1}$$

$$\rho\nabla\cdot(\mathbf{u}) = 0 \tag{2}$$

where ρ is density, $\mathbf{u}$ is velocity vector, p is pressure, $\mathbf{I}$ is unit matrix, μ is dynamic viscosity, and $\mathbf{F}$ is volume force vector.

A no-slip boundary condition was applied to the channel walls and the sphere surfaces. Laminar inflow with different volumetric flow rates of 25, 50, 100 and 500 μL/min was applied to the inlet boundary of the fluidic channel, and a pressure of 0 Pa was set to the outlet boundary. Due to the limitation of physical memory and time–cost in the numerical calculation, a predefined meshing process with normal element size was applied to all domains. Supplementary Table S1 lists the detailed parameters of element size set for meshing, such as maximum and minimum element size, and maximum element growth rate. Supplementary Figure S1 illustrates the translucent profiles of the microfluidic structures without and with spheres after meshing. The shear stress (τ) over the sphere surfaces was calculated by using the shear stress components in x-, y- and z-directions on the basis of the following equation:

$$\tau = \sqrt{\tau_x^2 + \tau_y^2 + \tau_z^2} \tag{3}$$

After simulations, contours and stream lines referring to the flow velocity in the fluidic channel and contours of the shear stress over the sphere surfaces were obtained.

2.4. Zebrafish Embryo Trapping, Culturing and Imaging

Embryos were obtained from natural spawning of wild type adult zebrafish and were collected in E3 medium (5 mM NaCl, 0.17 mM KCl, 0.33 mM $CaCl_2$, and 0.33 mM $MgSO_4$). After flushing the microfluidic channel with E3 medium, we used a pipette with a modified tip to transfer embryos into the embryo-culturing channel through the embryo inlet, which was then plugged up with a PTFE pillar. The microfluidic device was manually tilted in order to dock embryos in the traps, then fixed on the stage of a stereomicroscope (NSZ-608T, Novel Optics, Nanjing, China) with adhesive tape. For embryo culturing, E3 medium, initially loaded in a 100 mL syringe affixed on a syringe pump, was perfused into the embryo-culturing channel via PTFE tubing at a constant flow rate of 50 μL/min, and the room temperature was kept at 28.5 °C. In order to remove any bubbles from the embryo-culturing channel, the degassing chamber was connected to the house vacuum supplied by a vacuum pump (MPC 1201 T, Ilmvac, Germany), which can provide an ultimate pressure of below 2 mbar according to its datasheet. Time-lapse imaging of embryonic development was performed using the stereomicroscope equipped with a digital camera (E3CMOS, ToupTek Photonics, Hangzhou, China) and running its camera control software (ToupView, ToupTek Photonics) for image acquisition at a time interval of 150 s.

3. Results and Discussion

3.1. Degassing

During the long-term culturing and monitoring of zebrafish embryos, any air bubbles brought to the embryo-culturing channel and traps may squeeze the immobilized embryo out of its original position or obstruct the view of imaging, and thus affect the stability of embryo immobilization and the quality of time-lapse imaging. As such, we took advantage of the gas permeability of PDMS materials and removed air bubbles from the fluidic channel by applying a vacuum to the degassing chamber. Herein, we tested the degassing chamber, which was assigned to remove air bubbles attached to PDMS channel walls in the embryo-culturing channel. Once a low pressure (vacuum) was exerted over the degassing chamber via aspiration, a high-pressure difference was generated between the air in bubbles and the degassing chamber. Thus, air in the bubbles could be aspirated to pass through the gas-permeable PDMS and ultimately removed. Figure 4 shows the experimental results. Initially, there were some air bubbles attached to the surface of the embryo-culturing channel, as indicated by arrows in Figure 4A. After switching on the vacuum supply, bubbles started to reduce in size and number (Figure 4B,C) and finally disappeared within 3 s (Figure 4D). The result demonstrated that the degassing chamber was capable of removing air bubbles highly efficiently, thereby enabling the stable culturing and monitoring of zebrafish embryonic development.

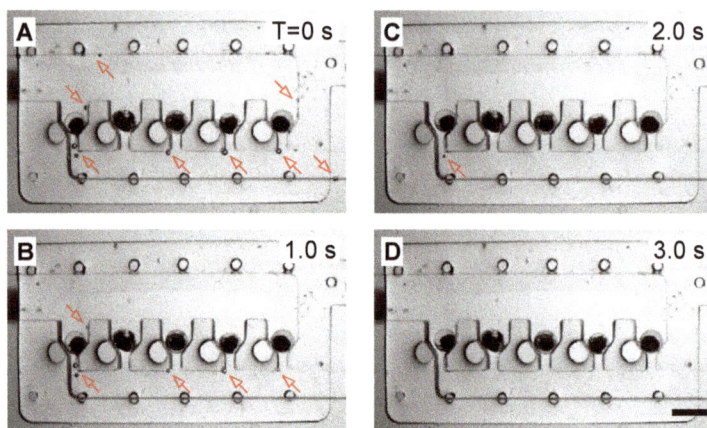

Figure 4. Fast evacuation of air bubbles (indicated by arrows) in the embryo-culturing channel within 3 s. (**A**) T = 0 s; (**B**) T = 1.0 s; (**C**) T = 2.0 s; (**D**) T = 3.0 s. Scale bar: 2 mm.

3.2. CFD Simulations

3D CFD simulations were performed to study the fluidic dynamics for embryo immobilization and the shear stress exerted on the immobilized embryos in the microfluidic device. Figure 5A shows the flow velocity across the embryo-culturing channel immobilized with five embryos. The stream lines indicate that the liquid flowed over each immobilized embryo and became trapped towards the outlet. The contour plots of vertical cross-sections indicate that the maximum value of flow velocity existed in the rightmost trap. As a result, the shear stress over the surface of the rightmost embryo exhibited a maximum value of 5.2×10^{-3} Pa among the five immobilized embryos perfused at 50 μL/min (Figure 5B). Based on the CFD simulations, we then obtained the maximum and mean shear stress over each embryo at different flow rates: 25, 50, 100 and 500 μL/min (Figure 5C,D). With the increase of flow rate, both maximum and mean values of shear stress showed an increase trend. When applying a flow rate of 100 μL/min, the maximum shear stress of the rightmost embryo exceeded 0.01 Pa. At a flow rate of 500 μL/min, the mean shear stress of the same embryo also exceeded 0.01 Pa. The CFD simulations thus indicated that a high flow rate resulted in high shear stress over the surfaces of embryos. Previous studies [15,18] have verified that the maximum shear stress in the order of 0.01 Pa has negligible influence on developing embryos due to the protection of the robust chorion membrane. In the experiment, we found that immobilized embryos during culturing were occasionally displaced from their original positions when a flow rate of 30 μL/min was applied for medium perfusion. This phenomenon can be elucidated by the fact that a flow rate lower than 30 μL/min cannot provide sufficient hydrodynamic forces to retain the embryos stably in their traps. Considering the required stable embryo immobilization and low shear stress over the embryos during long-term culturing and imaging, we, therefore, set the flow rate of medium perfusion to 50 μL/min in the experiment.

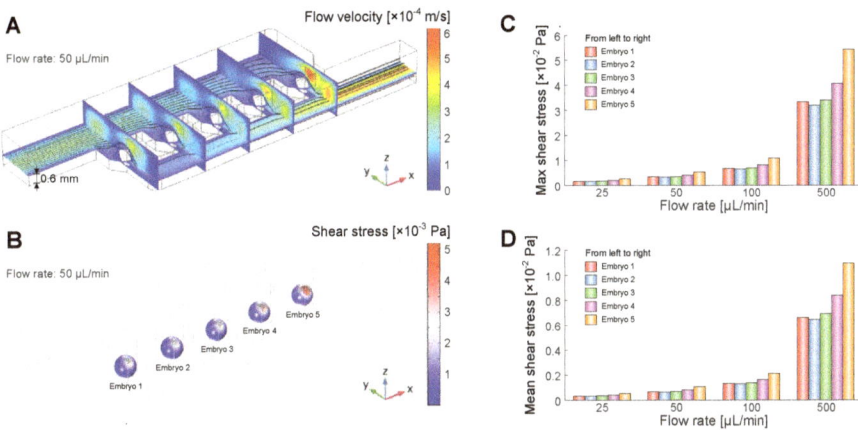

Figure 5. Computational fluid dynamics (CFD) simulations of flow velocity and shear stress for embryo immobilization in the microfluidic device. (**A**) Contour plots and stream lines of cross-sections in the *xy*-plane (across the horizontal middle of immobilized embryos, i.e., 0.6 mm above the channel bottom) and *yz*-plane (across the vertical middle of each immobilized embryo) showing the flow velocity across the embryo-culturing channel immobilized with five embryos. Flow rate was set to 50 μL/min in simulation. (**B**) Contour plot of shear stress over the surfaces of five immobilized embryos. Flow rate was set to 50 μL/min in simulation. (**C,D**) Simulated maximum and mean values of shear stress over each embryo at applied flow rates: 25, 50, 100 and 500 μL/min.

3.3. Zebrafish Embryonic Development

We then evaluated the microfluidic device as a stable platform to culture the immobilized zebrafish embryos and monitor their dynamic development. The immobilized embryos were perfused with E3 medium at a constant flow rate of 50 μL/min for a long-term culture (over 20 h). Meanwhile the embryonic development was traced by time-lapse imaging at a time interval of 150 s.

Figure 6 shows photographs every 2 h of the four recorded zebrafish embryos, which developed healthily and uniformly during the recording time course. Image acquisition started around 5 h post fertilization (T = 0 h) and the embryonic development had already entered the gastrula stage. The epiboly displaced the blastoderm margin and remained at 50%. The germ ring and the embryonic shield could be observed. Then, the epiboly continued until the yolk plug was completely covered by the blastoderm, and a tail bud became visible (4 h, bud stage). From this stage onward, the development of embryos entered the segmentation period. During this development period (6–16 h), somites appeared sequentially and developed in the trunk and tail, which elongated and became more prominent, and the early rudiments of primary organs and body movement could be observed. Afterwards, the embryos entered the pharyngula period for further morphogenetic development (18 h). Therefore, the aforementioned experimental results confirmed the capabilities of the microfluidic device for stable immobilization and long-term culturing of zebrafish embryos and monitoring their early-stage development without interfering in the intricate embryogenesis.

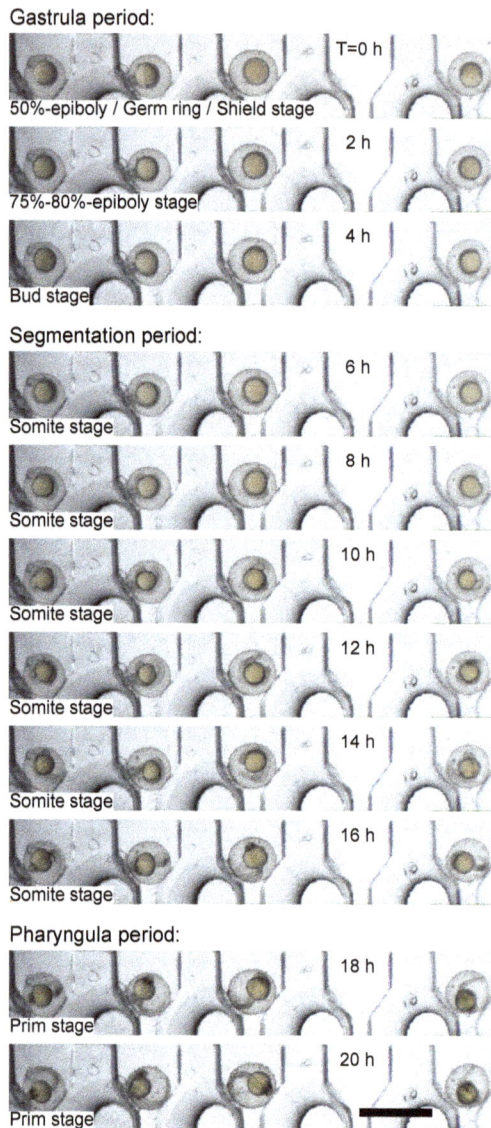

Figure 6. Twenty-hour live imaging of the early-stage development of zebrafish embryos immobilized in the microfluidic device. Scale bar: 2 mm.

4. Conclusions

We have presented a rapid-prototyping, low-cost and easy-to-operate microfluidic device for stable immobilization, long-term culturing and high-quality imaging of zebrafish embryos. The device has been simply fabricated by bonding a glass substrate with two layers of PDMS replicas from the 3D printed reusable masters. Experimental results have demonstrated that zebrafish embryos can be easily loaded into the embryo-culturing channel with a pipette and then rolled into the traps in turn under gravity by manually tilting the device. Suggested by the CFD simulations, we have

Micromachines **2019**, *10*, 168

optimized the embryo-trapping regime and diminished the potential shear stress exerted over the immobilized embryos. Continuous medium perfusion at a low flow rate of 50 μL/min provided adequate hydrodynamic forces to ensure the stable immobilization of the embryos. Air bubbles in the embryo-culturing channel and traps were rapidly and effectively removed by the degassing chamber on top. As a result, the zebrafish embryos were stably immobilized and underwent long-term time-lapse monitoring of their dynamic development. Culturing and monitoring over 20 h during the early stages of the embryonic development of the zebrafish successfully confirmed the functions of the microfluidic device. Therefore, the microfluidic platform promises to perform the stable immobilization, continuous medium perfusion, and long-term culturing of more zebrafish embryos with a potential scaling-up of the device, allowing for high-quality monitoring of dynamic embryonic development.

Supplementary Materials: The following are available online at http://www.mdpi.com/2072-666X/10/3/168/s1, Figure S1: Translucent meshing profiles of the microfluidic structures (A) without and (B) with spheres. Table S1: Element size parameters in meshing of 3D CFD simulation.

Author Contributions: Conceptualization, Z.Z. and D.P.; methodology, Z.Z., Y.G., Z.Y., and S.R.; validation, Z.Z., Y.G., Z.Y., and S.R.; formal analysis, Z.Z., Z.Y., and Z.M.; investigation, Z.Z., Z.Y., and S.R.; resources, Z.Z., M.L., and D.P.; writing—original draft preparation, Z.Z., Y.G., Z.M., and D.P.; writing—review and editing, Z.Z., Y.G., and D.P.; visualization, Z.Z., Y.G., and Z.Y.; supervision, Z.Z., M.L., and D.P.; project administration, Z.Z.; funding acquisition, Z.Z. and D.P.

Funding: This research was funded by the National Natural Science Foundation of China (grant numbers 61774036 and 31771299), the National Key Basic Research Program of China (grant number 2015CB352100), the Fundamental Research Funds for the Central Universities, and the National Center for International Research (grant number 2017B01012).

Acknowledgments: We would like to thank Hanyu Qin from ETH Zurich, Switzerland for critical reading of the manuscript, and Qingshun Zhao from Nanjing University, China for providing the zebrafish embryos.

Conflicts of Interest: The authors declare no conflict of interest.

References

1. MacRae, C.A.; Peterson, R.T. Zebrafish as tools for drug discovery. *Nat. Rev. Drug Discov.* **2015**, *14*, 721–731. [CrossRef] [PubMed]
2. Howe, K.; Clark, M.D.; Torroja, C.F.; Torrance, J.; Berthelot, C.; Muffato, M.; Collins, J.E.; Humphray, S.; McLaren, K.; Matthews, L.; et al. The zebrafish reference genome sequence and its relationship to the human genome. *Nature* **2013**, *496*, 498–503. [CrossRef] [PubMed]
3. Wheeler, G.N.; Brändli, A.W. Simple vertebrate models for chemical genetics and drug discovery screens: Lessons from zebrafish and Xenopus. *Dev. Dyn.* **2009**, *238*, 1287–1308. [CrossRef] [PubMed]
4. Basu, S.; Sachidanandan, C. Zebrafish: A multifaceted tool for chemical biologists. *Chem. Rev.* **2013**, *113*, 7952–7980. [CrossRef] [PubMed]
5. White, R.; Rose, K.; Zon, L. Zebrafish cancer: The state of the art and the path forward. *Nat. Rev. Cancer* **2013**, *13*, 624–636. [CrossRef] [PubMed]
6. Lammer, E.; Kamp, H.G.; Hisgen, V.; Koch, M.; Reinhard, D.; Salinas, E.R.; Wendler, K.; Zok, S.; Braunbeck, T. Development of a flow-through system for the fish embryo toxicity test (FET) with the zebrafish (*Danio rerio*). *Toxicol. In Vitro* **2009**, *23*, 1436–1442. [CrossRef] [PubMed]
7. Swain, J.E.; Lai, D.; Takayama, S.; Smith, G.D. Thinking big by thinking small: Application of microfluidic technology to improve ART. *Lab Chip* **2013**, *13*, 1213–1224. [CrossRef] [PubMed]
8. Kashaninejad, N.; Shiddiky, M.J.A.; Nguyen, N.-T. Advances in microfluidics-based assisted reproductive technology: From sperm sorter to reproductive system-on-a-chip. *Adv. Biosyst.* **2018**, *2*, 1700197. [CrossRef]
9. Hwang, H.; Lu, H. Microfluidic tools for developmental studies of small model organisms—Nematodes, fruit flies, and zebrafish. *Biotechnol. J.* **2013**, *8*, 192–205. [CrossRef] [PubMed]
10. Zhu, F.; Skommer, J.; Huang, Y.; Akagi, J.; Adams, D.; Levin, M.; Hall, C.J.; Crosier, P.S.; Wlodkowic, D. Fishing on chips: Up-and-coming technological advances in analysis of zebrafish and Xenopus embryos. *Cytom. Part A* **2014**, *85*, 921–932. [CrossRef] [PubMed]
11. Yang, F.; Gao, C.; Wang, P.; Zhang, G.-J.; Chen, Z. Fish-on-a-chip: Microfluidics for zebrafish research. *Lab Chip* **2016**, *16*, 1106–1125. [CrossRef] [PubMed]

12. Wielhouwer, E.M.; Ali, S.; Al-Afandi, A.; Blom, M.T.; Olde Riekerink, M.B.; Poelma, C.; Westerweel, J.; Oonk, J.; Vrouwe, E.X.; Buesink, W.; et al. Zebrafish embryo development in a microfluidic flow-through system. *Lab Chip* **2011**, *11*, 1815–1824. [CrossRef] [PubMed]

13. Yang, F.; Chen, Z.; Pan, J.; Li, X.; Feng, J.; Yang, H. An integrated microfluidic array system for evaluating toxicity and teratogenicity of drugs on embryonic zebrafish developmental dynamics. *Biomicrofluidics* **2011**, *5*, 024115. [CrossRef] [PubMed]

14. Choudhury, D.; van Noort, D.; Iliescu, C.; Zheng, B.; Poon, K.-L.; Korzh, S.; Korzh, V.; Yu, H. Fish and Chips: A microfluidic perfusion platform for monitoring zebrafish development. *Lab Chip* **2012**, *12*, 892–900. [CrossRef] [PubMed]

15. Akagi, J.; Khoshmanesh, K.; Evans, B.; Hall, C.J.; Crosier, K.E.; Cooper, J.M.; Crosier, P.S.; Wlodkowic, D. Miniaturized embryo array for automated trapping, immobilization and microperfusion of zebrafish embryos. *PLoS ONE* **2012**, *7*, e36630. [CrossRef] [PubMed]

16. Shi, W.; Qin, J.; Ye, N.; Lin, B. Droplet-based microfluidic system for individual *Caenorhabditis elegans* assay. *Lab Chip* **2008**, *8*, 1432–1435. [CrossRef] [PubMed]

17. Zhu, F.; Akagi, J.; Hall, C.J.; Crosier, K.E.; Crosier, P.S.; Delaage, P.; Wlodkowic, D. A high-throughput lab-on-a-chip interface for zebrafish embryo tests in drug discovery and ecotoxicology. *Proc. SPIE* **2013**, *8923*, 892345.

18. Akagi, J.; Khoshmanesh, K.; Hall, C.J.; Cooper, J.M.; Crosier, K.E.; Crosier, P.S.; Wlodkowic, D. Fish on chips: Microfluidic living embryo array for accelerated in vivo angiogenesis assays. *Sens. Actuators B-Chem.* **2013**, *189*, 11–20. [CrossRef]

19. Wang, K.I.K.; Salcic, Z.; Yeh, J.; Akagi, J.; Zhu, F.; Hall, C.J.; Crosier, K.E.; Crosier, P.S.; Wlodkowic, D. Toward embedded laboratory automation for smart lab-on-a-chip embryo arrays. *Biosens. Bioelectron.* **2013**, *48*, 188–196. [CrossRef] [PubMed]

20. Comina, G.; Suska, A.; Filippini, D. PDMS lab-on-a-chip fabrication using 3D printed templates. *Lab Chip* **2014**, *14*, 424–430. [CrossRef] [PubMed]

21. Shallan, A.I.; Smejkal, P.; Corban, M.; Guijt, R.M.; Breadmore, M.C. Cost-effective three-dimensional printing of visibly transparent microchips within minutes. *Anal. Chem.* **2014**, *86*, 3124–3130. [CrossRef] [PubMed]

22. Chan, H.N.; Chen, Y.; Shu, Y.; Chen, Y.; Tian, Q.; Wu, H. Direct, one-step molding of 3D-printed structures for convenient fabrication of truly 3D PDMS microfluidic chips. *Microfluid. Nanofluid.* **2015**, *19*, 9–18. [CrossRef]

23. Chen, Y.; Chan, H.N.; Michael, S.A.; Shen, Y.; Chen, Y.; Tian, Q.; Huang, L.; Wu, H. A microfluidic circulatory system integrated with capillary-assisted pressure sensors. *Lab Chip* **2017**, *17*, 653–662. [CrossRef] [PubMed]

micromachines

MDPI

Article

On-Chip Cell Incubator for Simultaneous Observation of Culture with and without Periodic Hydrostatic Pressure

Mitsuhiro Horade [1,*], Chia-Hung Dylan Tsai [2] and Makoto Kaneko [1]

[1]   Department of Mechanical Engineering, Osaka University, Suita 565-0871, Japan;
      mk@mech.eng.osaka-u.ac.jp
[2]   Department of Mechanical Engineering, National Chiao Tung University, Hsinchu 30010, Taiwan;
      dylantsai@nctu.edu.tw
*   Correspondence: horade@mech.eng.osaka-u.ac.jp; Tel.: +81-6-6879-7333

Received: 12 January 2019; Accepted: 15 February 2019; Published: 17 February 2019

Abstract: This paper proposes a microfluidic device which can perform simultaneous observation on cell growth with and without applying periodic hydrostatic pressure (Yokoyama et al. *Sci. Rep.* **2017**, *7*, 427). The device is called on-chip cell incubator. It is known that culture with periodic hydrostatic pressure benefits the elasticity of a cultured cell sheet based on the results in previous studies, but how the cells respond to such a stimulus during the culture is not yet clear. In this work, we focused on cell behavior under periodic hydrostatic pressure from the moment of cell seeding. The key advantage of the proposed device is that we can compare the results with and without periodic hydrostatic pressure while all other conditions were kept the same. According to the results, we found that cell sizes under periodic hydrostatic pressure increase faster than those under atmospheric pressure, and furthermore, a frequency-dependent fluctuation of cell size was found using Fourier analysis.

Keywords: on-chip cell incubator; periodic hydrostatic pressure; periodic pressure; time-lapse observation; cell growth; simultaneous multiple chamber observation

---

1. Introduction

There are many studies addressing cellular responses under mechanical and chemical stimulations [1–7]. Stimulation can be beneficial for cultured cells. For example, Di Carla et al. used caffeine as a xenobiotic stress-inducing agent and found the cell survival rate is promoted under such a chemical stress [8]. Some works used stimulations as a tool for determining cellular properties or physiologies. For example, Seo et al. showed the relationship between mechanical stimulation and the physiologies in a dystrophic heart [9]. Ito et al. applied mechanical stress to red blood cells (RBCs) for different durations and observed a 100 times difference of the time constant in their shape-recovery curves when the stress duration was just increased from 10 s to 180 s [7]. Sakuma et al. proposed a cell stress test by moving a RBC back and forth across a microfluidic constriction until the RBC eventually lost its deformability and used it as an index of RBC deformability [10]. There are also studies investigating cell alignment under stimulations. For example, Teramura et al. demonstrated that mechanical stimulation to human iPS cells altered the alignment of actin fibers as well as the expressions of the pluripotent related genes [11]. Subramony et al. investigated the role of nanofiber matrix alignment and mechanical stimulation on mesenchymal stem cell (MSC) differentiation [12]. While there are approaches using 3D printing for aligning cells [13–15], stimulation seems to be a more natural approach, since each cell can choose its comfortable position and orientation with fewer constraints.

Culture with periodic hydrostatic pressure here refers to a periodic stimulation of mechanical stress onto cells during cell culture. The concept of such a periodic hydrostatic pressure is can be explained with the analogy to the movement during a human workout. An interesting result was recently presented by Yokoyama group, who utilized periodic pressure during culturing smooth muscle (SMC) cells. They found an optimum recipe regarding culturing elastic cell sheet, and the applied frequency, the minimum pressure, and the maximum pressure of the recipe were 0.002 Hz, 110 kPa, and 180 kPa, respectively [16]. Under such an optimum recipe, they succeeded in growing a 10-layer cell sheet in 20 days. Furthermore, they made a vascular graft, with a diameter of 1 mm, from the cell sheet and transplanted the graft into a rat for reconnecting a cut artery. The results showed that the rat could continuously survive for 2.5 months after the transplantation, and new capillaries were found grown on the transplanted tissue. This is a sign showing that the body can successfully adapt to transplanted tissue and apply the necessary nutrition to the part.

The first prototype of on-chip cell incubator was recently presented with a result showing that the cells cultured with periodic hydrostatic pressure have a greater number of stress fibers [17]. While cell sheet cultured under periodic hydrostatic pressure is having a greater elasticity than that under atmospheric pressure [16], it is still not clear when and how this difference happens. The main goal of this work is to reveal the difference between cultured cells with and without periodic hydrostatic pressure during the culture. We focused on the cell behavior in the first one hour after cell seeding. The cell behavior was observed with time-lapse images taken from the observation window of a developed on-microscope incubation system. The proposed cell incubator was with two independent culture arrays in it. The key advantage is that we can compare both cell responses under the same conditions except periodic hydrostatic pressure. Through the experiments, we found that cells showed remarkable size growth under periodic hydrostatic pressure with the frequency-dependent fluctuation that the cell size changed with the applied pressure.

2. Materials and Method

2.1. Periodic Hydrostatic Pressure and Optimum Stimulus

Figure 1a illustrates the analogy of human exercise to cell culture with periodic hydrostatic pressure. The upper and lower figures in Figure 1a are examples of human exercise with weight lifting and cell culture with periodic hydrostatic pressure, respectively. Figure 1b,c explains how cells are cultured under atmospheric pressure and under periodic hydrostatic pressure. The rightmost pictures of Figure 1b,c are our preliminary results of cultured cells. In the results, stress fibers, the actin filaments, were stained with Acti-stain 488 Fluorescent phalloidin for the convenience of observation. It can be seen that the stress fibers of the cells under periodic hydrostatic pressure are thicker than those under atmospheric pressure. The results in Figure 1b,c show clear difference with and without periodic hydrostatic pressure.

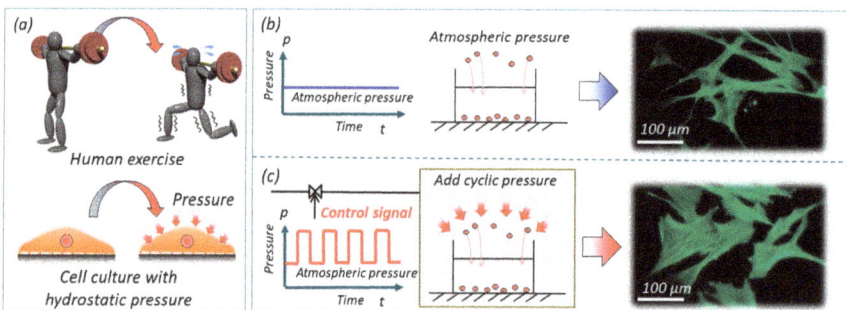

Figure 1. An overview of Periodic hydrostatic pressure. (**a**) The analogy of Periodic hydrostatic pressure to human exercise. (**b**) Conventional cell culture and a result showing the stress fibers of the cultured human smooth muscle cells (HSMCs) under atmospheric pressure. (**c**) Cell culture with Periodic hydrostatic pressure and a result showing the stress fibers of the cultured HSMCs under periodic pressure. The stress fibers, the actin filaments, were stained with Acti-stain 488 Fluorescent phalloidin.

The three most critical parameters of periodic hydrostatic pressure, are the maximum pressure, the minimum pressure, and frequency. These three parameters control the pressure pattern with respect to time during cell culture. In conventional works of pressure-based stimuli culture, the frequencies of periodic pressure are mostly around 1 Hz, just like heartbeat and breathing [16,18–21]. Different from those studies, we tested low frequency zone much less than 1 Hz and reached a new optimum frequency, which had not yet been found previously. As for the determination of the pressure pattern, we had to determine both the maximum and minimum pressures in addition to the frequency. It is well known that our blood pressure is slightly higher than atmospheric pressure, it is about 100 kPa. When the minimum value of blood pressure, which is also known as diastolic blood pressure, is roughly 80 mmHg, about 10 kPa. Thus, we set up the lower pressure is 110 kPa which corresponds to human minimum blood pressure. As for the maximum pressure, we selected it based on the gene expression, such as Fiburin and Lysyl [16]. Fiburin and Lysyl are two important components for stress fibers of human smooth muscle cells (HSMCs), where Fiburin is a key component of elastic fiber growth, and Lysyl helps cross-linking. After preliminary experiments, the maximum pressure of 180 kPa, roughly equivalent to 600 mmHg and about five times the blood pressure, was chosen [16]. Overall, the parameters of the maximum pressure, minimum pressure, and frequency were 180 kPa, 110 kPa, and 0.002 Hz, respectively. The time period of one cycle is about 8 min and 20 s. This means that the periodic pressure between 180 kPa and 110 kPa is given for every 250 s. Of course, we would like to make frequency lower than this, but it could not be done because the pH balance is broken due to pressurized CO_2. If pressurized continues for long time, CO_2 continues to dissolve into the culture solution, and pH becomes below 5.8, while the optimal cell culture pH is between 5.8 and 6.2.

2.2. Development of Cell Incubator and Pressure Control

We aimed to compare cell behaviors with and without periodic hydrostatic pressure during cell culture in this work. To do this, an important issue was how to simultaneously observe the cell groups. We propose the on-chip cell incubator where we can observe what happens in real-time on a single chip, as shown in Figure 2. The chamber arrays on the chip are placed in symmetry to the center line, and they are connected to two independent microfluidic channels. Through this device, we can impart pressure on cells with specified pressure patterns. Microfluidic devices have often been used for experiments under a microscope [22–24]. For example, Eyer et al. used a microchamber array for single cell isolation and analysis of intracellular biomolecules [25]. Reaction experiments and local irritation experiments on single cells can be performed using a microfluidic chip. The on-chip cell incubator in this method was fabricated using a standard photolithography approach with microelectromechanical systems (MEMS) technology and is made of polydimethylsiloxane (PDMS). The fabrications details of the PDMS chip can be found in Appendix A.

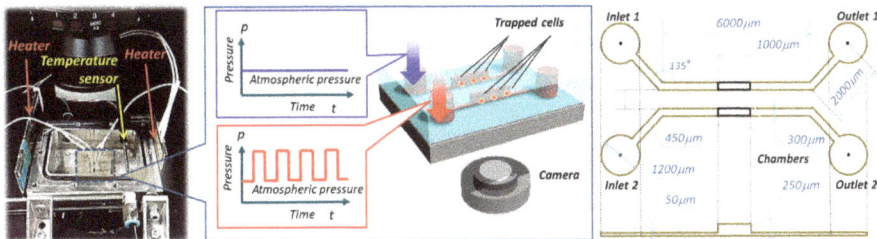

Figure 2. An overview of the experimental system. From left to right are the on-microscope incubation system, design of on-chip cell incubator and its dimensions, respectively. The system is mainly composed of two parallel chambers for simultaneously observation of cell culture with and without Periodic hydrostatic pressure.

The system is composed of a microscope (OLYMPUS: IX71, Olympus Co., Tokyo, Japan), a commercial incubator (SCA-80DS, ASTEC Inc., Fukuoka, Japan), a compressor (DOP-80SP, ULVAC Inc., Kanagawa, Japan), a flow regulator (ITV2030-212BL, SMC Co., Tokyo, Japan), a pressure sensor (HP101-C31-L50A*B/V1, Yokogawa Electric Co., Tokyo, Japan), a digital CMOS camera (C11440, Hamamatsu Photonics K.K., Shizuoka, Japan), and an on-microscope cell incubation system. The on-microscope incubation system is used for maintaining a culture environment as in the leftmost photo in Figure 2. The temperature around the chip is controlled by a feedback system using a heater and a temperature sensor to maintain an environment of 37 °C. The concentration of CO_2 is controlled at 5% by drawn gas from the commercial incubator to the on-microscope incubation system by the flow regulator. Both the inlet and outlet of the chip are connected with Teflon tubes to the pressure regulator, so that the inside of the micro chambers can be pressurized with a specified level at specified time sequences. As mentioned in the previous section, a pressure of 180 kPa and 110 kPa was applied as the periodic hydrostatic pressure in this work.

2.3. Chip Preparation and Experimental Procedure

In order to adhere cells to the PDMS surface for cell culture, 10% adhesion-assist protein fibronectin was coated on the surface of culture array by injecting 25 μL of it from the inlet of the device using a micropipette. The area inside of the PDMS flow channel and micro chamber arrays was filled with the solution and placed at 37 °C in the incubator for 60 min. Afterwards, the fibronectin solution was removed and rinsed with phosphate buffered salts (PBS).

Human smooth muscle cells (HSMCs) were used for all experiments in this work. Cells were injected into the chamber arrays for seeding from the inlets of the PDMS device. The concentration of HSMC was 2.5×10^5 cells/mL. Low cell density was used for the convenience of single cell analysis, particularly for avoiding overlaps of the projected area. The tube on the outlet side of the periodic hydrostatic pressure chamber was connected to the compressor for pressurized gas. Since the height of the micro chamber is higher than the other flow micro channel region, more cells can be trapped in the micro chamber due to inertial flow. After cell injection, the cells were gradually attached to the bottom of the glass by gravity force. For the culture with periodic hydrostatic pressure, the tube on the inlet side was sealed and periodic pressure was applied to the chamber. For the culture without periodic hydrostatic pressure, the tube was open to the atmosphere.

Time-lapse imaging was employed for comparing the difference between cell growth with and without periodic hydrostatic pressure. The camera takes a picture covering both chambers simultaneously every 10 s after seeding. We focused on how the cells grow during the first hour after seeding. Figure 3a shows an example of a captured time-lapse image. Both cells in the two micro chambers were recorded simultaneously, where the upper and lower chamber in Figure 3 are with and without periodic hydrostatic pressure mode, respectively. Only about 5–10 cells were trapped in the chamber of the on-chip cell incubator device for the convenience of single cell observation.

In order to evaluate the degree of cell growth, we introduced the projected area as an evaluation index where it was defined by the projected area of cell in the horizontal plane. An example of cell growth and image processing is demonstrated in Figure 3b,c, respectively. In Figure 3b, cell images were acquired for each record time, and images at 20 min, 40 min, and 60 min. Figure 3c shows the image at 60 min after seeding and the left shows the original image. The outline in the middle of Figure 3c allows us to determine the cell area, as shown in the rightmost image in Figure 3c. As a result, we can obtain the information of cell shape and area using image analysis software *Image J* (1. 50i, Wayne Rasband, National Institutes of Health, USA).

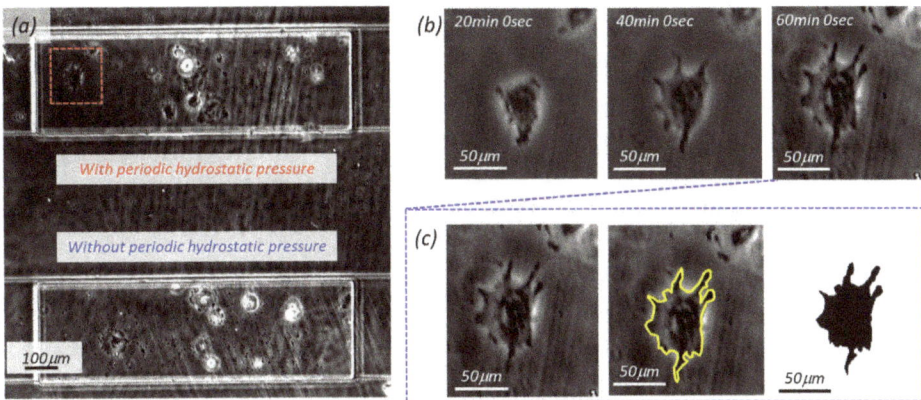

Figure 3. Captured pictures by time-lapse imaging. (**a**) Captured microscopic image where the chambers on the top and the bottom are the cells cultured with and without the proposed periodic hydrostatic pressure, respectively. An example of cell assessment is demonstrated using the highlighted cell. (**b**) Selected time-lapse images of the cell at the time of 20, 40, and 60 min. (**c**) The procedure to obtain the extracted cell area.

Figure 4 shows measured results of pressure control during the culture, and the pressure was cycling between 180 kPa and 110 kPa every 250 s. Figure 4a,b is close views of increasing and decreasing pressure periods, respectively. From Figure 4a,b, we can see a reasonably well controlled pressure where an error is roughly less than 1% with respect to the target value without overshoot. In addition, it can be seen that the time for pressurization to the target values is less than one second.

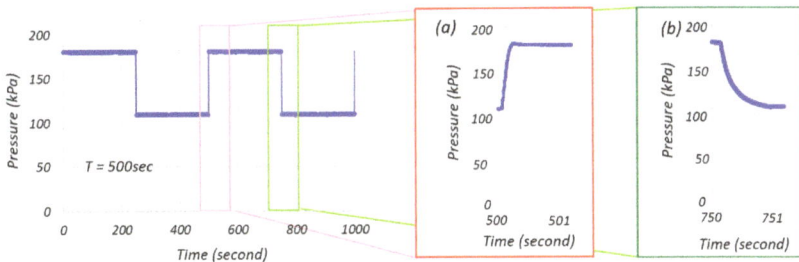

Figure 4. The performance of pressure control during periodic hydrostatic pressure. (**a**) A close view of the pressure increasing period. (**b**) A close view of the pressure decreasing period.

3. Results

3.1. Projected Area of Cells with and without Periodic Hydrostatic Pressure

Figure 5 shows the growth of the projected cell area with respect to time where Figure 5a,b denotes the projected areas and the average value among the six trapped cells, respectively. The cell groups with and without periodic hydrostatic pressure are indicated by red and blue marks and the original of the time axis is the starting time of pressurization. From Figure 5, no significant difference can be seen between the two groups with and without periodic hydrostatic pressure.

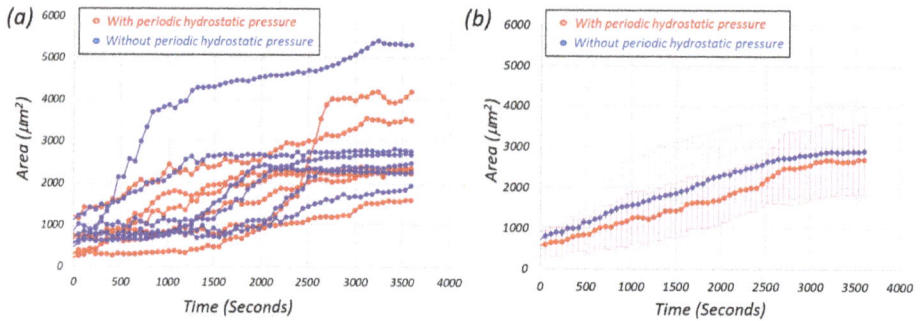

Figure 5. The growth of the projected area with respect to time, where the cell groups with and without periodic hydrostatic pressure are indicated by red and blue marks, respectively. (**a**) Measured area of six cells from the periodic hydrostatic pressure chamber (red) and 6 cells from control (blue), a culture chamber without periodic hydrostatic pressure (**b**) The average value and standard deviation from the six cells in each chamber are plotted. No significant difference between two.

An example of cell area changes under periodic hydrostatic pressure is shown in Figure 6, where there is a remarkable point (b). The projected cell area increased rapidly from (b) to (c), and was with nearly three times faster than the initial 15 min. After that, the cell entered another phase where the projected cell area increased with a slightly gentle slope from (c) to (d). It is interesting to know the tendency of the projected cell area with or without periodic hydrostatic pressure with such instances. A time-lapse cell behavior during cell culture can be found in the supplementary material, Videos S1 and S2. Video S1 shows the cell behavior from the point (a) to the point (d), and Video S2 shows the cell behavior after the point (c). After point (c), where rapid growth ended, interesting behavior was observed in which the cell periodically extends in terms of its size.

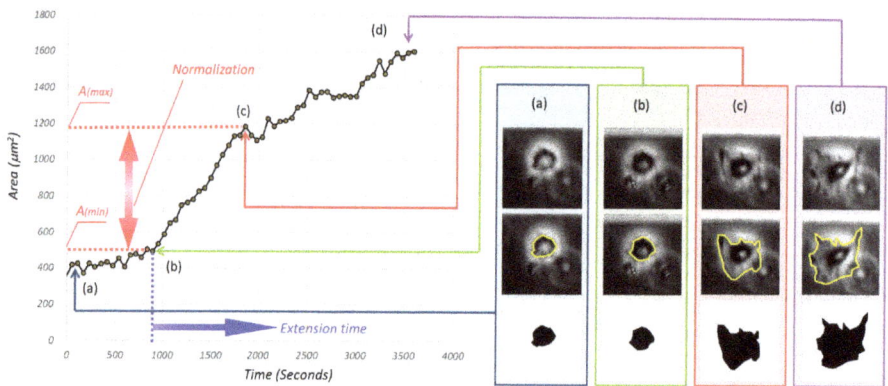

Figure 6. An example of the projected cell area with respect to time. (**a,b**) are the cell images at different instance, and from the top to the bottom are the original cell image, contour extraction, the projected cell area, respectively. It should be noted that the increase velocity of the projected cell area from (b) to (c) is larger than that of other phases, such as from (c) to (**d**).

For determining point (b) and point (c), we used S(*t*), defined by:

$$S(t) = \frac{A_{(t)} - A_{min}}{A_{max} - A_{min}} \qquad (1)$$

where A_{min}, A_{max}, and A(*t*) are the projected cell areas at the time in point (b), at the time in point (c), and at the time of t, respectively. Simply speaking, point (b) is the starting point when the cell size starts to increase rapidly and point (c) is the ending point when the rapid change of cell size is terminated.

Figure 7 shows the normalized S(*t*), where the cell groups with and without periodic hydrostatic pressure are indicated by red and blue marks, respectively. The origin of the horizontal axis is the time which corresponds to the point (b) in Figure 6. The velocity of the projected cell area increased dramatically after point (b). From Figure 7, both cell groups with and without periodic hydrostatic pressure resulted in 1.0 in about 20 min (1200 s), which matched well with the definition of A_{min} and A_{max}. On the other hand, the tendencies of S(*t*) shown by the two groups after 20 min (1200 s) were different. The cell group with periodic hydrostatic pressure continued to increase the projected cell area with a gentle slope, whereas the cell group without periodic hydrostatic pressure was only a tiny positive slope. The *p*-value of the T test for the cell size of two groups at 20 min (1200 s) was 0.173 and indicated no significant difference. However, when the time reached 30 min (1800 s), the *p*-value became less than 0.05 and demonstrated a significant difference between the growth rate with and without periodic hydrostatic pressure. In other words, the significant difference for the culture with and without periodic hydrostatic pressure happened after the time past 30 min in Figure 7, as the point (c) in Figure 6.

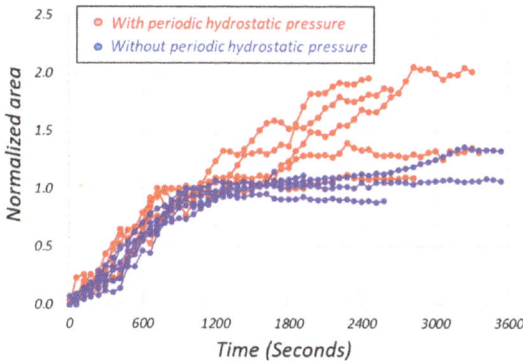

Figure 7. The normalized projected cell area with respect to the time where the origin of the time is redefined by the time corresponding to point (b) in Figure 6. The cell groups with and without periodic hydrostatic pressure are indicated by red and blue marks, respectively, and normalized for each of the six original data shown in Figure 5.

3.2. Periodic Characteristics in Cell Growth

Figure 7 shows two important results, one of which is that there is statistically meaningful difference at 30 min after cell extension, and the other is that the growth pattern is fluctuating with respect to time. To observe the growth pattern more qualitatively, let us rearrange the time domain, so that we can adjust the phase among all experiments under periodic hydrostatic pressure.

Figure 8 shows the normalized projected cell area S(*t*), where the origin of the horizontal axis is the time corresponding to (c) in Figure 6, more specifically when the pressure is switched from 180 kPa to 110 kPa. This means that A*max* in Equation (1) is the projected cell area at the time t when the pressure is switched from 180 kPa to 110 kPa in the nearest time around (c) in Figure 6.

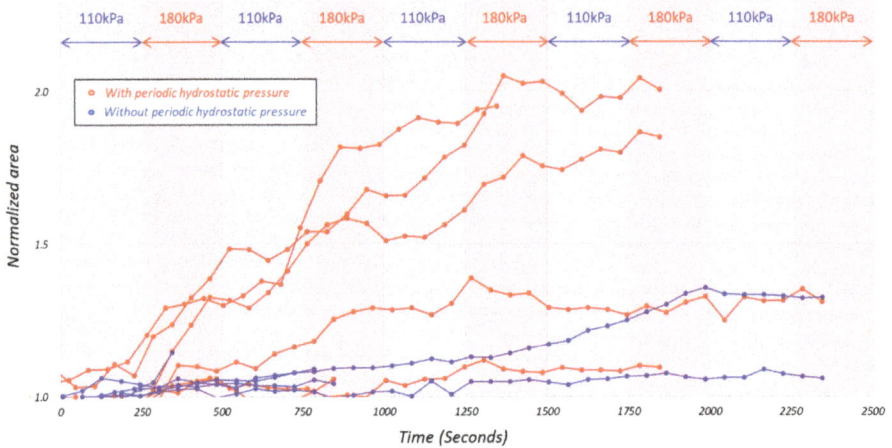

Figure 8. The normalized projected cell area S(*t*) where the origin of the horizontal axis is the time corresponding to (c) in Figure 6, more specifically when the pressure is switched from 180 kPa to 110 kPa in the nearest time around (c) in Figure 6, and normalized for each of the six original data shown in Figure 5.

3.3. Frequency Analysis on Projected Area of Cells

Figure 9 explains how to achieve the frequency analysis for one particular cell under periodic hydrostatic pressure, where Figure 9a–c denotes the normalized area with respect to time, the curve defined by ΔS(*t*) = S(*t*) − S(*t*)$_{\text{approximate curve}}$, and the frequency analysis, respectively. S(*t*)$_{\text{approximate curve}}$ is obtained by a linear fit with the-least-squares method as shown in Figure 9a. From Figure 9c, we can see an interesting observation, namely, frequency-dependent cell growth. The frequency of 0.002 Hz was a peak of the frequency analysis and it corresponds to the frequency of periodic hydrostatic pressure. Figure 9c–h shows three examples of frequency analysis where (c) through (e) is cultured under periodic hydrostatic pressure and (f) through (h) is cultured under atmospheric pressure. In the cell group cultured under periodic hydrostatic pressure, we can see clear amplitude in the frequency domain with 0.002 Hz, while we cannot see clear amplitude in the frequency domain with 0.002 Hz in the cell group cultured under atmospheric pressure.

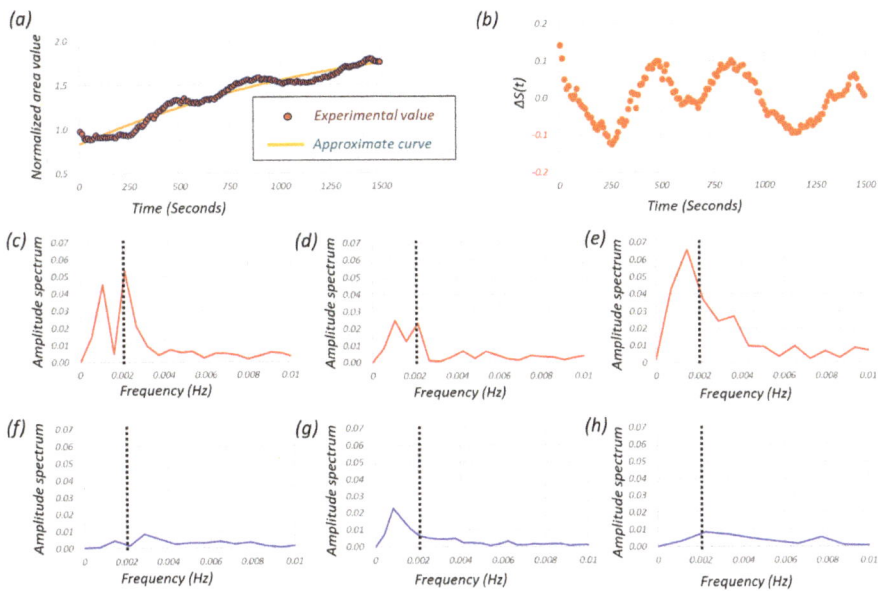

Figure 9. How to achieve frequency analysis, where (**a**) is the normalized area with respect to time where the approximate curve is shown by the line shown by red color, (**b**) the curve defined by S(*t*) − S(*t*)$_{approximate\ curve}$, and (**c**) the frequency analysis. The frequency analysis, where (**c–e**) are under periodic hydrostatic pressure, and (**f–h**) are cultured under atmospheric pressure. Three of each of the six original data were carried out.

4. Discussion

The on-chip cell incubator proposed in this paper includes two important key words, one of which is "same tme history" and the other is "simultaneous observation". "Same time history" means cells whose initial conditions are exactly the same, including the time for all cells in two chambers. It is a great advantage to completely avoid the influence for the results coming from the time difference among cells, and therefore, we can keep the culture condition the same in both chambers except either under periodic hydrostatic pressure or atmospheric pressure. "Simultaneous observation" allows us to visualize cell behaviors coming from only one parameter, in this work the effect of periodic hydrostatic pressure.

Using the on-chip cell incubator, we could observe how the cells grow by focusing in the first hour for both cultures with and without periodic hydrostatic pressure. The most interesting result under periodic hydrostatic pressure is that cells in growth period increase the projected cell area according to the pressure frequency imparted on the culture liquid. This effect is more enhanced for the cell whose size is bigger. To the best of the authors' knowledge, there was no such size fluctuation of cell size reported in literature under a periodic pressure stimulus. A natural question that comes up is why there has been no report on the frequency-dependent cell size fluctuation so far. Our work is based on an extremely low frequency, 0.002 Hz, while the frequencies of periodic pressure in former works are mostly around 1Hz, just like heartbeat and breathing [16,18–21]. We believe that several minutes are needed for cells to change size during periodic hydrostatic pressure and it is hard to change the size with a noticeable range under a pressure frequency with around 1 Hz.

5. Conclusions

In the same way that human muscle grows after exercise, it is known that an elastic cell sheet can be obtained by cell culture with periodic hydrostatic pressure. This paper presented simultaneous cell observation by an on-chip cell incubator with and without such a periodic hydrostatic pressure. The periodic pressure with an extremely low frequency of 0.002 Hz was imparted for one chamber and atmospheric pressure was given for the other one. The experiments were only focused on the first one hour after cell seeding, and significant difference of cell growth were observed. We also found an interesting phenomenon during periodic hydrostatic pressure where the projected areas of the cells increased at the same frequency as the pressure frequency imparted on them. For future work, we plan to test with difference frequencies and see in which frequency the frequency-dependent cell growth disappears.

Supplementary Materials: The following are available online at http://www.mdpi.com/2072-666X/10/2/133/s1, Video S1: the cell behavior from the point (a) to the point (d) during cell culture, and Video S2: the cell behavior against the pressure value after the point (c).

Author Contributions: All authors conceived and designed the experiments; M.H. performed the experiments; M.H. and C.T., contributed to the data analysis and interpretations; M.H., C.T., and M.K. wrote the paper.

Acknowledgments: This work was partially supported by JSPS KAKENHI Grant-in-Aid for Scientific Research(S) 15H05761, Challenging Exploratory Research 17K18854, Grant-in-Aid for Young Scientists (B) 17K14625, AMED (Japan Agency for Medical Research and Development) 18ek0109240h0002, Nanotech Career-up Alliance (Nanotech CUPAL), and the Ministry of Education via the "Nanotechnology Platform Undertaking". The microfluidic devices were fabricated at Nanotechnology Open Facilities in Osaka University [F-18-OS-0008 S-18-OS-0008].

Conflicts of Interest: The authors declare no conflict of interest.

Appendix A

In this research, PDMS (X-32-3094-2, Shin-Etsu Chemical Co., Ltd., and Tokyo, Japan) microfluidic chips for cell culture were fabricated by soft lithography technique [26]. The microfluidic device with different channel heights was designed in order to trap cells at the bottom of the chambers, and therefore, two-layer microstructures were fabricated with SU-8 photoresist (SU-8 3050, Microchem Corp., Massachusetts, USA), as shown in Figure A1. The width and length of the flow channels were 300 μm and 6000 μm, respectively, while the diameters of the inlet/outlet and the height of these structures were 1500 μm and 50 μm, respectively. After spin coating for the first layer of photoresist as shown in Figure A1, a baking step for 15 min at 95 °C was performed. Figure A1b shows that the SU-8 photoresist was exposed to 200 mJ/cm^2 of UV light with a mask aligner (MA-10, MIKASA CO., LTD, Tokyo, Japan), and then the first layer of photoresist was developed by 5 min baking at 95 °C, as shown in Figure A1c. Figure A1d shows the coating of the second layer of SU-8 photoresist for trapping the cells. The thickness of the second layer was designed as 250 μm and is baked for 300 min at 95 °C after the coating. The SU-8 photoresist was exposed to 500 mJ/cm^2 of UV light using the mask aligner, as shown in Figure A1e, and baked for another 20 min at 95 °C before the second-layer structure was developed, as shown in Figure A1f. The mold was fabricated through the process shown in Figure A1a–f. Afterwards, PDMS mixture with a weight ratio of 10:1 between the base and curing agent was poured onto the mold in a plastic dish. The PDMS was degassed for 30 min in vacuum desiccators, before being baked for 30 min at 90 °C. After the PDMS chip was completely cured, the chip was picked up by peeling off the PDMS replicas from the mold as shown in Figure A1g,h. The hardness of the PDMS chip was 80 and was important for suppressing the deformation of the chip for the periodic pressure application of cell culture. Next, holes of 1000 μm in diameter for inlet wells and outlet wells in the PDMS sheet were punched as shown in Figure A1i. Finally, a PDMS chip and glass plate were bonded by using a plasma bonding device (CUTE-MP, Femto Science Inc., Gyeonggi-Do, Korea) and baked for 10 min at 80 °C, as shown in Figure A1j. Figure A1k illustrates the fabricated microfluidic device.

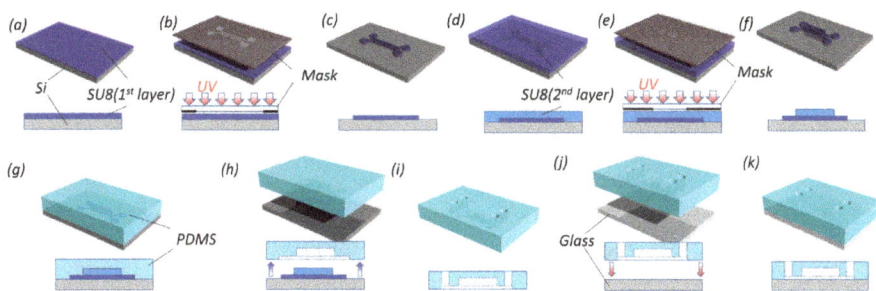

Figure A1. Fabrication of the microfluidic chip. The device consists of inlets, outlets, cell trap chambers. (**a–f**) Step-by-step photolithography for mold fabrication; (**g–k**) the step-by-step procedure for PDMS chip fabrication from the mold. PDMS and glass are integrated by plasma bonding.

References

1. Sies, H.; Berndt, C.; Jones, D.P. Oxidative stress. *Annu. Rev. Biochem.* **2017**, *86*, 715–748. [CrossRef] [PubMed]
2. Olagnier, D.; Peri, S.; Steel, C.; Van Montfoort, N.; Chiang, C.; Beljanski, V.; Slifker, M.; He, Z.; Nichols, C.N.; Lin, R.; et al. Cellular Oxidative Stress Response Controls the Antiviral and Apoptotic Programs in Dengue Virus-Infected Dendritic Cells. *PLoS Pathog* **2014**, *10*, e1004566. [CrossRef] [PubMed]
3. Veschgini, M.; Gebert, F.; Khangai, N.; Ito, H.; Suzuki, R.; Holstein, T.W.; Mae, Y.; Arai, T.; Tanaka, M. Tracking mechanical and morphological dynamics of regenerating Hydra tissue fragments using a two fingered micro-robotic hand. *Appl. Phys. Lett.* **2016**, *108*, 103702. [CrossRef]
4. Murakami, R.; Tsai, C.-H.D.; Ito, H.; Tanaka, M.; Sakuma, S.; Arai, F.; Kaneko, M. Catch, load and launch toward on-chip active cell evaluation. In Proceedings of the 2016 IEEE International Conference on Robotics and Automation (ICRA), Stockholm, Sweden, 16–21 May 2016; pp. 1713–1718.
5. Ohara, K.; Kawakami, D.; Takubo, T.; Mae, Y.; Tanikawa, T.; Honda, A.; Arai, T. Dextrous cell diagnosis using two-fingered microhand with micro force sensor. *J. Micro Nano Mechatron.* **2012**, *7*, 13–20. [CrossRef]
6. Horade, M.; Tsai, C.-H.D.; Ito, H.; Kaneko, M. Red Blood Cell Responses during a Long-Standing Load in a Microfluidic Constriction. *Micromachines* **2017**, *8*, 100. [CrossRef]
7. Ito, H.; Murakami, R.; Sakuma, S.; Tsai, C.-H.D.; Gutsmann, T.; Brandenburg, K.; Pöschl, J.M.B.; Arai, F.; Kaneko, M.; Tanaka, M. Mechanical diagnosis of human erythrocytes by ultra-high speed manipulation unraveled critical time window for global cytoskeletal remodeling. *Sci. Rep.* **2017**, *7*, 43134. [CrossRef] [PubMed]
8. Di Cara, F.; Maile, T.M.; Parsons, B.D.; Magico, A.; Basu, S.; Tapon, N.; King-Jones, K. The Hippo pathway promotes cell survival in response to chemical stress. *Cell Death Differ.* **2015**, *22*, 1526–1539. [CrossRef] [PubMed]
9. Seo, K.; Rainer, P.P.; Lee, D.I.; Hao, S.; Bedja, D.; Birnbaumer, L.; Cingolani, O.H.; Kass, D.A. Hyperactive Adverse Mechanical Stress Responses in Dystrophic Heart Are Coupled to Transient Receptor Potential Canonical 6 and Blocked by cGMP–Protein Kinase G Modulation. *Circ. Res.* **2014**, *114*, 823–832. [CrossRef] [PubMed]
10. Sakuma, S.; Kuroda, K.; Tsai, C.-H.D.; Fukui, W.; Arai, F.; Kaneko, M. Red blood cell fatigue evaluation based on the close-encountering point between extensibility and recoverability. *Lab Chip* **2014**, *14*, 1135. [CrossRef]
11. Teramura, T.; Takehara, T.; Onodera, Y.; Nakagawa, K.; Hamanishi, C.; Fukuda, K. Mechanical stimulation of cyclic tensile strain induces reduction of pluripotent related gene expressions via activation of Rho/ROCK and subsequent decreasing of AKT phosphorylation in human induced pluripotent stem cells. *Biochem. Biophys. Res. Commun.* **2012**, *417*, 836–841. [CrossRef]
12. Subramony, S.D.; Dargis, B.R.; Castillo, M.; Azeloglu, E.U.; Tracey, M.S.; Su, A.; Lu, H.H. The guidance of stem cell differentiation by substrate alignment and mechanical stimulation. *Biomaterials* **2013**, *34*, 1942–1953. [CrossRef] [PubMed]

13. Itoh, M.; Nakayama, K.; Noguchi, R.; Kamohara, K.; Furukawa, K.; Uchihashi, K.; Toda, S.; Oyama, J.-I.; Node, K.; Morita, S. Scaffold-Free Tubular Tissues Created by a Bio-3D Printer Undergo Remodeling and Endothelialization when Implanted in Rat Aortae. *PLoS ONE* **2015**, *10*, e0136681.

14. Lee, V.K.; Kim, D.Y.; Ngo, H.; Lee, Y.; Seo, L.; Yoo, S.-S.; Vincent, P.A.; Dai, G. Creating perfused functional vascular channels using 3D bio-printing technology. *Biomaterials* **2014**, *35*, 8092–8102. [CrossRef]

15. Murphy, S.V.; Atala, A. 3D bioprinting of tissues and organs. *Nat. Biotechnol.* **2014**, *32*, 773–785. [CrossRef] [PubMed]

16. Yokoyama, U.; Tonooka, Y.; Koretake, R.; Akimoto, T.; Gonda, Y.; Saito, J.; Umemura, M.; Fujita, T.; Sakuma, S.; Arai, F.; et al. Arterial graft with elastic layer structure grown from cells. *Sci. Rep.* **2017**, *7*, 427. [CrossRef] [PubMed]

17. Horade, M.; Kaneko, M.; Tsai, C.D.; Ito, H.; Higashino, N.; Akai, T.; Yokoyma, U.; Ishikawa, Y.; Sakuma, S.; Arai, F. On-Chip Cell Gym. In Proceedings of the 30th IEEE Conference on Micro Electro Mechanical Systems (MEMS2017), Las Vegas, NV, USA, 22–26 January 2017; pp. 603–604.

18. Buschmann, M.D.; Gluzband, Y.A.; Grodzinsky, A.J.; Hunziker, E.B. Mechanical compression modulates matrix biosynthesis in chondrocytelagarose culture. *J. Cell Sci.* **1995**, *108*, 1497–1508. [PubMed]

19. Carver, S.E.; Heath, C.A. Increasing extracellular matrix production in regenerating cartilage with intermittent physiological pressure. *Biotechnol. Bioeng.* **1999**, *62*, 166–174. [CrossRef]

20. Hall, A.C.; Urban, J.P.G.; Gehl, K.A. The effects of hydrostatic pressure on matrix synthesis in articular cartilage. *J. Orthop. Res.* **1991**, *9*, 1–10. [CrossRef]

21. Suh, J.K.; Baek, G.H.; Årøen, A.; Malin, C.M.; Niyibizi, C.; Evans, C.H.; Westerhausen-Larson, A. Intermittent sub-ambient interstitial hydrostatic pressure as a potential mechanical stimulator for chondrocyte metabolism. *Osteoarthr. Cartil.* **1999**, *7*, 71–80. [CrossRef]

22. Horade, M.; Mizuta, Y.; Kaji, N.; Higashiyama, T.; Arata, H. Plant-on-a-chip microfluidic-system for quantitative analysis of pollen tube guidance by signaling molecule: Towards cell-to-cell communication study. In Proceedings of the 16th International Conference on Miniaturized Systems for Chemistry and Life Sciences (MicroTAS 2012), Okinawa, Japan, 28 October–1 November 2012; pp. 1027–1029.

23. Yasaki, H.; Yasui, T.; Yanagida, T.; Kaji, N.; Kanaki, M.; Fukuyama, M.; Nagashima, K.; Kawai, T.; Baba, Y. Microfluidic long-pore-channel to highlight bacteria contents. In Proceedings of the 20th International Conference on Miniaturized Systems for Chemistry and Life Sciences (MicroTAS 2016), Dublin, Ireland, 9–13 October 2016; pp. 101–102.

24. Shimizu, K.; Shunori, A.; Morimoto, K.; Hashida, M.; Konishi, S. Development of a biochip with serially connected pneumatic balloons for cell-stretching culture. *Sens. Actuators B Chem.* **2011**, *156*, 486–493. [CrossRef]

25. Eyer, K.; Kuhn, P.; Hanke, C.; Dittrich, P.S. A microchamber array for single cell isolation and analysis of intracellular biomolecules. *Lab Chip* **2012**, *12*, 765–772. [CrossRef] [PubMed]

26. Anderson, J.R.; Chiu, D.T.; Jackman, R.J.; Cherniavskaya, O.; McDonald, J.C.; Wu, H.; Whitesides, S.H.; Whitesides, G.M. Fabrication of Topologically Complex Three-Dimensional Microfluidic Systems in PDMS by Rapid Prototyping. *Anal. Chem.* **2000**, *72*, 3158–3164. [CrossRef] [PubMed]

micromachines

MDPI

Article

A Silicon-based Coral-like Nanostructured Microfluidics to Isolate Rare Cells in Human Circulation: Validation by SK-BR-3 Cancer Cell Line and Its Utility in Circulating Fetal Nucleated Red Blood Cells

Gwo-Chin Ma [1,2], Wen-Hsiang Lin [1], Chung-Er Huang [3,4], Ting-Yu Chang [1], Jia-Yun Liu [1], Ya-Jun Yang [1], Mei-Hui Lee [1], Wan-Ju Wu [5], Yun-Shiang Chang [6,*] and Ming Chen [1,5,6,7,8,9,*]

[1] Department of Genomic Medicine and Center for Medical Genetics, Changhua Christian Hospital; and Department of Genomic Science and Technology, Changhua Christian Hospital Healthcare System, Changhua 50046, Taiwan; 128729@cch.org.tw (G.-C.M.); 397620cch@gmail.com (W.-H.L.); taiwanbird@gmail.com (T.-Y.C.); 182011@cch.org.tw (J.-Y.L.); 157097@cch.org.tw (Y.-J.Y.); 29561@cch.org.tw (M.-H.L.)

[2] Department of Medical Laboratory Science and Biotechnology, Central Taiwan University of Science and Technology, Taichung 40601, Taiwan

[3] International College of Semiconductor Technology, National Chiao Tung University, Hsinchu 30010, Taiwan; cehuang5858@gmail.com

[4] Cytoaurora Biotechnologies, Inc. Hsinchu Science Park, Hsinchu 30016, Taiwan

[5] Department of Obstetrics and Gynecology, Changhua Christian Hospital, Changhua 50006, Taiwan; crystalwu835@gmail.com

[6] Department of Molecular Biotechnology, Da-Yeh University, Changhua 51591, Taiwan

[7] Department of Obstetrics and Gynecology, College of Medicine, National Taiwan University, Taipei 10041, Taiwan

[8] Department of Medical Genetics, National Taiwan University Hospital, Taipei 10041, Taiwan

[9] Department of Life Science, Tunghai University, Taichung 40704, Taiwan

* Correspondence: yschang@mail.dyu.edu.tw (Y.-S.C.); mingchenmd@gmail.com (M.C.); Tel.: +886-4-851-1888 (ext. 4265) (Y.-S.C.); +886-4-723-8595 (ext. 2323) (M.C.)

Received: 24 January 2019; Accepted: 14 February 2019; Published: 17 February 2019

Abstract: Circulating fetal cells (CFCs) in maternal blood are rare but have a strong potential to be the target for noninvasive prenatal diagnosis (NIPD). "Cell Reveal[TM] system" is a silicon-based microfluidic platform capable to capture rare cell populations in human circulation. The platform is recently optimized to enhance the capture efficiency and system automation. In this study, spiking tests of SK-BR-3 breast cancer cells were used for the evaluation of capture efficiency. Then, peripheral bloods from 14 pregnant women whose fetuses have evidenced non-maternal genomic markers (e.g., de novo pathogenic copy number changes) were tested for the capture of circulating fetal nucleated red blood cells (fnRBCs). Captured cells were subjected to fluorescent in situ hybridization (FISH) on chip or recovered by an automated cell picker for molecular genetic analyses. The capture rate for the spiking tests is estimated as 88.1%. For the prenatal study, 2–71 fnRBCs were successfully captured from 2 mL of maternal blood in all pregnant women. The captured fnRBCs were verified to be from fetal origin. Our results demonstrated that the Cell Reveal[TM] system has a high capture efficiency and can be used for fnRBC capture that is feasible for the genetic diagnosis of fetuses without invasive procedures.

Keywords: cbNIPD; fnRBC; capture efficiency; microfluidics; nanostructure

1. Introduction

Since the first report of circulating fetal cells (CFCs) in maternal blood in 1959 [1], CFCs have been expected as the potential target of noninvasive prenatal diagnosis (NIPD). However, the isolation of CFCs for genetic analysis is always a challenge because of the scarcity of the cells in maternal circulation (1/10,000–1,000,000) [2]. Recently, by advances in knowledge about CFCs and in technology at single-cell genetic analyses, cell-based NIPD (cbNIPD) have again been in focus [3]. In contrast to the popular noninvasive prenatal testing (NIPT) based on cell-free fetal DNA (cffDNA) [4–13], which mainly reflects the genetic complement of placental trophoblasts and cannot recognize the condition of fetoplacental mosaicism (a situation where there is a discrepancy between the genomic makeup of the fetus and placenta) [14,15], cell-based technology had been reported to be able to capture not only trophoblasts but also fetal nucleated red blood cells (fnRBC, which can truly reflect the fetal genome). Nevertheless, most previous reports regarding cbNIPD focused on capturing trophoblasts from placenta that prohibited a definite diagnosis of fetuses and thus were not superior to cffDNA testing [16–18]. One major criticism of the previous studies is that very few fetal specific antigens are available since nucleated red blood cells (nRBCs) in maternal circulation can be of both maternal and fetal origin [19,20]. It is mandatory to verify that the captured nRBCs are indeed of fetal origin.

Only a few research groups published study results on capturing fnRBCs [19,21–24]. In our previous report, we have verified our captured circulating nRBCs were indeed of fetal origin using whole genome amplification (WGA) followed by subsequent short tandem repeat (STR) analyses, with a limited sample size (*n* = 5) [19]. There are two directions to solve this hurdle: one is to explore more fetal specific antigens to undoubtedly identify fnRBCs [25–27] and the other is to optimize the efficiency of the cell capture platform used. In this study, we adopted the latter strategy to overcome this difficulty by demonstrating that at least a significant proportion of the captured nRBCs are fetal origin, in contrast to most previous reports that showed a rarity of fnRBCs (one in 30 mL maternal blood) by their capturing methodologies [3,28,29].

Rare cell populations in human circulation (i.e., CFCs and circulating tumor cells (CTCs)) can be isolated by different methodologies [30–36], including (1) immunoaffinity-based positive/negative enrichment; (2) biophysical-based selections by density gradient, size, electrical signature, or acoustophoretic mobility; (3) direct image modalities either by improving the efficiency of imaging or by replacing the enrichment through high-speed fluorescent imaging [37]; and (4) functional assays based on the bioactivity of cells such as protein secretion or cell adhesion [33]. Our platform (named Cell Reveal™ system) is classified as an immunoaffinity-based positive enrichment system coupled with a proprietary direct imaging modality which can accurately map the coordinates of the cells captured, followed by the subsequent recovery of the captured cells by an automated cell picker upgraded from a manual micropipetting system [19]. The microfluidics we used was named as "Coral Chip", an upgraded version of the PicoBioChip [19], for its coral-like nanostructure clearly visible under the scanning electronic microscope (SEM).

In this study, we evaluate the capture efficiency of the Cell Reveal™ system by spiking tests of SK-BR-3 breast cancer cells. Both array comparative genomic hybridization (aCGH) and next generation sequencing (NGS) were used to elucidate the characteristic molecular signatures of such cancer cells. Then, we validate the use of the platform for a series of prenatal cases in which at least one undisputable non-maternal genomic marker is present in the fetuses, for example, in those women who carried male fetus (Y chromosome will be the non-maternal marker) and in those women with de novo genomic imbalances such as trisomies or chromosome copy number changes. Genetic analyses, including fluorescence in situ hybridization (FISH), aCGH, and STR analyses, were directly performed for the captured cells, which confirm the captured nRBC are indeed from fetuses (i.e., fnRBCs). Our results demonstrated that by capturing fnRBCs and using the subsequent well-established comprehensive genomic approaches, a true NIPD with resolutions similar to the invasive sampling is closer to reality.

2. Materials and Methods

2.1. Materials

Two cell lines were used to create artificial cell mixtures in the cell spiking test: (1) SK-BR-3 (human breast cancer cells, HTB-30, ATCC, Manassas, VA, USA), which expresses the cell markers of epithelial cell adhesion molecule (EpCAM) and cytokeratin (CK) and lacks the leukocyte common antigen (CD45). SK-BR-3 cancer cells were maintained in McCoy's 5A medium (BioConcept, Allschwil, Switzerland), supplemented with 10% fetal bovine serum (FBS) and 100 units/mL antibiotic-antimycotic (Gibco, Grand island, NY, USA). The other cell line was (2) Jurkat (immortalized human T lymphocyte cells), which expresses the cell marker of CD45 and lacks EpCAM and CK. Jurkat cells were maintained in an RPMI-1640 medium (BioConcept, Allschwil, Switzerland), supplemented with 10% FBS and 100 units/mL antibiotic-antimycotic (Gibco, Grand island, NY, USA). Prior to be mixed, both cell lines were incubated with anti-EpCAM antibody at 37 °C for 45 min and then spun at 300× g for 10 min to collect the cell pellets. The cell mixture was prepared by spiking 5×10^3 SK-BR-3 cells into 10^6 Jurkat cells and was resuspended in 200 µL Dulbecco's phosphate-buffered saline (DPBS), which was used as the model sample for the evaluation of the capture efficiency of the Cell Reveal^{TM} system.

Blood samples collected from pregnant women were then used for the cbNIPD study. The fnRBCs which have distinct cell markers, such as the cluster of differentiation 71 (CD71), glycophorin A (GPA), the cluster of differentiation 36 (CD36), and epsilon hemoglobin, permitting to be isolated from the maternal blood [38–41] were chosen as the target for genetic analysis. The cluster of differentiation 45 (CD45) expressed on all white blood cells (WBCs) but not on fnRBCs was used as a negative selection marker for the fnRBC capture. Fourteen pregnant women with singleton pregnancies (at gestational age (GA) of 13^{+4}–27^{+5} week^{+day} who have received invasive procedures (chorionic villus sampling or amniocentesis) and with confirmed fetuses that have evident non-maternal genomic markers, including 4 cases with de novo pathogenic copy number changes (9p24.2p23 deletion, $n = 1$; 10q25.2q26.12 deletion, $n = 1$; 21q22.11q22.3 deletion, $n = 1$; and 22q11.21 deletion, $n = 1$), 4 cases with trisomic chromosomes (48,XXY,+18, $n = 1$; 47,XY,+18, $n = 2$; and 47,XY,+21, $n = 1$), and 6 euploid cases with a male karyotype (46,XY, $n = 6$) were recruited in this study. For each case, approximately 8 mL of venous blood were collected and stored in the BD vacutainer ®with acid citrate dextrose (ACD) solution A (Becton, Dickinson and Company, Franklin Lakes, NJ, USA). This study was approved by the Ethics Committee of the Changhua Christian Hospital, Changhua, Taiwan (project ID: CCH-IRB-171215). All participants gave written informed consent before the study began.

2.2. Coral Chip Manufacture

The Coral Chip is a silicon (Si)-based chip with a porous morphology on the inside of microfluidic chambers that are capable to capture targeted cells from a cell mixture. The chip is fabricated using the metal-assisted chemical etching technology as previously described [19], with minor modifications (Figure 1). Briefly, 5 instead of 3 microfluidic chambers were created in this chip to extend the surface area for cell capture. Moreover, the fabrication sequence was revised. The starting materials of p-type (100) Si wafers followed the standard cleaning to remove the environmental contaminants. Then, the plasma-enhanced chemical vapor deposition deposited a SiNx layer for a hard mask on Si wafers. The chip's pattern was defined by using the standard photolithography technique and the inductively coupled plasma etched SiNx hard mask pattern. A 20 nm Ag film was evaporated onto surfaces of wafers and was lift off the metal caps on the photoresist. The wafers were etched in a HF/H_2O_2 mixture solution, with a concentration of 4.8 M and 0.3 M, respectively. After finishing the Ag removal, the hard mask SiNx was etched in a 125 °C H_3PO_4 and a Si nanostructure with porous morphology was formed. The wafers were cut into a standard-sized Coral Chip to fit the microfluidic component of the Cell Reveal^{TM} system. The surface of the chip was finally modified by silane deposition and coated with biotinylated PLL-g-PEG + streptavidin.

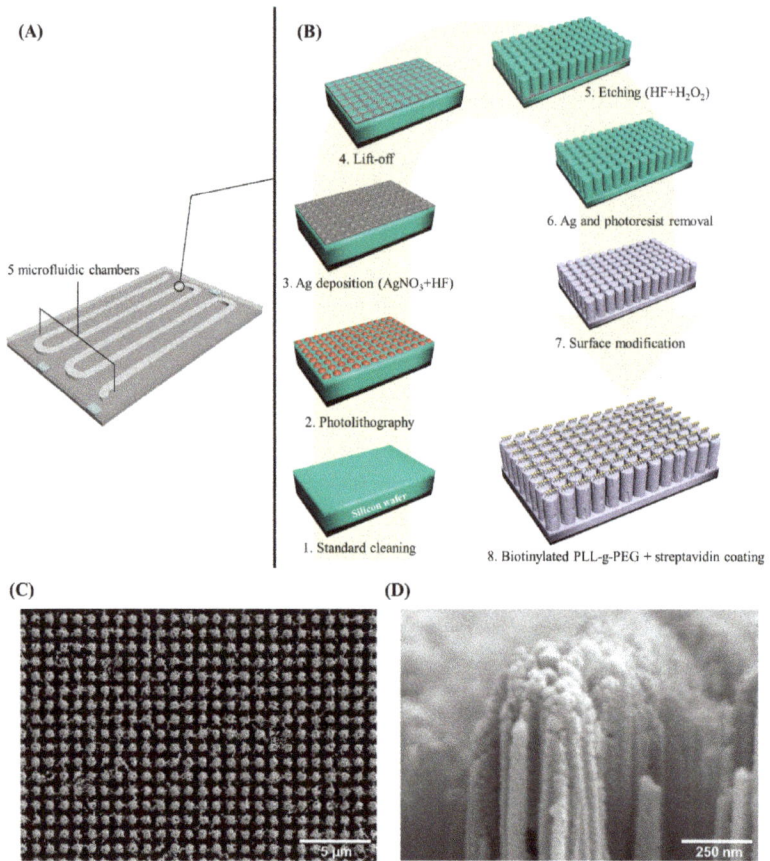

Figure 1. The silicon-based microfluidic Coral Chip. (**A**) An exemplified Coral Chip with 5 microfluidic chambers. (**B**) The manufacturing follow chart of the Coral Chip chamber surface: 1. standard cleaning, 2. photolithography, 3. Ag deposition, 4. liftoff, 5. etching, 6. Ag and photoresist removal, 7. surface modification, and 8. biotinylated PLL-g-PEG + streptavidin coating. (**C**) Top view and (**D**) lateral view of a scanning electron microscope (SEM) image of the Coral Chip.

When the chip was used for cell capture, the potential targeted cells are pre-labeled with biotin-conjugated monoclonal antibodies. The strong interaction between streptavidin and biotin enables a high efficiency for cell capture by the chip.

2.3. Cell Spiking Test

The mixed cell suspension of SK-BR-3 and Jurkat cells was injected into the Cell Reveal™ system for the evaluation of the capture efficiency. The subsequent procedures were automatically carried out by the system with a cell flow rate of 0.6 mL/h. The inputted cell suspensions were fixed in 4% paraformaldehyde. Then, Triton X-100 (0.1%) and 2% BSA (bovine serum albumin) were added to increase the cellular permeability and to prevent nonspecific binding sites. The antibody used for the primary capture of SK-BR-3 cells is anti-EpCAM (Figure 2A). Then, the captured cells were treated with anti-CK and anti-CD45 antibodies. Finally, fluorescence-labeled secondary antibodies were used to stain the targeted cells. The chips are examined using a fluorescence microscope equipped with a built-in automated inspection and image analysis system to filter out images of Jurkat cells for further analyses. The SK-BR-3 cells can therefore be targeted, identified, and enumerated. Image analyses with the count-in/filter-out criteria for SK-BR-3 and Jurkat cells are EpCAM+/CK+/CD45−/Hoechst+ and EpCAM-/CK-/CD45/Hoechst+, respectively. Data for the test were repeated in quadruplicate. The capture efficiency and false capture rate were determined as the number of captured SK-BR-3 cells divided by the total number of spiked SK-BR-3 cells and the number of captured Jurkat cells divided by the total number of background Jurkat cells, respectively.

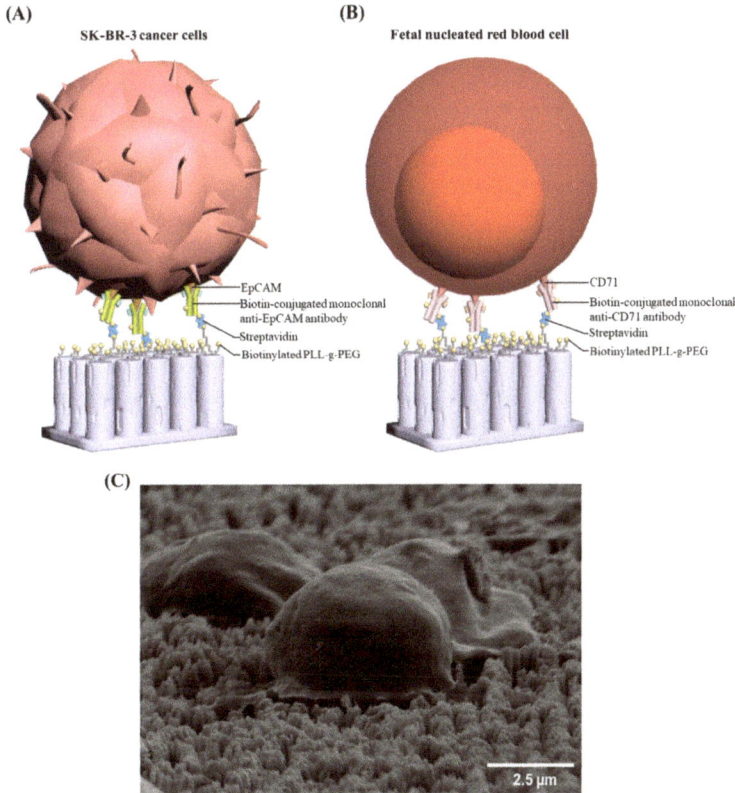

Figure 2. Rare cell captures by the Coral Chip: The Coral Chip surface is coated with biotinylated PLL-g-PEG + streptavidin, and the potential targeted cells are pre-labeled with biotinylated antibodies. The strong interaction between streptavidin and biotin enhances the capturing effect. (**A**) A schematic diagram of the SK-BR-3 cancer cells captured from an artificial cell mixture with a large amount of Jurkat cells as the background. (**B**) A schematic diagram of the fetal nucleated red blood cells (fnRBCs) captured from the peripheral blood of pregnant women. (**C**) Scanning electron microscope (SEM) micrographs of the targeted cells captured on Coral Chip.

2.4. Fetal Nucleated Red Blood Cells (fnRBCs) Capture

The whole blood sample was flown through the automated Cell Reveal™ system (with a rate of 0.6 mL/h) and then fnRBCs were captured by Coral Chip. For each test run, 4 Coral Chips can be used simultaneously to analyze 8 mL blood (2 mL blood per chip). The antibody used for primary capture of fnRBCs is anti-CD71 (Figure 2B). The captured cells were then treated with anti-GPA and anti-CD45 antibodies and stained by fluorescence-labeled secondary antibodies. As a result, the fnRBCs can be automatically targeted, identified, and enumerated by image analyses with the count-in/filter-out criteria CD71+/GPA+/CD45−/Hoechst+ for fnRBCs and CD71-/GPA-/CD45+/Hoechst+ for maternal WBCs.

2.5. Fluorescence in Situ Hybridization (FISH)

FISH was performed directly on Coral Chip capturing for fnRBCs. Prior to hybridization, the formaldehyde on Coral Chip were treated by a 10 mM sodium citrate at 90 °C for 20 min; followed by being immersed in 0.1% Triton-X at room temperature for 10 min; and then followed by serial washes of 0.2 N HCl at 25 °C for 20 min, double distilled water at 25 °C for 3 min, 2× saline-sodium citrate (SSC) at 25 °C for 3 min, and an immersion of Vysis pretreatment solution (1 N NaSCN) (Abbott, Lake Bluff, IL, USA) at 25 °C overnight. Then, the Coral Chip was deposited in purified water at 25 °C for 1 min, 2× SSC at 25 °C for 5 min (repeated two times), pepsin solution (10 µL 10% Pepsin/40 ml 0.01 N HCl) at 37 °C for 3 min, and 2× SSC at 25 °C for 5 min (repeated two times). Finally, the Coral Chip was immersed in 70% ethanol at 4 °C for 1 min, 85% ethanol at 4 °C for 1 min, and 100% ethanol at 4 °C for 1 min and dried at 50 °C for 5 min. The interphase FISH for chromosomes 18, 21, or Y was conducted on captured fnRBCs. For the hybridization experiment, the Coral Chips were dehydrated in an ethanol series and hybridized overnight in a moist chamber at 37 °C. The chips were washed for 2 min in 0.4× SSC at 70 °C and for 5 min in 4× SSC, 0.1% Tween 20 at room temperature and blocked in 4× SSC, 3% bovine serum albumin (BSA), 0.1% Tween 20 at 37 °C for 30 min. The hybridization signal was detected with a Nikon-Ni-E microscope system (Nikon, Tokyo, Japan). The chromosomes were counterstained with 0.125 µg/mL 4′,6-diamidino-2-phenylindole (DAPI) in Antifade (Vysis, Downers Grove, IL, USA). The FISH analyses were performed using the Aquarius®FAST FISH Prenatal kit (Cytocell, Cambridge, UK) for trisomy 18 and 21 fetuses (the chromosome 18 probe for the centromere of chromosome 18 (D18Z1) and the chromosome 21 probe for D21S270, D21S1867, D21S337, D21S1425, and D21S1444 were labeled with aqua and orange fluorophores, respectively) and using the centromeric enumeration probe (CEP) X SpectrumOrange/Y SpectrumGreen DNA probe kit (Vysis, Downers Grove, IL, USA) for euploid male fetuses (the chromosome X probe for Xp11.1q11.1 alpha satellite DNA and the chromosome Y probe for Yq12 satellite III were labeled with orange and green fluorophores, respectively).

2.6. Captured Cells Recovery

The cells captured on Coral Chip (i.e., SK-BR-3 cells and fnRBCs) (Figure 2C) can be recovered by an automated cell picker which is upgraded from the manual micropipetting system that we previously reported [19]. That is, the target cell location coupled with the coordinates were acquired by the Cell Reveal™ system. Then, the Cell Reveal™ system removed a computer lid covering the Coral Chip during the cell capture process and exposed the microfluidic chamber to the cell picker. Finally, the in-house developed software coordinates the fluorescent microscope and the pipetting system to recover the target cells (Figure 3).

(A) (B)

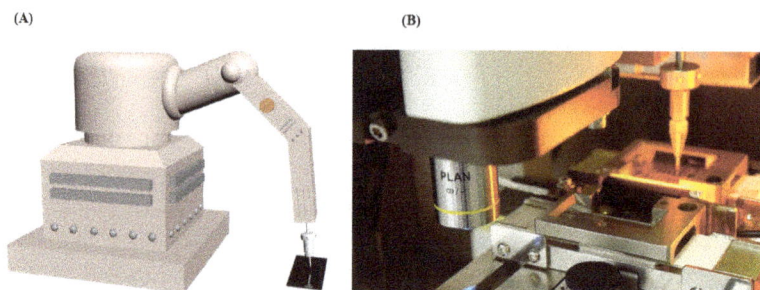

Figure 3. An automated cell picker. (**A**) A schematic diagram of the cell picker machine. (**B**) The targeted cell enriched on the Coral Chip can be automatically recovered by the cell picker.

2.7. Whole Genome Amplification (WGA)

Five to 15 captured cells recovered from the Coral Chip were pooled in a single 0.2 mL PCR tube. The recovered cells were subjected to whole genome amplification (WGA) using the PicoPLEX Single Cell WGA Kit (Takara Bio, Mountain View, CA, USA) following the manufacturer's instructions. Amplified DNA was purified using the QIAquick PCR purification kit (Qiagen, Hilden, Germany). The DNA purities and concentrations were examined by a Nanodrop 2000 spectrophotometer (Thermo Fisher Scientific, DE, USA).

2.8. Array Comparative Genomic Hybridization (aCGH)

Approximately 1 μg of purified WGA DNA was fluorescently labeled with Cy3 d-CTP or Cy5-dCTP using the SureTag DNA Labeling Kit (Agilent Technologies, CA, USA) and then cleaned up by a Microcon YM-30 centrifugal filter unit (Millipore, MA, USA). The yield DNA was hybridized with a CytoScan 60 × 8K microarray chip (Agilent customer array, Changhua Christian Hospital, Changhua, Taiwan) following the manufacturer's instructions. The image on a chip was acquired with a G4900DA SureScan microarray scanner (Agilent Technologies) and analyzed with Agilent Genomic Workbench software (Agilent Technologies) for chromosome gain or loss across all 24 chromosomes. Aberrations were detected using the default setting with the z-score algorithm conjugated with a filter of a minimum of 5 Mb aberrations.

2.9. Next Generation Sequencing (NGS)

Approximately 1 μg of purified WGA DNA was used for library construction by the Ion Xpress Plus gDNA Fragment Library Preparation Kit Set (Life technologies, Carlsbad, CA, USA) following the manufacturer's instructions. The quantity of library was determined using Qubit dsDNA HS assay kits (Life technologies) with Qubit fluorometers (Life technologies). The template-positive Ion Sphere Particles were generated using Ion PGM Hi-Q Template Kits (Life technologies) with the Ion OneTouch 2 Instrument (Life technologies) and then enriched with the Ion OneTouch ES Instrument (Life technologies). Sequencing was performed on the Ion Torrent PGM Instrument (Life technologies) platform with the Ion PGM Hi-Q Sequencing Kit and Ion 316 chip (Life technologies). The sequencing data analysis was performed by using the cloud-based the Ion Reporter™ Server System (https://ionreporter.thermofisher.com/ir/).

2.10. Short Tandem Repeat (STR) Analysis

The STR analysis was performed for gender determination in order to confirm that the circulating cells captured are indeed from male fetuses instead of maternal origin. The GenomeLab™ Human STR Primer Set kit (Beckman Coulter, Brea, CA, USA) containing the primer pair of gender-specific AMEL locus was used to analyze the STR pattern on the GenomeLab™ GeXP Genetic Analysis System (Beckman Coulter). The data were then analyzed by the FRAGMENTS application program (Beckman Coulter).

3. Results

3.1. Capture Efficiency Estimated by Cell Spiking Test

Four model samples, each prepared by spiking SK-BR-3 cells into background Jurkat cells, were used to evaluate the capture efficiency of the Cell Reveal™ system. The cell capture experiment was carried out according to the procedure mentioned above (Section Cell Spiking Test). The mean of the capture rate is 88.17% (range: 80.24%–94.56%). The mean of the false capture rate is close to 0% (range: 0%–0.0007%) (Table 1).

Table 1. A summary of the cell spiking test: In each sample, 5×10^3 SK-BR-3 breast cancer cells were mixed with 10^6 Jurkat cells in 200 µL Dulbecco's phosphate-buffered saline (DPBS) and subjected to Cell Reveal™ system to examine the capture efficiency.

Sample No.	No. of Captured SK-BR-3 cells	No. of Falsely Captured Jurkat Cells	Capture Rate for SK-BR-3 Cells (%)	False Capture Rate for Jurkat Cells (%)
1	4012	7	80.24	0.0007
2	4241	0	84.82	0
3	4683	0	93.06	0
4	4728	1	94.56	0.0001
Mean	4405	2	88.17	0.0002

3.2. Circulating fnRBC Captured by Coral Chip

In every 2 mL of the maternal blood being tested on 1 Coral Chip, the circulating fnRBCs were always captured in all the 14 pregnant women examined (Table 2). The fnRBCs were enriched on the chip (Figure 4) and identified based on the count-in/filter-out criteria of CD71+/GPA+/CD45−/Hoechst+ by a fluorescence microscope equipped with a built-in automated inspection and image analysis system (Figure 5). The cells automatically captured by system were rechecked manually. All the captured cells passed the count-in/filter-out criteria of fnRBCs, suggesting a low false capture rate. The number of captured fnRBCs were 2–71 cells per 2 mL of maternal blood. The total numbers of fnRBCs captured were 273 cells. As a result, the overall capture rate is estimated as 9.75 fnRBCs per ml maternal blood per individual (Table 2).

Table 2. The validation of the cell-based noninvasive prenatal diagnosis (cbNIPD) in 14 pregnant women.

Case No.	MA (Year)	GA (Week+day)	Pre-acquired Fetal Genetic Condition	cbNIPD		Validated * Method
				No. of fnRBCs Captured (in 2 mL Maternal Blood)	Non-maternal Genomic Markers Used to Confirm the Fetal Origin of Captured Cells	
1	30	27+5	arr[GRCh37] 9p24.2p23 (2267812_13374304) × 1 dn	10	1. 9p24.2p23 deletion 2. Chr Y	aCGH [pooled 8]
2	38	20+6	arr[GRCh37] 10q25.2q26.12 (114393625_121720948) × 1 dn	47	1. 10q25.2q26.12 deletion 2. Chr Y	aCGH [pooled 13]
3	31	25	arr[GRCh37] 21q22.11q22.3 (35703384_48056450) × 1 dn	47	1. 21q22.11q22.3 deletion 2. Chr Y	aCGH [pooled 15]
4	40	18	arr[GRCh37] 22q11.21 (18894835_21505417) × 1 dn	18	22q11.21 deletion	aCGH [pooled 10]
5	28	15+6	48,XXY,+18	7	T18	FISH [4]
6	37	13+4	47,XY,+18	25	T18	FISH [10]
7	29	16	47,XY,+18	3	T18	FISH [3]
8	34	20+6	47,XY,+21	14	T21	FISH [6]
9	43	25+6	46,XY	3	Chr Y	FISH [3]
10	32	19	46,XY	2	Chr Y	FISH [2]
11	29	24+6	46,XY	10	Chr Y	FISH [6]
12	37	15	46,XY	10	Chr Y	FISH [4]
13	28	24	46,XY	71	Chr Y	FISH [22]
14	42	24	46,XY	6	Chr Y	STR analysis [pooled 5]

* The number in the bracket indicates the number or pooled number of captured cells used for validation. MA, maternal age; GA, gestational age; fnRBC, fetal nucleated red blood cell; Chr, chromosome; T18, trisomy 18; T21, trisomy 21; aCGH, array-based comparative genomic hybridization; FISH, fluorescence in situ hybridization; and STR, short tandem repeat.

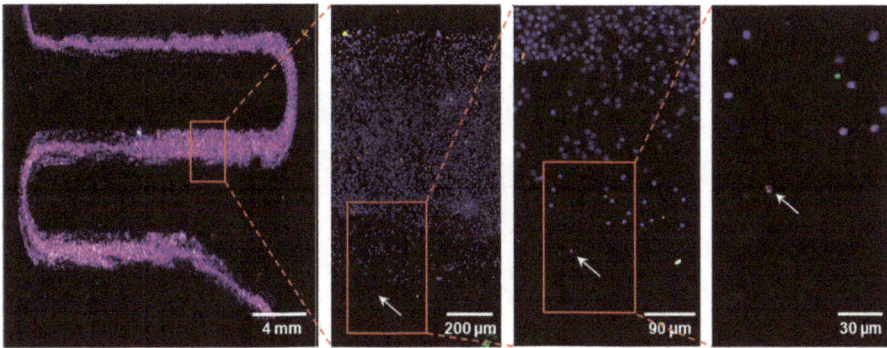

Figure 4. Images of fetal nucleated red blood cells (fnRBCs) enrichment on chambers of a Coral Chip. An fnRBC identified by immunocytochemistry is indicated by an arrow.

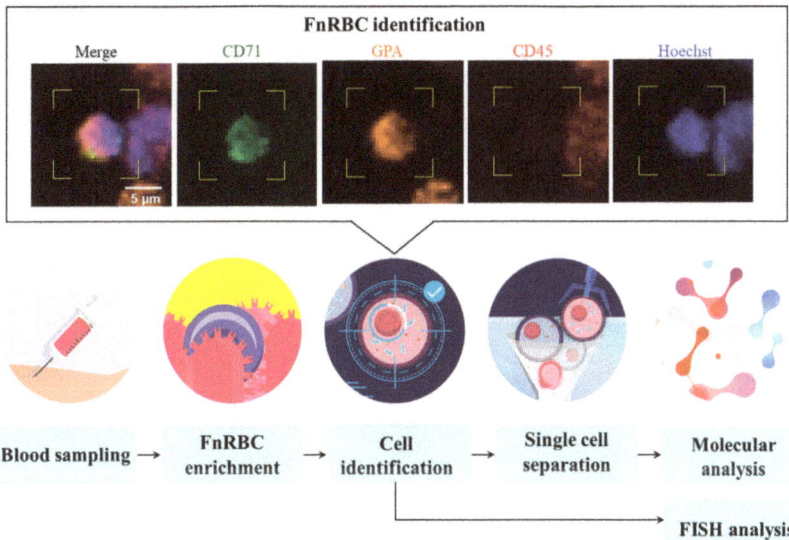

Figure 5. The process flow diagram of a cell-based noninvasive prenatal diagnosis (cbNIPD) by the fetal nucleated red blood cells (fnRBCs) enrichment strategy. The fnRBCs were identified based on the count-in/filter-out criteria of CD71+/GPA+/CD45−/Hoechst+.

3.3. FISH

Interphase FISH for the captured fnRBCs from the blood of the 4 pregnant women with a fetus of trisomy 18 or trisomy 21 (cases 5–8 in Table 2) and for 5 of the 6 pregnant women with euploid male fetuses (cases 9–13 in Table 2) revealed correct diagnoses in all cases. For each case, at least 2 fnRBCs were examined on the chip. Figure 6 exemplified a FISH result using the CEP X SpectrumOrange/Y SpectrumGreen DNA probe kit (Vysis, Downers Grove, IL, USA) for a pregnant women with a euploid male fetus (case 13 in Table 2). The fnRBC can be distinguished from the maternal WBC by the signals of chromosome X and Y: the fnRBC has 1 orange and 1 green signal, and the maternal WBC has 2 orange signals (Figure 6).

(A)

(B)

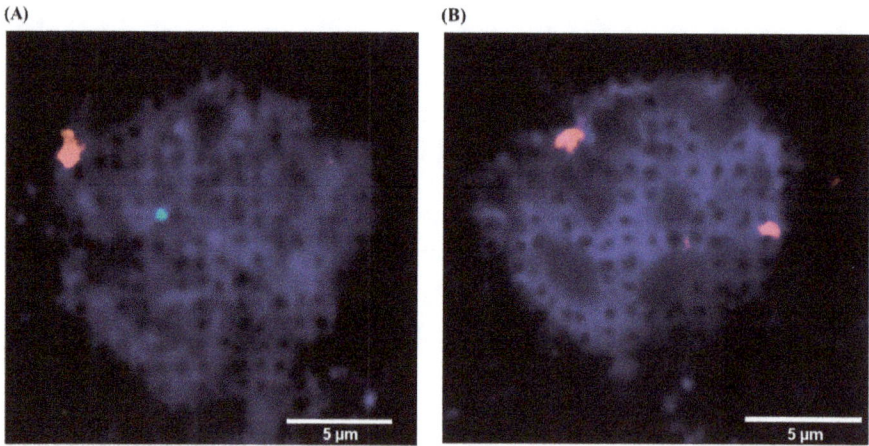

Figure 6. The fluorescent in situ hybridization (FISH) for cells on a Coral Chip. The cells are from the blood of a pregnant women with an euploid male fetus (case 13 in Table 2). The FISH was directly performed on the chip using the CEP X SpectrumOrange/Y SpectrumGreen DNA probe kit (Vysis, Downers Grove, IL, USA). (**A**) The fetal nucleated red blood cell (fnRBC) can be distinguished from (**B**) the maternal white blood cell (WBC) by the signals of chromosome X and Y: the fnRBC has 1 orange and 1 green signal, and the maternal WBC has 2 orange signals.

3.4. Captured Cells Recovery

The recovery rate for the targeted cells is estimated to be 90%. About 10% of cells were lost when they were pulled out from the chip by the micropipetting system.

3.5. WGA

All pooled captured cells underwent WGA successfully except those with a total number of cells that were too few (namely, less than 5 cells) to reach the amplified threshold for subsequent molecular genetic analyses by aCGH, NGS, or STR analyses. Overall, the SK-BR-3 cell WGA DNA as well as the fnRBC WGA DNA from 11 prenatal cases (cases 1–6, 8, 11–14 in Table 2) were obtained. The WGA products were 30 uL in total, with a concentration ranged from 150–275 ng/uL.

3.6. aCGH and NGS Analysis

For SK-BR-3 cells, both aCGH and NGS analyses were performed, and the recognizable genomic features of the SK-BR-3 cell line [42] were identified (Figure 7A,B). For the four prenatal cases with de novo pathogenic copy number changes (cases 1–4 in Table 2), aCGH were performed and the results are consistent with the fetal genetic features pre-acquired by aCGH of amniotic fluid. An exemplified aCGH result for the captured fnRBCs (case 2 in Table 2) is showed in Figure 7C.

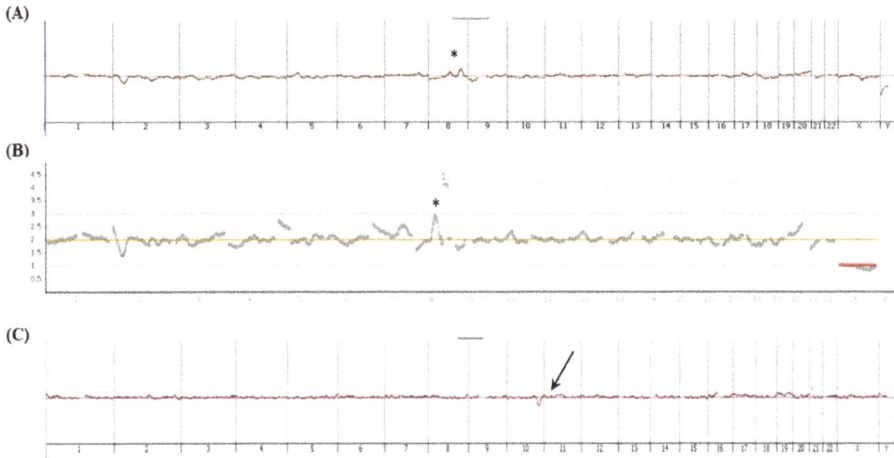

Figure 7. The molecular analyses for targeted cells enriched on and then captured from the Coral Chip. (**A**) The array comparative genomic hybridization (aCGH) and (**B**) the next generation sequencing (NGS) for the SK-BR-3 cancer cells. The recognizable genomic imbalance [42] in chromosome 8 was denoted by a star. (**C**) The aCGH for the circulating fetal cells with a de novo deletion in chromosome 10q25.2q26.12 (i.e., the case 2 in Table 2). The 10q25.2q26.12 deletion is indicated by an arrow. The DNA used for the molecular analyses was extracted from 4–5 captured cells and then amplified by the PicoPLEX Single Cell WGA Kit (Takara Bio, Mountain View, CA, USA).

3.7. STR Analysis

An STR analysis was performed in fnRBCs of 1 prenatal euploid male case (case 14 in Table 2) for gender determination. The results demonstrated the captured fnRBCs contain the informative STR marker on chromosome Y and are indeed of fetal origin (data not shown).

4. Discussion

Although a number of research groups had made tremendous efforts on isolating CFCs (especially fnRBC) and tried to apply the technology to clinical utility, very few had actually reported successful results [19,21–24]. Meanwhile, some of the published studies have the potential to become a laboratory developed test, but the laborious experimental steps made those published reports questionable if these tests can truly turn into a reliable and stable system being adopted by clinical cytogenetics laboratories. Our cell capture system is nearly automated in both processes of the cell capture and recovery. Moreover, the Coral Chip used is manufactured by photolithography and etching, a process easy to achieve standardization and production compared with other nanostructure wet etching methods. As a result, this system has the scalability and potential to become an in vitro diagnostic which may change the landscape now that it has been dramatically reshaped by the popular cffDNA testing. Currently, most of the emerging platforms are targeted at trophoblasts [16,28,31]. It is reasonable since trophoblasts are much larger than the background of maternal WBCs and adds another useful determinant to differentiate trophoblasts from maternal cells, and having an intact trophoblast can still provide much more information than the fragmented cffDNA degraded from trophoblasts. However, fnRBC is indeed representative of the true fetal genome, and therefore, its priority of cbNIPD should be higher than trophoblasts. In our previous proof-of-principle pilot [19], we demonstrated the feasibility of our platform to detect fetal aneuploidy by using common trisomies (i.e., trisomy 13, 18, and 21). Here, we further expand our case series into those with de novo chromosome copy number changes and those carrying male fetuses, and we demonstrated the cells being captured are indeed of fetal origin by different genetic analyses including FISH, aCGH, and STR.

It is now well-known the trend of the variation of the fetal DNA fractions during the whole gestation, as well as the possible confounding factors (e.g., material body mass index, fetoplacental mosaicism, anticoagulation therapy, vanishing twin syndrome, and genetic chimerism caused by blood transfusion or maternal malignancy) that may cause false-positives or false-negatives by the cffDNA testing [7,15,43]. On the contrary, a recent report showed that the maternal body mass index (BMI) has no effect on the number of CFCs being captured [28], a fact hinting that cbNIPD may have much fewer limitations than cffDNA testing and a greater potential to achieve a true NIPD in the future. However, it should be highlighted that any cbNIPD platforms must be able to demonstrate its feasibility through prospective, double-blinded, large-scale clinical trials to convince the clinical communities, hopefully in the near future, that indeed it is a workable solution and can then compete with the now very successful cffDNA NIPT. It can also be anticipated that new confounding factors may affect the accuracy of cbNIPD theoretically, such as fetomaternal hemorrhage, a not uncommon complication during the gestation [44].

5. Conclusions

Our results demonstrated that the Cell Reveal^{TM} system has a high capture efficiency and can be used for fnRBC capture and recovery that is feasible for the genetic diagnosis of fetuses without invasive procedures. However, to convince its clinical utility in cbNIPD, a prospective, large-scale, randomized study is needed.

Author Contributions: Conceptualization, G.-C.M., Y.-S.C., and M.C.; methodology, G.-C.M., C.-E.H., and M.C.; validation, G.-C.M., W.-H.L., C.-E.H., and M.C.; formal analysis, G.-C.M., W.-H.L., C.-E.H., and J.-Y.L.; investigation, W.-H.L., and M.-H.L.; resources, M.C. and W.-J.W.; data curation, G.-C.M., W.-H.L., C.-E.H., and M.C.; writing—original draft preparation, G.-C.M., T.-Y.C., Y.-S.C., and M.C.; writing—review and editing, G.-C.M., Y.-S.C., and M.C.; visualization, G.-C.M., Y.-S.C., M.C., and Y.-J.Y.; supervision, G.-C.M., Y.-S.C., and M.C.; project administration, G.-C.M., Y.-S.C., and M.C.; funding acquisition, G.-C.M., Y.-S.C., and M.C.

Funding: Part of this research was funded by the Ministry of Science and Technology, Executive Yuan, Taiwan (grant number MOST 107-2314-B-371-009-MY3) and by Changhua Christian Hospital, Taiwan (grant number SMI-2018-07).

Acknowledgments: The authors would like to thank Wei-Cheng Hsu, Sheng-Wen Chen, and Hsin-Cheng Ho of CytoAurora Biotechnologies Inc. for their assistance in the manuscript preparation and Yu-Hsiang Tang and Nien-Nan Chu of the Instrument Technology Research Center, National Applied Research Laboratories, Hsinchu, Taiwan for their assistance with the SEM imaging.

Conflicts of Interest: C.-E.H. is the CEO and holds equity of the CytoAurora Biotechnologies Inc., Hsinchu Science Park, Hsinchu, Taiwan. M.C. is the Honorary Cofounder of CytoAurora Biotechnologies Inc., Hsinchu Science Park, Hsinchu, Taiwan. All other authors declare no conflict of interests.

References

1. Zipursky, A.; Hull, A.; White, F.D.; Israels, L.G. Foetal erythrocytes in the maternal circulation. *Lancet* **1959**, *1*, 451–452. [CrossRef]
2. Norwitz, E.R.; Levy, B. Noninvasive prenatal testing: The future is now. *Rev. Obstet. Gynecol.* **2013**, *6*, 48–62. [CrossRef] [PubMed]
3. Beaudet, A.L. Using fetal cells for prenatal diagnosis: History and recent progress. *Am. J. Med. Genet. C Semin. Med. Genet.* **2016**, *172*, 123–127. [CrossRef] [PubMed]
4. Wu, W.J.; Ma, G.C.; Lee, M.H.; Chen, Y.C.; Chen, M. Ultrasonography for prognosis in case of trisomy 14 confined placental mosaicism developing after preimplantation genetic screening. *Ultrasound Obstet. Gynecol.* **2017**, *50*, 128–130. [CrossRef] [PubMed]
5. Wu, W.J.; Ma, G.C.; Lin, Y.S.; Yeang, C.H.; Ni, Y.H.; Li, W.C.; Tsai, H.D.; Shur-Fen Gau, S.; Chen, M. Detection of 22q11.2 microduplication by cell-free DNA screening and chromosomal microarray in fetus with multiple anomalies. *Ultrasound Obstet. Gynecol.* **2016**, *48*, 530–532. [CrossRef] [PubMed]
6. Cheng, H.H.; Ma, G.C.; Tsai, C.C.; Wu, W.J.; Lan, K.C.; Hsu, T.Y.; Yang, C.W.; Chen, M. Confined placental mosaicism of double trisomies 9 and 21: Discrepancy between noninvasive prenatal testing.; chorionic villus sampling and postnatal confirmation. *Ultrasound Obstet. Gynecol.* **2016**, *48*, 251–253. [CrossRef] [PubMed]

7. Gregg, A.R.; Skotko, B.G.; Benkendorf, J.L.; Monaghan, K.G.; Bajaj, K.; Best, R.G.; Klugman, S.; Watson, M.S. Noninvasive prenatal screening for fetal aneuploidy.; 2016 update: A position statement of the American College of Medical Genetics and Genomics. *Genet. Med.* **2016**, *18*, 1056–1065. [CrossRef]
8. Sinkey, R.G.; Odibo, A.O. Cost-effectiveness of old and new technology for aneuploidy screening. *Clin. Lab Med.* **2016**, *36*, 237–248. [CrossRef]
9. Chitty, L.S.; Lo, Y.M. Noninvasive Prenatal Screening for Genetic Diseases Using Massively Parallel Sequencing of Maternal Plasma DNA. *Cold Spring Harb Perspect. Med.* **2015**, *5*, a023085. [CrossRef]
10. Norton, M.E.; Jacobsson, B.; Swamy, G.K.; Laurent, L.C.; Ranzini, A.C.; Brar, H.; Tomlinson, M.W.; Pereira, L.; Spitz, J.L.; Hollemon, D.; et al. Cell-free DNA analysis for noninvasive examination of trisomy. *N. Engl. J. Med.* **2015**, *372*, 1589–1597. [CrossRef]
11. Wapner, R.J.; Babiarz, J.E.; Levy, B.; Stosic, M.; Zimmermann, B.; Sigurjonsson, S.; Wayham, N.; Ryan, A.; Banjevic, M.; Lacroute, P.; et al. Expanding the scope of noninvasive prenatal testing: detection of fetal microdeletion syndromes. *Am. J. Obstet. Gynecol.* **2015**, *212*, 332.e1-9. [CrossRef]
12. Yeang, C.H.; Ma, G.C.; Hsu, H.W.; Lin, Y.S.; Chang, S.M.; Cheng, P.J.; Chen, C.A.; Ni, Y.H.; Chen, M. Genome-wide normalized score: a novel algorithm to detect fetal trisomy 21 during noninvasive prenatal testing. *Ultrasound Obstet. Gynecol.* **2014**, *44*, 25–30. [CrossRef] [PubMed]
13. Agarwal, A.; Sayres, L.C.; Cho, M.K.; Cook-Deegan, R.; Chandrasekharan, S. Commercial landscape of noninvasive prenatal testing in the United States. *Prenat. Diagn.* **2013**, *33*, 521–531. [CrossRef] [PubMed]
14. Grati, F.R.; Malvestiti, F.; Ferreira, J.C.; Bajaj, K.; Gaetani, E.; Agrati, C.; Grimi, B.; Dulcetti, F.; Ruggeri, A.M.; De Toffol, S.; et al. Fetoplacental mosaicism: potential implications for false-positive and false-negative noninvasive prenatal screening results. *Genet. Med.* **2014**, *16*, 620–624. [CrossRef] [PubMed]
15. Meck, J.M.; Dugan, E.K.; Matyakhina, L.; Aviram, A.; Trunca, C.; Pineda-Alvarez, D.; Aradhya, S.; Klein, R.T.; Cherry, A.M. Noninvasive prenatal screening for aneuploidy: positive predictive values based on cytogenetic findings. *Am. J. Obstet. Gynecol.* **2015**, *213*, 214.e1-5. [CrossRef] [PubMed]
16. Hou, S.; Chen, J.F.; Song, M.; Zhu, Y.; Jan, Y.J.; Chen, S.H.; Weng, T.H.; Ling, D.A.; Chen, S.F.; Ro, T.; et al. Imprinted nanoVelcro microchips for isolation and characterization of circulating fetal trophoblasts: toward noninvasive prenatal diagnostics. *ACS Nano* **2017**, *11*, 8167–8177. [CrossRef] [PubMed]
17. Breman, A.M.; Chow, J.C.; U'Ren, L.; Normand, E.A.; Qdaisat, S.; Zhao, L.; Henke, D.M.; Chen, R.; Shaw, C.A.; Jackson, L.; et al. Evidence for feasibility of fetal trophoblastic cell-based noninvasive prenatal testing. *Prenat. Diagn.* **2016**, *36*, 1009–1019. [CrossRef] [PubMed]
18. Kølvraa, S.; Singh, R.; Normand, E.A.; Qdaisat, S.; van den Veyver, I.B.; Jackson, L.; Hatt, L.; Schelde, P.; Uldbjerg, N.; Vestergaard, E.M.; et al. Genome-wide copy number analysis on DNA from fetal cells isolated from the blood of pregnant women. *Prenat. Diagn.* **2016**, *36*, 1127–1134. [CrossRef]
19. Huang, C.E.; Ma, G.C.; Jou, H.J.; Lin, W.H.; Lee, D.J.; Lin, Y.S.; Ginsberg, N.A.; Chen, H.F.; Chang, F.M.; Chen, M. Noninvasive prenatal diagnosis of fetal aneuploidy by circulating fetal nucleated red blood cellsand extravillous trophoblasts using silicon-based nanostructured microfluidics. *Mol. Cytogenet.* **2017**, *10*, 44. [CrossRef]
20. Calabrese, G.; Fantasia, D.; Alfonsi, M.; Morizio, E.; Celentano, C.; Guanciali Franchi, P.; Sabbatinelli, G.; Palka, C.; Benn, P.; Sitar, G. Aneuploidy screening using circulating fetal cells in maternal blood by dual-probe FISH protocol: a prospective feasibility study on a series of 172 pregnant women. *Mol. Genet. Genomic Med.* **2016**, *4*, 634–640. [CrossRef]
21. Feng, C.; He, Z.; Cai, B.; Peng, J.; Song, J.; Yu, X.; Sun, Y.; Yuan, J.; Zhao, X.; Zhang, Y. Noninvasive Prenatal Diagnosis of Chromosomal Aneuploidies and Microdeletion Syndrome Using Fetal Nucleated Red Blood Cells Isolated by Nanostructure Microchips. *Theranostics* **2018**, *8*, 1301–1311. [CrossRef] [PubMed]
22. Wei, X.; Ao, Z.; Cheng, L.; He, Z.; Huang, Q.; Cai, B.; Rao, L.; Meng, Q.; Wang, Z.; Sun, Y.; et al. Highly sensitive and rapid isolation of fetal nucleated red blood cells with microbead-based selective sedimentation for noninvasive prenatal diagnostics. *Nanotechnology* **2018**, *29*, 43401. [CrossRef] [PubMed]
23. Zhang, H.; Yang, Y.; Li, X.; Shi, Y.; Hu, B.; An, Y.; Zhu, Z.; Hong, G.; Yang, C.J. Frequency-enhanced transferrin receptor antibody-labelled microfluidic chip (FETAL-Chip) enables efficient enrichment of circulating nucleated red blood cells for noninvasive prenatal diagnosis. *Lab Chip* **2018**, *18*, 2749–2756. [CrossRef] [PubMed]
24. Byeon, Y.; Ki, C.S.; Han, K.H. Isolation of nucleated red blood cells in maternal blood for Noninvasive prenatal diagnosis. *Biomed Microdevices* **2015**, *17*, 118. [CrossRef] [PubMed]

25. Kanda, E.; Yura, H.; Kitagawa, M. Practicability of prenatal testing using lectin-based enrichment of fetal erythroblasts. *J. Obstet. Gynecol. Res.* **2016**, *42*, 918–926. [CrossRef] [PubMed]

26. Hatt, L.; Brinch, M.; Singh, R.; Møller, K.; Lauridsen, R.H.; Schlütter, J.M.; Uldbjerg, N.; Christensen, B.; Kølvraa, S. A new marker set that identifies fetal cells in maternal circulation with high specificity. *Prenat. Diagn.* **2014**, *34*, 1066–1072. [CrossRef] [PubMed]

27. Zimmermann, S.; Hollmann, C.; Stachelhaus, S.A. Unique monoclonal antibodies specifically bind surface structures on human fetal erythroid blood cells. *Exp. Cell Res.* **2013**, *319*, 2700–2707. [CrossRef]

28. Kruckow, S.; Schelde, P.; Hatt, L.; Ravn, K.; Petersen, O.B.; Uldbjerg, N.; Vogel, I.; Singh, R. Does maternal body mass index affect the quantity of circulating fetal cells available to use for cell-based noninvasive prenatal test in high-risk pregnancies? *Fetal Diagn. Ther.* **2018**, 1–4. [CrossRef]

29. Schlütter, J.M.; Kirkegaard, I.; Petersen, O.B.; Larsen, N.; Christensen, B.; Hougaard, D.M.; Kølvraa, S.; Uldbjerg, N. Fetal gender and several cytokines are associated with the number of fetal cells in maternal blood–an observational study. *PLoS ONE* **2014**, *9*, e106934. [CrossRef]

30. Jan, Y.J.; Chen, J.F.; Zhu, Y.; Lu, Y.T.; Chen, S.H.; Chung, H.; Smalley, M.; Huang, Y.W.; Dong, J.; Chen, L.C.; et al. NanoVelcro rare-cell assays for detection and characterization of circulating tumor cells. *Adv. Drug Deliv. Rev.* **2018**, *125*, 78–93. [CrossRef]

31. Laget, S.; Broncy, L.; Hormigos, K.; Dhingra, D.M.; BenMohamed, F.; Capiod, T.; Osteras, M.; Farinelli, L.; Jackson, S.; Paterlini-Br échot, P. Technical Insights into Highly Sensitive Isolation and Molecular Characterization of Fixed and Live Circulating Tumor Cells for Early Detection of Tumor Invasion. *PLoS ONE* **2017**, *12*, e0169427. [CrossRef] [PubMed]

32. Schreier, S.; Sawaison, P.; Udomsanpetch, R.; Triampo, W. Advances in rare cell isolation: an optimization and evaluation study. *J. Transl. Med.* **2017**, *15*, 6. [CrossRef]

33. Ferreira, M.M.; Ramani, V.C.; Jeffrey, S.S. Circulating tumor cell technologies. *Mol. Oncol.* **2016**, *10*, 374–394. [CrossRef] [PubMed]

34. Gross, A.; Schoendube, J.; Zimmermann, S.; Steeb, M.; Zengerle, R.; Koltay, P. Technologies for single-cell isolation. *Int. J. Mol. Sci.* **2015**, *16*, 16897–16919. [CrossRef] [PubMed]

35. Sahmani, M.; Vatanmakanian, M.; Goudarzi, M.; Mobarra, N.; Azad, M. Microchips and their significance in isolation of circulating tumor cells and monitoring of cancers. *Asian Pac. J. Cancer. Prev.* **2016**, *17*, 879–894. [CrossRef] [PubMed]

36. Ramirez, J.M.; Fehm, T.; Orsini, M.; Cayrefourcq, L.; Maudelonde, T.; Pantel, K.; Alix-Panabières, C. Prognostic relevance of viable circulating tumor cells detected by EPISPOT in metastatic breast cancer patients. *Clin. Chem.* **2014**, *60*, 214–221. [CrossRef] [PubMed]

37. Seppo, A.; Frisova, V.; Ichetovkin, I.; Kim, Y.; Evans, M.I.; Antsaklis, A.; Nicolaides, K.H.; Tafas, T.; Tsipouras, P.; Kilpatrick, M.W. Detection of circulating fetal cells utilizing automated microscopy: potential for noninvasiveprenatal diagnosis of chromosomal aneuploidies. *Prenat. Diagn.* **2008**, *28*, 815–821. [CrossRef] [PubMed]

38. Al-Mufti, R.; Hambley, H.; Farzaneh, F.; Nicolaides, K.H. Fetal erythroblasts in maternal blood in relation to gestational age. *J. Matern. Fetal Neonatal. Med.* **2003**, *14*, 392–397. [CrossRef]

39. Bianchi, D.W.; Zickwolf, G.K.; Yih, M.C.; Flint, A.F.; Geifman, O.H.; Erikson, M.S.; Williams, J.M. Erythroid-specific antibodies enhance detection of fetal nucleated erythrocytes in maternal blood. *Prenat. Diagn.* **1993**, *13*, 293–300. [CrossRef]

40. Ziegler, B.L.; Müller, R.; Valtieri, M.; Lamping, C.P.; Thomas, C.A.; Gabbianelli, M.; Giesert, C.; Bühring, H.J.; Kanz, L.; Peschle, C. Unicellular-unilineage erythropoietic cultures: Molecular analysis of regulatory gene expression at sibling cell level. *Blood* **1999**, *93*, 3355–3368.

41. Sørensen, M.D.; Gonzalez Dosal, R.; Jensen, K.B.; Christensen, B.; Kølvraa, S.; Jensen, U.B.; Kristensen, P. Epsilon haemoglobin specific antibodies with applications in noninvasive prenatal diagnosis. *J. Biomed. Biotechnol.* **2009**, *2009*, 659219. [CrossRef] [PubMed]

42. Babayan, A.; Alawi, M.; Gormley, M.; Müller, V.; Wikman, H.; Mcmullin, R.P.; Smirnov, D.A.; Li, W.; Geffken, M.; Pantel, K.; Joosse, S.A. Comparative study of whole genome amplification and next generation sequencing performanceof single cancer cells. *Oncotarget* **2016**, *8*, 56066–56080. [CrossRef]

43. Ma, G.C.; Wu, W.J.; Lee, M.H.; Lin, Y.S.; Chen, M. The use of low molecular weight heparin reduced the fetal fraction and rendered the cell-free DNA testing for fetal trisomy 21 false negative. *Ultrasound Obstet. Gynecol.* **2018**, *51*, 276–277. [CrossRef] [PubMed]

44. Trola, L.; Al-Kouatly, H.B.; McKurdy, R.; Konchak, P.S.; Weiner, S.; Berghella, V. The recurrent risk of fetomaternal hemorrhage. *Fetal Diagn. Ther.* **2019**, *45*, 1–12. [CrossRef]

![micromachines logo] *micromachines*

MDPI

Article

A Microfluidic Micropipette Aspiration Device to Study Single-Cell Mechanics Inspired by the Principle of Wheatstone Bridge

Yong-Jiang Li [1,†], Yu-Nong Yang [2,†], Hai-Jun Zhang [2], Chun-Dong Xue [1], De-Pei Zeng [2], Tun Cao [1,*] and Kai-Rong Qin [1,*]

[1] School of Optoelectronic Engineering and Instrumentation Science, Dalian University of Technology, Dalian 116024, Liaoning, China; yongjiangli@dlut.edu.cn (Y.-J.L.); xuechundong@dlut.edu.cn (C.-D.X.)

[2] School of Biomedical Engineering, Dalian University of Technology, Dalian 116024, Liaoning, China; yangyunong@mail.dlut.edu.cn (Y.-N.Y.); haijunzhang@mail.dlut.edu.cn (H.-J.Z.); 1925044995@mail.dlut.edu.cn (D.-P.Z.)

* Correspondence: caotun1806@dlut.edu.cn (T.C.); krqin@dlut.edu.cn (K.-R.Q.); Tel.: +86-0411-8470-9690 (K.-R.Q.)

† These authors contributed equally to this work.

Received: 10 January 2019; Accepted: 11 February 2019; Published: 16 February 2019

Abstract: The biomechanical properties of single cells show great potential for early disease diagnosis and effective treatments. In this study, a microfluidic device was developed for quantifying the mechanical properties of a single cell. Micropipette aspiration was integrated into a microfluidic device that mimics a classical Wheatstone bridge circuit. This technique allows us not only to effectively alter the flow direction for single-cell trapping, but also to precisely control the pressure exerted on the aspirated cells, analogous to the feature of the Wheatstone bridge that can precisely control bridge voltage and current. By combining the micropipette aspiration technique into the microfluidic device, we can effectively trap the microparticles and Hela cells as well as measure the deformability of cells. The Young's modulus of Hela cells was evaluated to be 387 ± 77 Pa, which is consistent with previous micropipette aspiration studies. The simplicity, precision, and usability of our device show good potential for biomechanical trials in clinical diagnosis and cell biology research.

Keywords: micropipette aspiration; microfluidics; single-cell mechanics; Wheatstone bridge

1. Introduction

The biomechanical properties of single cells serve as critical factors in directing the physiological functions of cells, such as cell growth, proliferation, and migration, which ultimately contribute to pathophysiological progression [1–3]. Typically, cancer cells are more deformable than healthy ones, which facilitates their metastatic journey into the blood stream [1]. Cell adhesion results in the mechanical scaffold for cell cortex tension to drive cell sorting during gastrulation [4]. Intrinsic mechanical changes in cell and tissue structure lead to the development of malignancy and metastasis [5]. Not only cell mechanical properties affect cell functions. On the contrary, biological processes also change the cell mechanics. The stiffness increases as cells enter mitosis [6], as tumor cells transit to premalignant stage [7], and when red blood cells are infected with malaria [8,9]. In this context, the characterization of cellular biomechanics could provide novel insight in understanding the development of human diseases such as tumor and cancer, showing a good potential in early disease diagnosis and effective treatments. Therefore, considerable interest have been aroused in determining the biomechanical properties of single cells.

To date, numerous quantitative micromanipulation techniques have been developed to evaluate the mechanical properties of single cells, such as micropipette aspiration (MPA), optical tweezers, magnetic twisting cytometry, and atomic force microscopy [10]. Among these methods, MPA provides a non-invasive, simple, and direct approach to measure cell mechanics at the single-cell level [11]. The classical MPA experiment consists of partial or complete suction of single cells into a glass micropipette using negative pressure. By recording the cell deformation to applied pressure, the intrinsic mechanical properties of individual cells can be determined using various models [12–16]. Yet, the small pressure control and manual trapping of target cells in suspension or attached cells make MPA operation challenging and time-consuming. Continuous evaporation loss results in a drifting baseline of the aspiration pressure, leading to measurement inaccuracy [14,17]. Recently, Shojaei-Baghini et al. [18] reported an automated MPA. Yet, a proportional–integral–derivative (PID) position controller, motorized pressure system, and real-time visual tracking system are necessary. Although the MPA technique is theoretically straightforward, it requires not only specialized equipment to precisely control the small pressure, but also highly delicate manipulation to manually target the cells at the single-cell level [19].

Recent developments in microfabrication and microfluidic techniques can solve the problems discussed above. Microfluidics have advantages in single-cell manipulation and precise mechanical stimuli loading, which are challenging for traditional MPA. Moreover, microfluidic devices are inherently matched with the individual cell in scale. Therefore, microfluidics is an effective technique for single-cell mechanics studies [20,21]. A variety of microfluidic forms and techniques have been developed for single-cell mechanical characterization, including constriction channel [9,22,23], fluid stress [24,25], optical stretcher [26], electro-deformation, and electroporation. A few researchers have also applied the MPA technique to microfluidic devices. Chen et al. [27] combined an impedance analyzer and MPA for the simultaneous mechanical and electrical characterization of single cells. Guo et al. [9,19] demonstrated a microfluidic micropipette aspiration with two-layer microstructure and membrane microvalves, which takes advantage of fluidic circuitry to attenuate exerted pressure on cells within a funnel constriction channel for mechanical characterization. However, these microfluidic devices mentioned above employed either complicated microfluidic structure design or sophisticated peripheral systems for single-cell manipulation and exerting forces. An easy-to-use microfluidic device is thus required to characterize single-cell mechanics.

Herein, a microfluidic device is proposed for quantifying cell mechanics at the single-cell level by combining the micropipette aspiration technique and the Wheatstone bridge principle. The microfluidic analog of the Wheatstone bridge allows effective trapping of single cells and precise control of the suction pressure on aspirated cells. The combination of MPA can quantitatively measure the cell deformability, revealing the advantages of simplicity in implementation, ease of use, and reduction of sample consumption. The simplicity, precision, and usability of our device show its potential for biomechanical trials in clinical diagnosis and cell biology research.

2. Materials and Methods

2.1. The Principle of the Microfluidic Wheatstone Bridge

The microfluidic device was designed based on the analog of the classical Wheatstone bridge (Figure 1). We applied the principle to quantitatively regulate the flow rate and pressure difference (equivalent to current and voltage in an electric circuit) through the bridge channel by adjusting flow resistances (Figure 1d).

Following the Darcy–Weisbach equation, the flow resistance of a rectangular microchannel is expressed as [28]:

$$R = \frac{\Delta p}{Q} = \frac{C(\alpha)}{32} \frac{\eta L P^2}{A^3}, \tag{1}$$

where

$$C(\alpha) = 96(1 - 1.3553\alpha + 1.9467\alpha^2 - 1.7012\alpha^3 + 0.9564\alpha^4 - 0.2537\alpha^5), \tag{2}$$

in which $\alpha = H/W$ is the aspect ratio, η is the viscosity, L is the microchannel length, Δp is the pressure difference, Q is the flow rate, and P and A are the perimeter and the area of the rectangle cross section, respectively. From Equations (1) and (2), it can be determined that flow resistance only depends on geometry and dimensions for a given solution.

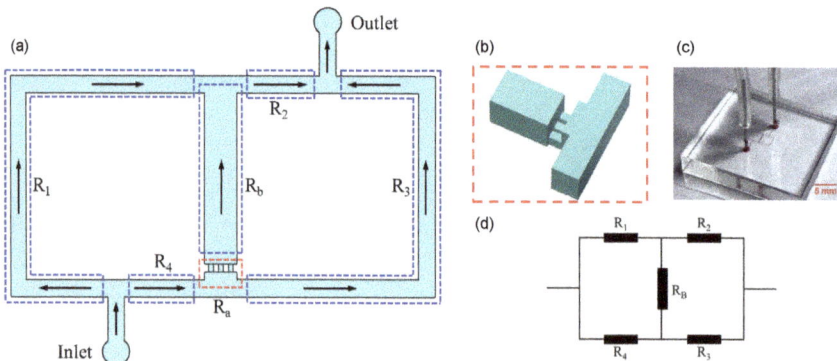

Figure 1. (**a**) Schematic of the Wheatstone bridge microchannel. The microchannel is divided into four segments with distinct flow resistances (R_1, R_2, R_3, R_4) by the inlet, outlet, and bridge microchannels. (**b**) Micropipette aspiration microchannels. (**c**) A picture of the device. (**d**) The equivalent circuit of the Wheatstone bridge microchannel.

According to the equivalent circuit of the microfluidic Wheatstone bridge (Figure 1d), the flow rate through the bridge channel can be given by:

$$Q_B = Q_t \frac{R_1 R_3 - R_2 R_4}{(R_1 + R_4)(R_2 + R_3) + R_B(R_1 + R_2 + R_3 + R_4)}, \tag{3}$$

where R_1, R_2, R_3, R_4 are flow resistances, Q_t is the total flow rate, and R_B is the total resistance of the bridge channel, which is expressed by:

$$R_B = R_b + R_a/N, \tag{4}$$

where R_a and R_b are the flow resistances of the single aspiration channel and the fractional bridge channel, respectively (Figure 1a). N denotes the number of open micropipette aspiration channels (Figure 1b), herein three micropipette aspiration channels were designed. Consequently, the pressure difference of the micropipette aspiration channel can be written as:

$$\Delta p_A = \frac{Q_B R_a}{N}. \tag{5}$$

In the above expressions (Equations (3) and (5)), Q_B and Δp_A are functions of total flow rate and flow resistances, which can be thus quantitatively controlled by regulating Q_t and microchannel structures.

To enhance the trapping efficiency, we regulate the flow direction through the bridge channel ($Q_B > 0$) and ensure its flow rate is larger than that through microchannel R_3, which should satisfy the conditions:

$$R_1 R_3 > R_2 R_4 \tag{6}$$

and

$$Q_B > Q_3 = \frac{Q_B R_B + R_2 Q_t}{R_2 + R_3}, \tag{7}$$

where Q_3 denotes the flow rate within the branch R_3.

2.2. Fabrication and Operation of the Microfluidic Device

The microfluidic MPA device based on the principle of the Wheatstone bridge consists of a polydimethylsiloxane (PDMS)–glass chip fabricated by standard soft-lithography techniques (Figure 1c). The microchannels were patterned in PDMS (Sylgard 184, Dow Corning, Midland, MI, USA) by replica molding. The mold was prepared by spin-coating a thin layer of negative photoresist (SU8-2050, MicroChem, Newton, MA, USA) onto silicon wafers polished on one side (111 N-type, Lijing Ltd., Quzhou, China) and patterned with UV mask aligner (URE-2000/35, Chinese Academy of Sciences, Beijing, China). Next, the micro-channel layer was obtained by pouring PDMS with 10:1 (w/w) base:crosslinker ratio onto the mold, yielding a thickness of approximately 3 mm. After curing the elastomer for 2 h at 80 °C, the PDMS slab was peeled from the mold, punched, and hermetically bonded to a coverslip by plasma oxidation. In our device, all the microchannels consist of rectangular cross sections. According to the requirements in Equations (6) and (7), the dimensions of microchannels were determined as shown in Table 1.

Table 1. Dimensions and resistances of microchannels in the microfluidic Wheatstone bridge. Microchannels are denoted by their flow resistances.

Microchannel	Length (μm)	Width (μm)	Height (μm)	Resistance ($\times 10^{14}$ N·s/m^5)
R_1	5000	30	30	1.76
R_2	1000	30	30	0.35
R_3	5000	30	30	1.76
R_4	1000	30	30	0.35
R_b	1990	88	30	0.13
R_a	10	8	8	0.69

For the operation of the microfluidic micropipette aspiration device (see Figure 2), the inlet was connected to a syringe pump (Pump 11 Elite, Harvard Apparatus, MA, USA) filled with the cell suspension. During the cell trapping, the cell suspension was pumped at the volume flow rate Q_t increased from 20 μL/h to 120 μL/h at a step of 10 μL/h. In each case, we waited at least 1 min to observe the cell deformation. Images of cell deformation were captured only when the cell maintained its shape for at least 1 min (i.e., the stable state). The cell deformation due to the micropipette aspiration was observed under an optical microscope (CKX41, Olympus, Tokyo, Japan) with a CCD camera. The recorded images were further applied to measure the mechanical properties of single cells.

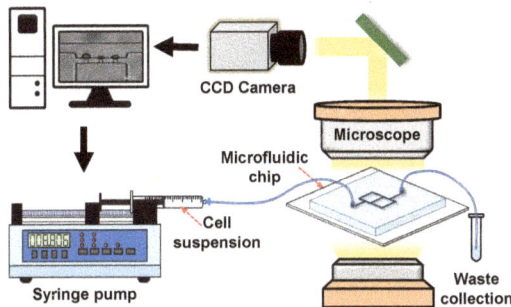

Figure 2. Schematic of the microfluidic micropipette aspiration system.

2.3. Cell Culture and Sample Preparation

The Hela cell line was purchased from the Cell Resource Center in the Shanghai Institutes for Biological Sciences (SIBS, Shanghai, China). Dulbecco's Modified Eagle Medium

(DMEM/high glucose), fetal bovine serum (FBS), phosphate buffered saline (PBS), trypsin-EDTA, and penicillin/streptomycin were provided by Hyclone (Thermo Scientific, Waltham, MA, USA). The Hela cell line was cultured in standard culture flasks using DMEM supplemented with 10% FBS, 1% penicillin/streptomycin, and 1% 1-glutamine (Sigma-Aldrich, St. Louis, MO, USA). After the third generation, the cells were collected via trypsin, and then a cell suspension at a density of $\sim 10^6$ cells/mL was made.

2.4. Measurement of Cell Mechanics

In general, the two most popular models for analyzing single-cell mechanics treat the cell either as a homogeneous elastic solid or a drop of liquid encapsulated by an elastic solid shell [14]. Here, we adopt the elastic solid model of Theret et al. [12]. Figure 3 presents a schematic diagram of a spherical cell aspirated into an MPA channel. The Young's modulus of single cells to pressure is thus expressed as:

$$E = \frac{\frac{3\Delta p_A \Phi}{2\pi}}{\left(\frac{L_p}{R_p}\right)},$$ (8)

where E is Young's modulus, Δp_A is the suction pressure indicated in Equation (5), and Φ is a term that depends on the geometry of the micropipette. A typical value for Φ is $\Phi = 2.1$. L_p denotes the extension length of the surface of the cell into the micropipette (see Figure 3). R_p is the hydraulic radius of the micropipette aspiration channel [29], which can be given as:

$$R_p = \frac{W_a H_a}{W_a + H_a},$$ (9)

where W_a and H_a are the width and height of the micropipette aspiration channel, respectively (see Figure 3).

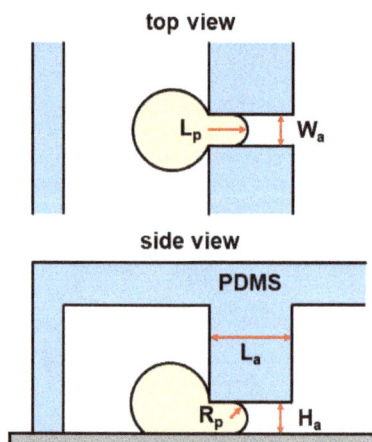

Figure 3. Schematic of a spherical cell aspirated into an micropipette aspiration (MPA) channel with a suction pressure Δp_A. PDMS: polydimethylsiloxane.

In the measurement procedure, we measured the protrusion length L_p into the micropipette aspiration channel at the stable state, where no significant deformation occurred for at least 1 min. The suction pressure Δp_A was calculated with the analytical results in Equation (5) according to the corresponding volume flow rate Q_t. The Young's modulus of a Hela cell was then evaluated with Equation (8). All the measurements were calibrated by measuring the bridge channel width (88 µm) in pixels. Note that cell mechanics were only characterized when the cell behaved as a solid for the protrusion length $L_p \leq L_a/2$ at the stable state. The instances where cells flowed entirely into or passed through the MPA channels were not considered.

2.5. COMSOL Simulation

The velocity and pressure fields were numerically studied using COMSOL Multiphysics. A 3-D simulation was conducted with the dimensions indicated in Table 1. Using the linear flow module (spf), the velocity and pressure distributions were measured at the flow rate Q_t increased from 20 µL/h to 120 µL/h at the step of 20 µL/h. The simulation results of aspiration pressure Δp_A were calculated by averaging the pressure drops along the centerline of the micropipette aspiration channels. In addition, the particle tracing module (fpt) was applied to track the microparticle movements within the microchannel, which was used to evaluate the trapping efficiency of the micropipette channels.

3. Results

3.1. Quantitative Control of Aspiration Pressure

Micropipette aspiration relies on the suction pressure exerted on a single cell to study its biomechanical properties. Firstly, the pressure difference exerted on trapped cells was investigated both analytically and numerically. When single cells were trapped by the micropipette aspiration channels ($N = 0$), it showed a uniform pressure field at both ends of the micropipette aspiration channels (see Figure 4). When N ($N = 1, 2,$ or 3) micropipette aspiration channels were open, the pressure decreased linearly along the centerline of the open micropipette aspiration channel, in which the pressure difference between its two ends was measured at different flow rates Q_t. The results were compared with analytical ones calculated with Equation (5). The pressure difference was linearly proportional to the volume flow rate Q_t (Figure 5). These two results showed a discrepancy of 10% at a maximum flow rate ($Q_t = 120$ µL/h), revealing that Equation (5) is reliable for the calculation of the pressure difference across the micropipette aspiration channel.

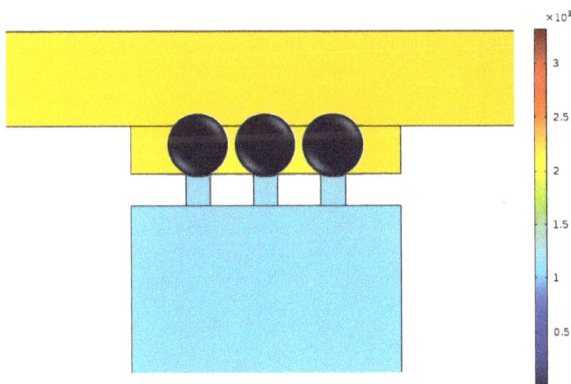

Figure 4. Pressure distribution around the micropipette aspiration channels.

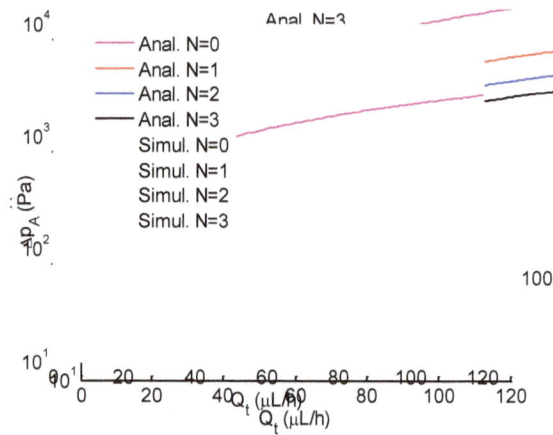

Figure 5. Comparison between analytical (solid lines) and simulation (circle markers) results of the pressure difference across the micropipette aspiration channel versus the total flow rate. N is the number of open aspiration channels.

3.2. Effective Trapping of Microparticles and Single Cells

The hydrodynamic trapping efficiency of the micropipette aspiration channels was validated both numerically and experimentally. Figure 6a illustrates the numerical simulation result of velocity distribution and streamlines in the region of micropipette aspiration channels. When a microparticle suspension was introduced into the inlet at a velocity of 0.01 m/s, the microparticle close to the side wall flowed along the streamlines and ultimately entered a micropipette aspiration channel (Figure 6b). The trapping efficiency was validated experimentally by introducing microparticle and cell suspension. In both cases, either microparticles or single cells were feasibly trapped by the micropipette aspiration channels (Figure 7), revealing the trapping/aspiration effectiveness of the microfluidic device. In particular, the same cell population showed a different mechanical property, indicated by the variations in protrusion length into the aspiration channels (Figure 7b).

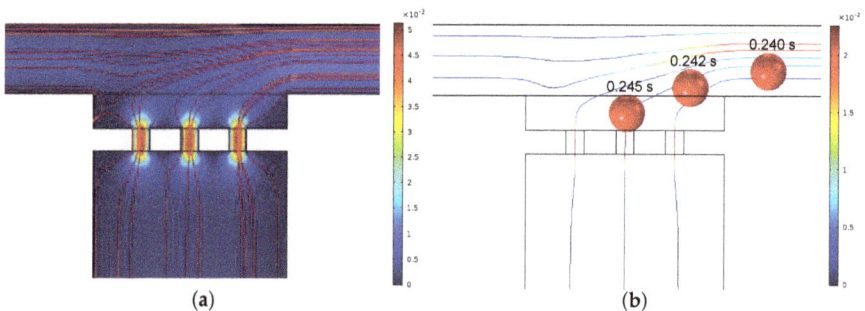

(a)   (b)

Figure 6. (a) Velocity distribution and streamlines around the micropipette aspiration channels for $N = 3$. (b) Tracking trajectory of a particle at different time intervals.

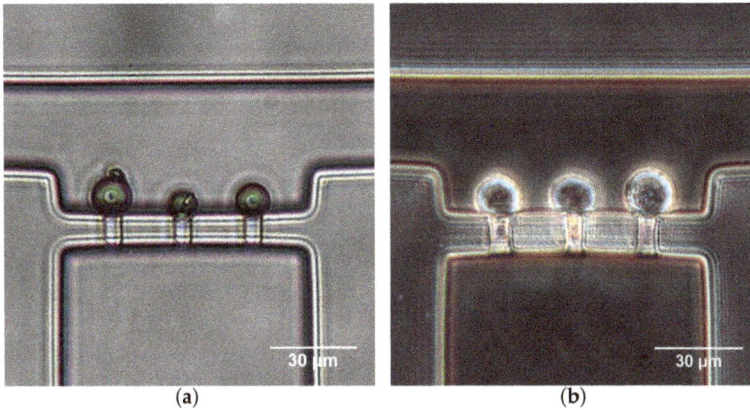

Figure 7. Microparticle (**a**) and Hela cell (**b**) trapping with micropipette aspiration channels at $Q_t = 40$ µL/h.

3.3. Biomechanical Properties of Single Cells

Figure 8 shows a demonstration of Hela cells aspirated into the micropipette aspiration channels. To the same applied suction pressures, cells presented different changes in shape, revealing the heterogeneous mechanical properties of cell populations (Figure 8). Additionally, the cell deformation mainly included two forms. In one case, cells showed hemispherical projections into the MPA channels (Figure 8a). In another case, cell membranes extended completely into or even passed through the MPA channels. The two cases respectively demonstrate the solid-like and liquid-like behaviors of cells.

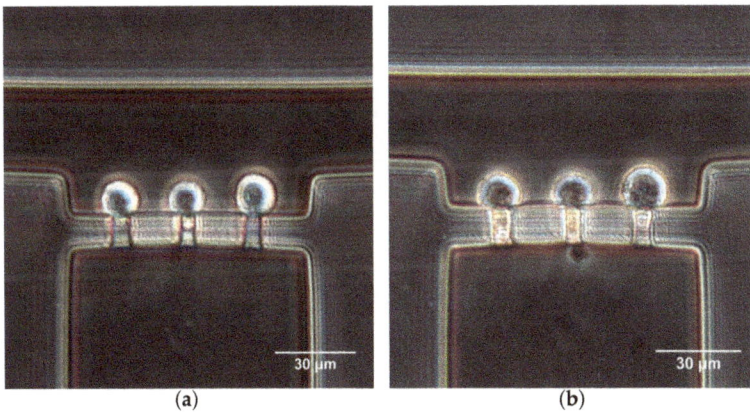

Figure 8. Hela cells aspirated into the micropipette aspiration channels at (**a**) $Q_t = 40$ µL/h and (**b**) $Q_t = 30$ µL/h.

The Young's modulus of Hela cells was evaluated for solid-like cells at the stable state. The results are shown in Figure 9. The average Young's modulus of Hela cells was 387 ± 77 Pa. This value is comparable to previous studies using magnetic tweezer [30] or micropipette aspiration technique [31], but it differs with that measured by atomic force microscopy [3,32–35]. It is clear that most measurements were at low flow rate Q_t (20–40 µL/h), where the aspiration pressure is low. As Q_t increased, only four measurements included one or two open micropipette channels ($N = 1$ or $N = 2$), indicating a relatively higher Young's modulus than those at low flow rates.

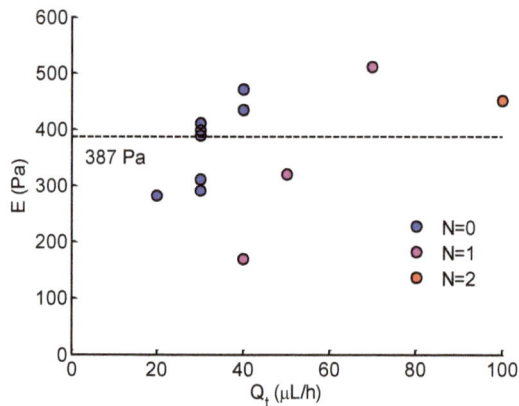

Figure 9. Young's modulus of Hela cells calculated at different Q_t. N denotes the number of open micropipette aspiration channels.

4. Discussion

The micropipette aspiration technique has been widely used in recent cell biology research, such as cell mechanical properties [19,29], molecular mechanics [17], cell response to mechanic stimuli, and single cell manipulation. Conventional MPA involves delicate manipulations conducted with specialized equipment by highly skilled technicians. Significant evaporation in the chamber leads to a drift in the null setting for the pressure [14]. Few studies have applied a microfluidic platform to conduct MPA for single-cell mechanics characterization, yet complicated microfluidic structures and sophisticated peripheral systems are required for implementation [19,27,36]. In this study, we presented a novel and easy-to-use microfluidic device by coupling the MPA technique and the principle of the Wheatstone bridge circuit. Typically, a classical Wheatstone bridge is an electrical circuit used to measure an unknown electrical resistance by balancing two legs of a bridge circuit, one leg of which includes a variable resistance. It reveals the advantages of high measurement accuracy and simple operation. Owing to these advantages and the comparability between the electrical field and flow field, the principle of the Wheatstone bridge circuit has been employed to control the flow in microfluidic devices [37,38]. Although a microfluidic Wheatstone bridge allows control of the bridge pressure and flow direction by a variable fluidic resistance, it causes difficulties in fabrication and problems in quantifying the resistance through the membrane deformation [38], leading to a quantitative control of the bridge pressure. In order to simplify the problem, in this study, we used fixed resistances and regulated the bridge pressure by controlling the flow rate. The designed device enables effective alteration of the flow direction for single-cell trapping and precise control of suction pressure in the micropipette aspiration channels. The combination of the MPA technique in the microfluidic Wheatstone bridge further improved the trapping efficiency (Figure 7) and provided a quantitative measurement of cell deformability (Figure 8). After image capture, by withdrawing the cell suspension at the inlet, aspirated cells can be released for further cell mechanics measurements. Thus, this device can be used for the long-term study of single cells' mechanical analysis.

The intrinsic mechanical properties of single cells is closely related to cell adhesion, migration, and motility [1–3]. Typically, Young's modulus is regarded as a biomarker of cell motility, especially in estimating the metastasis of cancer cells. Based on the distinctions between healthy and diseased cells [3,35], Young's modulus is suggested to be a diagnostic marker in the clinical setting. For the MPA technique, Young's modulus is determined by measuring deformation to applied force in conjunction with theoretical models [12–16]. The homogeneous half-space model [12,17] applied in this study is based on the assumption of small deformation. The model is employed best to problems in which the displacements and velocities are small. From the Young's modulus results in Figure 9, most measurements were conducted at low flow rate Q_t (20–40 µL/h) due to the small deformations. As the flow rate increased (corresponding to high pressure), the protrusion edge reached or extended the length of the micropipette channels (see Figure 8b). In these cases, the deformation was too large, such that the calculated Young's modulus may have been underestimated. From this viewpoint, the measurements were conducted only when the protrusion length was less than half of the micropipette channel length at the stable state. This is why only three measurements were included for high flow rate ($Q_t > 40$ µL/h). Note that the analytical aspiration pressure (see Figure 5) was calculated with the average dimensions indicated in Table 1, which was further applied to evaluate Young's modulus. In fact, due to the fabrication error, the actual dimensions were slightly different from the set values, which may have had a slight impact on measurement accuracy.

In Figure 9, it is shown that the measurement number markedly declined at a high flow rate Q_t. The measurement could only be obtained in the case of 1 or 2 open micropipette channels ($N = 1$ or $N = 2$). According to the analytical results of pressure (see Figure 5), the increase in the number of open micropipette channels resulted in the decrease in suction pressure. From this point, more open micropipette channels ($N \geq 1$) will benefit the measurement at a high flow rate, corresponding to high pressure. Thus, to improve the device and make it work effectively in a wide force range, properly increasing the number of micropipette channels may extend the measurement to a high applied force. In addition, an alternative enabling the device to be available for high force is to improve or develop a theoretical model for large deformations to applied force.

The discussion above reveals that the present microfluidic device combines the advantages of the MPA technique and the Wheatstone bridge principle, which shows the potential for the biomechanical characterization of single cells. This technique also has limitations. One is the realization of high throughput. Although the device can trap and release cells for repetitive and long-term study, repetitive operations lead to inconvenience and changes in cell mechanics. Future improvements will focus on improving the high-throughput capacity of the device by optimizing the structures (e.g., distributing the designed chip in a starlike disposition with the same cell suspension inlet). In this way, more measurements could be achieved simultaneously.

5. Conclusions

In this study, we developed a novel microfluidic device for quantifying cell mechanics at the single-cell level. The designed device combines the micropipette aspiration technique and the Wheatstone bridge principle, which allows single cells to be trapped effectively, precise control of the suction pressure, and quantitative measurement of the deformability, revealing the advantages of simplicity in implementation, ease-of-use, and reduction of sample consumption. The simplicity, precision, and usability of our device show great potential for biomechanical trials in clinical diagnosis and cell biology research.

Author Contributions: Y.-J.L. and K.-R.Q. conceived and designed the experiments; Y.-N.Y. performed the experiments; D.-P.Z. and H.-J.Z. analyzed the data; C.-D.X. and T.C. gave scientific support and conceptual advices; Y.-J.L. and K.-R.Q. wrote the draft. All authors discussed the results and commented on the manuscript.

Funding: This research was funded by the National Natural Science Foundation of China (No. 11672065) and the Fundamental Research Fund of Dalian University of Technology (DUT18RC(3)030).

Acknowledgments: The authors would like to warmly thank Jin-Yu Shao (Department of Biomedical Engineering, Washington University in Saint Louis) for the scientific discussions.

Conflicts of Interest: The authors declare no conflict of interest.

References

1. Cross, S.E.; Jin, Y.S.; Rao, J.; Gimzewski, J.K. Nanomechanical analysis of cells from cancer patients. *Nat. Nanotechnol.* **2007**, *2*, 780–783. [CrossRef] [PubMed]
2. Lee, G.Y.H.; Lim, C.T. Biomechanics approaches to studying human diseases. *Trends Biotechnol.* **2007**, *25*, 111–118. [CrossRef] [PubMed]
3. Luo, Q.; Kuang, D.D.; Zhang, B.Y.; Song, G.B. Cell stiffness determined by atomic force microscopy and its correlation with cell motility. *Biochim. Biophys. Acta* **2016**, *1860*, 1953–1960. [CrossRef] [PubMed]
4. Maître, J.L.; Berthoumieux, H.; Krens, S.F.G.; Salbreux, G.; Julicher, F.; Paluch, E.; Heisenberg, C.P. Adhesion functions in cell sorting by mechanically coupling the cortices of adhering cells. *Science* **2012**, *338*, 253–256. [CrossRef] [PubMed]
5. Kumar, S.; Weaver, V.M. Mechanics, malignancy, and metastasis: The force journey of a tumor cell. *Cancer Metastasis Rev.* **2009**, *28*, 113–127. [CrossRef] [PubMed]
6. Théry, M.; Bornens M. Get round and stiff for mitosis. *HFSP J.* **2008**, *2*, 65–71. [CrossRef] [PubMed]
7. Tavares, S.; Vieira, A.F.; Taubenberger, A.V.; Araújo, M.; Martins, N.P.; Brás-Pereira, C.; Polónia, A.; Herbig, M.; Barreto, C.; Otto, O.; et al. Actin stress fiber organization promotes cell stiffening and proliferation of pre-invasive breast cancer cells. *Nat. Commun.* **2017**, *8*, 15237. [CrossRef] [PubMed]
8. Suresh, S.; Spatz, J.; Mills, J.P.; Micoulet, A.; Dao, M.; Lim, C.T.; Beil, M.; Seufferlein, T. Connections between single-cell biomechanics and human disease states: gastrointestinal cancer and malaria. *Acta Biomater.* **2005**, *1*, 15–30. [CrossRef] [PubMed]
9. Guo, Q.; Reiling, S.J.; Rohrbach, P.; Ma, H. Microfluidic biomechanical assay for red blood cells parasitized by plasmodium falciparum. *Lab Chip* **2012**, *12*, 1143–1150. [CrossRef] [PubMed]
10. Lim, C.T.; Zhou, E.H.; Li, A.; Vedula, S.R.K.; Fu, H.X. Experimental techniques for single cell and single molecule biomechanics. *Mater. Sci. Eng. C* **2006**, *26*, 1278–1288. [CrossRef]
11. Guevorkian, K.; Maître, J.L. Micropipette aspiration: A unique tool for exploring cell and tissue mechanics in vivo. *Methods Cell Biol.* **2017**, *139*, 187–201. [PubMed]
12. Theret, D.P.; Levesque, M.J.; Sato, M.; Nerem. R.M.; Wheeler, L.T. The application of a homogeneous half-space model in the analysis of endothelial cell micropipette measurements. *J. Biomech. Eng.* **1988**, *110*, 190–199. [CrossRef] [PubMed]
13. Shao, J.Y.; Hochmuth, R.M. Micropipette suction for measuring piconewton forces of adhesion and tether formation from neutrophil membranes. *Biophys. J.* **1996**, *71*, 2892–2901. [CrossRef]
14. Hochmuth, R.M. Micropipette aspiration of living cells. *J. Biomech.* **2000**, *33*, 15–22. [CrossRef]
15. He, J.H.; Xu, W.; Zhu, L. Analytical model for extracting mechanical properties of a single cell in a tapered micropipette. *Appl. Phys. Lett.* **2007**, *90*, 023901. [CrossRef]
16. Khalilian, M.; Navidbakhsh, M.; Valojerdi, M.R.; Chizari, M.; Yazdi, P.E. Estimating Young's modulus of zona pellucida by micropipette aspiration in combination with theoretical models of ovum. *J. R. Soc. Interface* **2010**, *7*, 687–694. [CrossRef] [PubMed]
17. Lee, L.M.; Liu, A.P. The Application of Micropipette Aspiration in Molecular Mechanics of Single Cells. *J. Nanotechnol. Eng. Med.* **2014**, *5*, 040902. [CrossRef]
18. Shojaei-Baghini, E.; Zheng, Y.; Sun, Y. Automated micropipette aspiration of single cells. *Annals Biomed. Eng.* **2013**, *41*, 1208–1206. [CrossRef]
19. Guo, Q.; Park, S.; Ma, H.S. Microfluidic micropipette aspiration for measuring the deformability of single cells. *Lab Chip* **2012**, *12*, 2687–2695. [CrossRef]
20. Zheng, Y.; Sun, Y. Microfluidic devices for mechanical characterisation of single cells in suspension. *Micro Nano Lett.* **2011**, *6*, 327. [CrossRef]
21. Murphy, T.; Zhang, Q.; Naler, L.; Ma, S.; Lu, C. Recent advances on microfluidic technologies for single cell analysis. *Analyst* **2018**, *143*, 60–80. [CrossRef] [PubMed]
22. Rosenbluth, M.J.; Lam, W.A.; Fletcher, D.A. Analyzing cell mechanics in hematologic diseases with microfluidic biophysical flow cytometry. *Lab Chip* **2008**, *8*, 1062–1070. [CrossRef] [PubMed]
23. Gabriele, S.; Versaevel, M.; Preira, P.; Theodoly, O. A simple microfluidic method to select, isolate, and manipulate single cells in mechanical and biochemical assays. *Lab Chip* **2010**, *10*, 1459–1467. [CrossRef] [PubMed]

24. Lee, S.S.; Yim, Y.; Ahn, K.H.; Lee, S.J. Extensional flow-based assessment of red blood cell deformability using hyperbolic converging microchannel. *Biomed. Microdevices* **2009**, *11*, 1021–1027. [CrossRef] [PubMed]

25. Forsyth, A.M.; Wan, J.D.; Ristenpart W.D.; Stone H.A. The dynamic behavior of chemically 'stiffened' red blood cells in microchannel flows. *Microvasc. Res.* **2010**, *80*, 37–43. [CrossRef] [PubMed]

26. Guck, J.; Schinkinger, S.; Lincoln, B.; Wottawah, F.; Ebert, S.; Romeyke, M.; Lenz, D.; Erickson, H.M.; Ananthakrishnan, R.; Mitchell, D.; et al. Optical deformability as an inherent cell marker for testing malignant transformation and metastatic competence. *Biophys. J.* **2005**, *88*, 3689–3698. [CrossRef] [PubMed]

27. Chen, J.; Zheng, Y.; Tan, Q.Y.; Zhang, Y.L.; Li, J.; Geddie, W.R.; Jewett, M.A.S.; Sun, Y. A microfluidic device for simultaneous electrical and mechanical measurements on single cells. *Biomicrofluidics* **2011**, *5*, 014113. [CrossRef] [PubMed]

28. Tan, W.H.; Takeuchi, S. A trap-and-release integrated microfluidic system for dynamic microarray applications. *Biomicrofluidics* **2007**, *104*, 1146–1151. [CrossRef]

29. Lee, L.M.; Liu, A.P. A microfluidic pipette array for mechanophenotyping of cancer cells and mechanical gating of mechanosensitive channels. *Lab Chip* **2015**, *15*, 264–273. [CrossRef]

30. De Vries, A.H.B.; Krenn, B.E.; Van Driel, R.; Subramaniam, V.; Kanger, J.S. Direct Observation of Nanomechanical Properties of Chromatin in Living Cells. *Nano Lett.* **2007**, *7*, 1424–1427. [CrossRef]

31. Kim, W. A micro-aspirator chip using vacuum expanded microchannels for high-throughput mechanical characterization of biological cells. Master Thesis, Texas A & M University, College Station, TX, USA, August 2010; pp. 47–49.

32. Yokokawa, M.; Takeyasu, K.; Yoshimura, S.H. Mechanical properties of plasma membrane and nuclear envelope measured by scanning probe microscope. *J. Microsc.* **2008**, *232*, 82–90. [CrossRef] [PubMed]

33. Leporatti, S.; Vergara, D.; Zacheo, A.; Vergaro, V.; Maruccio, G.; Cingolani, R.; Rinaldi, R. Cytomechanical and topological investigation of MCF-7 cells by scanning force microscopy. *Nanotechnology* **2009**, *20*, 055103. [CrossRef]

34. Hayashi, K.; Iwata, M. Stiffness of cancer cells measured with an AFM indentation method. *J. Mech. Behav. Biomed. Mater.* **2015**, *49*, 105–111. [CrossRef]

35. Hao, S.J.; Wan, Y.; Xia, Y.Q.; Zou, X.; Zheng, S.Y. Size-based separation methods of circulating tumor cells. *Adv. Drug Deliv. Rev.* **2018**, *125*, 3–20. [CrossRef] [PubMed]

36. Lee L.M.; Lee J.W.; Chase D.; Gebrezgiabhier D.; Liu A.P. Development of an advanced microfluidic micropipette aspiration device for single cell mechanics studies. *Biomicrofluidics* **2016**, *10*, 054105. [CrossRef] [PubMed]

37. Plecis, A.; Chen, Y. Microfluidic analogy of the Wheatstone bridge for systematic investigations of electro-osmotic flows. *Anal. Chem.* **2008**, *80*, 3736–3742. [CrossRef]

38. Tanyeri, M.; Ranka, M.; Sittipolkul, N.; Schroeder, C.M. Microfluidic Wheatstone bridge for rapid sample analysis. *Lab Chip* **2011**, *11*, 4181. [CrossRef]

micromachines

MDPI

Review

Microfluidic Single-Cell Manipulation and Analysis: Methods and Applications

Tao Luo [1,†]**, Lei Fan** [1,†]**, Rong Zhu** [2] **and Dong Sun** [1,3,*]

1   Department of Biomedical Engineering, City University of Hong Kong, Hong Kong, China;
    taoluo4-c@my.cityu.edu.hk (T.L.); leifan-c@my.cityu.edu.hk (L.F.)
2   State Key Laboratory of Precision Measurement Technology and Instruments, Department of Precision
    Instrument, Tsinghua University, Beijing 100084, China; zr_gloria@mail.tsinghua.edu.cn
3   Shenzhen Research Institute of City University of Hong Kong, Shenzhen 518057, China
*   Correspondence: medsun@cityu.edu.hk; Tel.: +852-3442-8405
†   These authors contributed equally to this work.

Received: 19 December 2018; Accepted: 30 January 2019; Published: 1 February 2019

Abstract: In a forest of a hundred thousand trees, no two leaves are alike. Similarly, no two cells in a genetically identical group are the same. This heterogeneity at the single-cell level has been recognized to be vital for the correct interpretation of diagnostic and therapeutic results of diseases, but has been masked for a long time by studying average responses from a population. To comprehensively understand cell heterogeneity, diverse manipulation and comprehensive analysis of cells at the single-cell level are demanded. However, using traditional biological tools, such as petri-dishes and well-plates, is technically challengeable for manipulating and analyzing single-cells with small size and low concentration of target biomolecules. With the development of microfluidics, which is a technology of manipulating and controlling fluids in the range of micro- to pico-liters in networks of channels with dimensions from tens to hundreds of microns, single-cell study has been blooming for almost two decades. Comparing to conventional petri-dish or well-plate experiments, microfluidic single-cell analysis offers advantages of higher throughput, smaller sample volume, automatic sample processing, and lower contamination risk, etc., which made microfluidics an ideal technology for conducting statically meaningful single-cell research. In this review, we will summarize the advances of microfluidics for single-cell manipulation and analysis from the aspects of methods and applications. First, various methods, such as hydrodynamic and electrical approaches, for microfluidic single-cell manipulation will be summarized. Second, single-cell analysis ranging from cellular to genetic level by using microfluidic technology is summarized. Last, we will also discuss the advantages and disadvantages of various microfluidic methods for single-cell manipulation, and then outlook the trend of microfluidic single-cell analysis.

Keywords: microfluidics; single-cell manipulation; single-cell analysis

---

1. Introduction

Over the past few decades, cellular heterogeneity has gradually been emphasized on fundamental biological and clinical research as numerous novel tools/methods for single-cell analysis have emerged [1]. Phenotype heterogeneity between genetically identical cells plays an important role in tumor metastasis [2], drug resistance [3], and stem cell differentiation [4]. For instance, different responses of individual cells to drugs cause the emergence of drug-resistant cells, but only a small percentage (0.3%) of these cells have the ability for tumor recurrence [5]. However, cellular heterogeneity has been masked for a long time because previous biological studies are mainly based on manipulating and analyzing cells at the bulk-scale, which interpreted all phenomena by using average results. Until today, single-cell study still has been recognized as the most straightforward way to performance comprehensive heterogeneity study from the aspects of cellular behavior to genetic expression. Comprehensive single-cell study heavily relies on the use of high-throughput and efficient tools for manipulating and analyzing cells at the single-cell level.

Single-cell analysis is technically more difficult than bulk-cell analysis in terms of the sizes of cells and the concentrations of cellular components. The majority of cells, such as mammalian and bacteria cells, have sizes at the scale of microns. Therefore, manipulation of those cells at the single-cell level becomes difficult when using traditional biological tools, such as petri-dishes and well-plates. Additionally, most of the intracellular, extracellular components are presented in very small concentrations and have a wide range of concentrations, which demand highly sensitive and specific detection methods. Many single-cell analysis applications require a single-cell isolation first, and multi-well plates are commonly used in most biological labs for single-cell isolation, which is low in efficiency and labor-intensive [6]. While the use of robotic liquid handling workstation reduces the labor intensity, it is very expensive for some labs to afford it [7]. Flow cytometry or laser scanning cytometry, which rapidly screens fluorescently labeled cells in a flow, has been developed and recognized as a golden standard for single-cell analysis for a long time [8]. Taking flow cytometry as an example, although they are automatic, capable of multiple detections, and efficient in single-cell sorting, they are bulky, mechanically complicated, expensive, and demanding for relatively large sample volumes. Besides, they can only be used for analyzing cells at one time-point. Hence, it is impossible to use flow cytometry for continuously monitoring cell dynamics. Owing to the capability of manipulating and controlling fluids in the range of micro to pico-liters, microfluidics has been developed as a platform-level and continuously evolving technology for single-cell manipulation and analysis for about two decades.

Microfluidics has many incomparable advantages over conventional techniques. Firstly, the microfluidic chip can be flexibly designed to fulfill the demands of diverse single-cell manipulation and analysis tasks. For instance, single-cell manipulation can be achieved by using either passive [9–11] or active [12,13] method, and single-cell analysis can be achieved by implementing either optical [14,15] or electrochemical [16,17] method. Secondly, miniaturized microfluidic systems work can work with very small volume (down to pL level) of liquid, which helps to reduce sample loss and decrease dilution, resulting in highly sensitive detections. Hence, numerous microfluidics-based biosensors have been developed. Thirdly, microfluidics allows for high-throughput parallel manipulation and analysis of the sample, which is beneficial for the statistically meaningful single-cell analysis. Fourthly, multiple functionalities are easily integrated on the same chip, which allows for automation, and can also avoid contamination and errors introduced by manual operations. Many single-cell studies require single-cell capture/isolation, and different microfluidic methods, such as hydrodynamic [11,18,19], electrical [20], optical [21], magnetic [22], and acoustic [23] methods, have been developed. Various detection methods, such as fluorescence microscopy, fluorometry and mass spectroscopy, can be combined with microfluidic systems for single-cell analysis from cell morphology to secreted proteins. As for either single-cell manipulation or single-cell analysis, it is hard to obtain a comprehensive result by merely using one method. Therefore, two or more methods are usually combined into a microfluidics system for various single-cell studies [24,25].

While reviews about single-cell manipulation and analysis by using microfluidics are reported almost every year, systematic summarization of this area can give valuable references to both academic and industrial fields. In this review, we mainly focus on microfluidic technologies for single-cell manipulation analysis from the aspects of methods and applications. We highlight methods that are promising for future development, which are discussed in terms of single-cell manipulation including hydrodynamic, electrical, optical, magnetic, acoustic, and micro-robots assisted methods. We also highlight applications that are accepted by academic and industrial fields, which are discussed in terms of single-cell analysis from cellular to protein analysis. Last, we also discuss the technology and application trend for microfluidics based single-cell analysis.

2. Microfluidic Single-Cell Manipulation

With the development of Microelectromechanical Systems (MEMS) technology, many micro-scale devices have been fabricated for bioanalysis at single-cell resolution. As a powerful technology to perform precise fluidic control, microfluidics has attracted great interests for various single-cell manipulations, such as single-cell encapsulation and single-cell trapping (Table 1). Nowadays, various methods, including hydrodynamic, electrical, optical, acoustic, magnetic, and micro-robotic method, were used for diverse microfluidic single-cell manipulations.

Table 1. Various single-cell manipulations.

Manipulations	Descriptions
Single-cell encapsulation [26–30]	Entrapping single cells in isolated microenvironments
Single-cell sorting [12,31–33]	Separating homogenous populations of cells from heterogeneous populations at the single-cell resolution
Single-cell trapping [34–36]	Immobilizing single cells from bulk cells on the designated positions.
Single-cell isolation [37,38]	Pick or isolate single cells from bulk populations
Single-cell rotation [39]	Rotating targeted single cells
Single-cell pairing [20,40]	Positioning two homo- or heterotypic cells in proximity or contact
Single-cell patterning [23]	Positioning single cells on a substrate with defined spatial selection
Single-cell stretching [41]	Using external forces to deform targeted single cells
Single-cell transportation [42]	Moving cells at the single-cell level
Single-cell lysis [43,44]	Breaking down the targeted single cells
Single-cell stimulation [19,45,46]	Applying external physical/chemical/biological cues to stimulate targeted single cells

2.1. Hydrodynamic Method

Compared with other single-cell manipulation methods, the hydrodynamic method is much simpler and higher throughput. Hydrodynamic manipulation mainly relies on the interaction among microstructure, fluid, and cells. This technique has the advantages of high throughput, less damage to cells, mature chip fabrication crafts, and easy integration with other analysis functionalities. Based on the used mechanisms, the hydrodynamic method can be categorized into droplet microfluidics, inertial microfluidics, vortex, and mechanical methods.

2.1.1. Droplet Microfluidics

Droplet microfluidics has attracted more and more interests for its capability to encapsulate cells and many reagents in a microscale environment [47]. It is also a powerful technique for high-throughput single-cell encapsulation. The size, shape, and uniformity of droplets can be precisely controlled. This method usually requires two immiscible fluids to create monodispersed water-in-oil (w/o) microdroplets with sizes that range from the submicron to several hundreds of microns [48]. As shown in Figure 1, three types of microfluidic droplet generation approach usually used: T-junction, flow focusing, and co-flow [28]. However, the basic principles of these three approaches are the same. One fluid becomes the dispersed phase to form the droplets and the other fluid becomes the continuous phase to separate the droplets. As illustrated in Figure 1a, Zhang et al. integrated T-junction structure with droplet inspection, single-cell droplet sorting and exporting on one chip to analyze DNA and RNA at both gene-specific and whole-genome levels [49]. Zilionis et al. utilized flow focusing structure, shown in Figure 1b, to conduct single-cell barcoding and sequencing [27]. Adams et al. encapsulated multi-component in one droplet using co-flow method, as shown in Figure 1c [50]. All three approaches require to control each fluid phase precisely, which usually make the system a little bit complex. Khoshmanesh et al. proposed a novel mechanism for generating microscale droplets of aqueous solutions in oil using a highly porous PDMS sponge [51]. Compared to the existing microfluidic droplet generation approach, the sponge-based approach is a self-sufficient, stand-alone device, which can be operated without using pumps, tubes, and microfluidic skills. Single-cell is randomly encapsulated in each droplet, and the number of cells in each droplet follows nonuniform poisson distribution. Therefore, the cell suspension usually requires to be highly diluted before encapsulation to ensure only one single cell in one droplet. However, this procedure leads to reagent waste and degrades throughput because most of the generated droplets do not contain single cells. Several methods have been developed to overcome this limitation. For example, post-sorting based on property differences can be applied to enhance single-cell encapsulation efficiency. Moreover, this technique is hampered by some other drawbacks. For instance, the cell culture in droplet is suspended, which means that the culture of adherent cell in droplet is difficult. Introducing/picking components in/out of droplets without risking cross-contamination among cell-contained droplets is difficult.

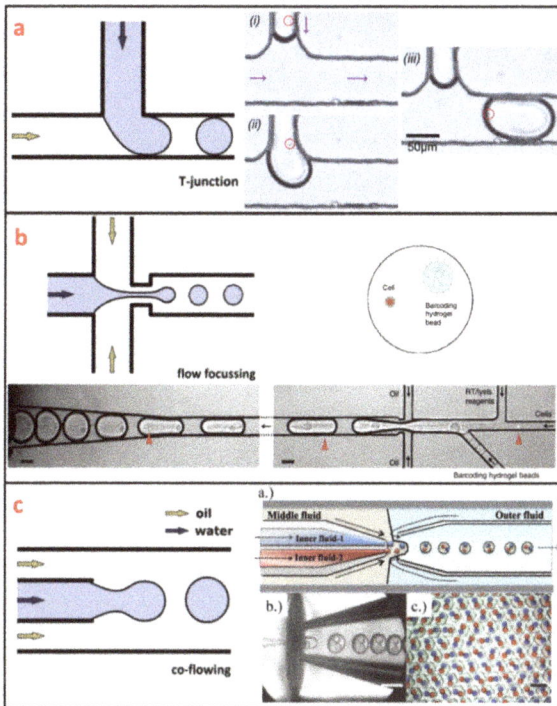

Figure 1. Three types of methods and designs for microfluidic single-cell droplet generation. Adapted by permission from Reference [28], copyright Royal Society of Chemistry 2015. (**a**) T-junction microchannel based droplet generation for encapsulating single cells for cultivation and genomic analysis. Adapted by permission from reference [49], under the Creative Commons Attribution License 2017. (**b**) Flow-focusing for encapsulating a single cell and a barcoding hydrogel bead in a droplet for single-cell barcoding and sequencing. Adapted by permission from Reference [27], copyright Nature Publishing Group 2017. (**c**) Co-flow for preparing multiple component double emulsions. Adapted by permission from Reference [50], copyright Royal Society of Chemistry 2012.

2.1.2. Inertial Microfluidics

As a cross-streamline cell manipulation method, inertial microfluidics is based on using inertial forces under certain high flow rates to continuously focus and sort cells with different sizes and shapes, as illustrated in Figure 2a [52]. The inertial microfluidics has the advantages of simple chip structure, ultra-high throughput, and less damage to cells which is beneficial to maintain high cell survival rate for downstream culture. However, intercellular interaction can greatly reduce the efficiency of cell manipulation by using inertial microfluidics. Therefore, it is only applicable to work under a certain cell concentration. This method is usually used for separating single circulating cancer cells from blood cells, and it usually requires pre-dilution of blood samples [53,54]. Nathamgari et al. used inertial microfluidics to separate the single neural cells and clusters from a population of chemically dissociated neurospheres shown in Figure 2a [55]. In contrast to previous sorting technologies that require operating at high flow rates, they implemented a spiral microfluidic channel in a novel-focusing regime that occurs at relative lower flow rates. The curvature-induced Dean's force focused the smaller single cells towards the inner wall and the larger clusters towards the center of channel.

2.1.3. Vortex

Vortex based method is based on generating a time-averaged secondary flow known as steady streaming eddies [56,57], which is generated by the interaction between frequency oscillations of the fluid medium and fixed cylinder in a microchannel to trap single cells. Therefore, the vortex approach is also called single-cell hydrodynamic tweezer. In 2006, Lutz et al. used motile phytoplankton cells to measure the trapping location and trapping force [34]. They proved that each eddy traps a single-cell near the eddy center, precisely at the channel midplane, and the trapped cell is completely suspended by the fluid without touching any solid surface. Furthermore, the trapping force are comparable to dielectrophoretic force and optical tweezer force, whereas the trap environment is within physiological limits of shear in arterial blood flow. Thus, the hydrodynamic tweezer does not limit the cell type, shape, density, or composition of the fluid medium. As shown in Figure 2b, Hayakawa et al. adopted three micropillars arranged in a triangular configuration and an xyz piezoelectric actuator to apply the circular vibration to generated vortex for trapping and 3D rotating mouse oocytes single-cell [58]. Additionally, they measured the rotational speeds in the focal and vertical planes as $63.7 \pm 4.0° \cdot s^{-1}$ and $3.5 \pm 2.1° \cdot s^{-1}$, respectively.

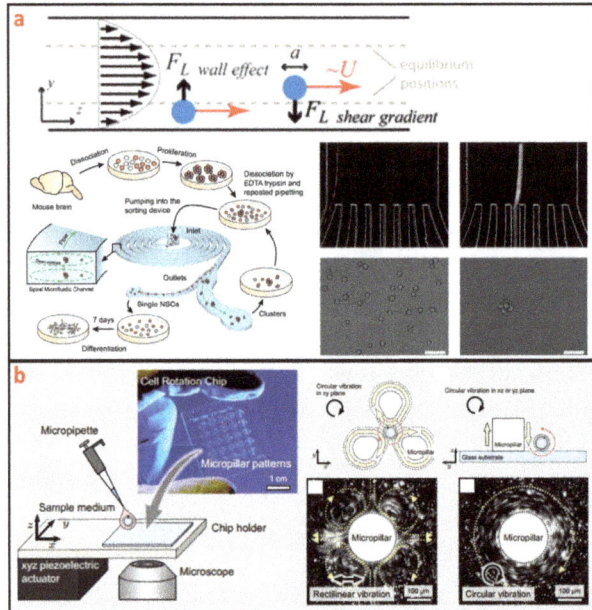

Figure 2. Inertial and vortex microfluidic single-cell manipulation and some designs. (**a**) Inertial microchannel to separate single-cell from cell cluster. Adapted by permission from References [52,55], copyright Royal Society of Chemistry 2009 and 2015. (**b**) Vortex generated by micropillar to rotate single-cell. Adapted by permission from Reference [58], under the Creative Commons Attribution License 2015.

2.1.4. Mechanical Method

Mechanical method for single-cell manipulation mainly refers to the use of membrane pump, microvalve, and microstructures.

- Microvalve

Having the advantages of small size, fast response, and simple fabrication, the active microvalve is widely used in the manipulation of single cells [29]. Multilayer soft lithography, which is the basement of microvalve, was first developed by Quake's group [59]. They used this technique to build active microfluidic systems containing on-off valves and switching valves. The invention of these valves made it possible to realize high density fluid control and large-scale functional integration on a chip, which is a milestone in the development of microfluidic technology. Thus, this kind of microvalve is called "Quake valve". The structure of "Quake valve" consists of a vertically crossed flow layer and a control layer, between which a deformable film is formed. When the pressure is applied through the control layer, the film deforms and sticks to the bottom surface of flow channel to block the flow in the flow layer. On the contrary, when the control layer does not exert pressure or apply a relatively small pressure, the flow layer appears to be open or semi-open. Fluidigm Company of United States developed a single-cell automatic pretreatment system called C1 Single-Cell Auto Prep System based on this technology, which can automatically separate 96 suspended cells at one time. At present, the system has been used in many universities and research institutions in the world for single-cell genomics research. Shalek et al. uses this system to isolate single-cell and conduct RNA-seq libraries prepared from over 1700 primary mouse bone-marrow-derived dendritic cells shown in Figure 3 [60]. While microvalves offer many advantages, their external control devices are extremely complex and cumbersome. Thus, the external operation of large-scale integrated microfluidic chips based on microvalves must be simplified and adapted to the working habits of biological researchers.

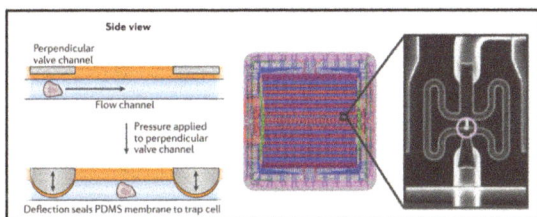

Figure 3. Microfluidic valve single-cell isolation and a design Adapted by permission from References [29,60], copyright Nature Publishing Group 2017 and 2014.

- Microstructure

The method based on microstructures, such as microtraps and microwells, only requires researcher to design microstructures whose sizes are similar to the size of a single cell. After injecting cells into microfluidic chip, cells flow along the streamlines of the laminar flow, and can be trapped by shearing force generated by microstructure. Current microstructure for single-cell trap include U-shaped, S-shaped, and microwell based traps.

U-shaped microtrap was demonstrated firstly by Di Carlo et al. to trap and culture single Hela cells [61]. The U-shaped structure was fabricated by polydimethylsiloxane (PDMS), bonded with glass and there was a gap between the trap and the glass substrate for increasing the single-cell trap efficiency. Then, the U-shaped microtrap was modified by Wlodkowic et al. to add three gaps in the edge of U-shaped groove for increasing the cell viability [19]. In the last years, Tran et al. designed U-shaped micro sieve comprised semicircular arcs spaced at specific offsets and distance as shown in Figure 4a [62]. Using this micro sieve, they realized label-free and rapid human breast cancer single-cell isolation with up to 100% trapping yield and >95% sequential isolation efficiency. Luo et al. fabricated a high throughput single-cell trap and culture platform to investigate clonal growth of arrayed single-cells under chemical/electrical stimuli for the week-scale period [63]. To achieve deterministic single-cell capture in large-sized microchambers, a U-shaped sieve with a 5μm-thick bottom microchannel was used to capture a single-cell, and two stream focusing arms were placed in front of the sieve to enhance capture efficiency. Zhang et al. demonstrated a handheld single-cell pipette, which allows for rapid single-cell isolation from low concentration cell suspension, by using a U-shaped microtrap, as shown in Figure 4b [37].

S-shaped microstructure is based on the different fluidic resistance in different position of microchannel. The S-shaped microstructure was firstly reported by Tan et al. to trap and release microbeads [64]. After microbead trapping, one objective bead could be release by optical microbubbles. In recent years, S-shaped microstructure was also further optimized. Kim. et al. proposed a simple, efficient S-shaped microfluidic array chip integrated with a size-based cell bandpass filter [65]. The key advancement to this chip is not the optimization of single-cell trap, but the capability of trapping cells within a specific range of sizes. Mi et al. combined U-shape and S-shape, which is named m-by-n trap units, to pattern single Hela cells, as shown in Figure 4c [10]. Each unit has two roundabout channels and one capture channel. Different from previous S-shaped microchannels, this structure enables each trap unit to be treated equally and independently. Therefore, any unit can be selected for finalizing the geometric parameters of the fluidic channels to satisfy the capture condition.

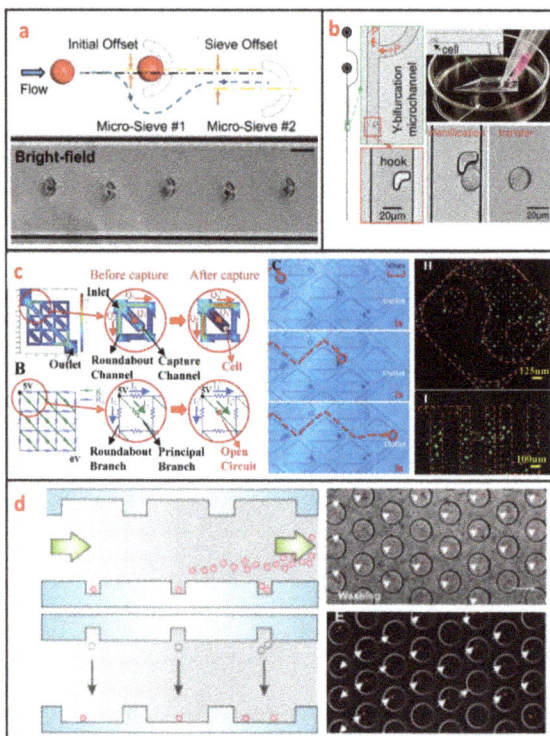

Figure 4. Microstructure based single-cell manipulation methods and some typical designs. (**a**) Micro-sieve to isolate floating single cancer cell under continuous flow. Adapted by permission from Reference [62], copyright Royal Society of Chemistry 2016. (**b**) A microfluidic pipette with a micro-hook for trapping and releasing a single cell. Adapted by permission from Reference [37], copyright Royal Society of Chemistry 2016. (**c**) Microchannel to trap single-cell based on fluidic circuit. Adapted by permission from Reference [10], copyright Royal Society of Chemistry 2016. (**d**) A dual-microwell design for the trap and culture of single cells. Adapted by permission from Reference [66], copyright Royal Society of Chemistry 2015.

Microwell is another passive microstructure for single-cell trapping. This structure is mainly based on the size match of the single cell and microwell. When the diameter of microwell approaches the diameter of a single cell, redundant cells will be wash out and one single cell settled down to one microwell can be trapped. After injecting cell suspension to microfluidic channel, cells will sediment down and go into microwells. When the depth of the microwell is deep enough, vortex generates inside the microwell and single cells can be trapped firmly. Most of current microwells were fabricated by soft lithography. In 2004, Revzin et al. developed a cytometry platform for characterization and sorting of individual leukocytes [32]. Poly (ethylene glycol) (PEG) was employed to fabricate arrays of microwells composed of PEG hydrogel walls and glass substrates. PEG micropatterned glass surfaces were further modified with cell-adhesive ligands, poly-L-lysine, anti-CD5, and antiCD19 antibodies. Cell occupancy reached 94.6 ± 2.3% for microwells decorated with T-cell specific anti-CD5 antibodies. Later on, Rettig et al. fabricated tens of thousands of microwells on a glass substrate [9]. And they characterized microwell occupancy for a range of dimensions and seeding concentration using different cells. For culturing single-cell in one chamber, Lin et al. fabricated a dual-well (DW) device which allows for highly efficient loading of single-cells into large microwells for single-cell culture, as shown in Figure 4d [66]. Single-cell loading in large microwells is achieved by utilizing small microwells to trap single cells followed by using gravity to transfer the trapped single cells into large microwells for single-cell culture.

2.2. Electrical Method

The electrical methods have been widely used to trap and pattern single-cell because it usually imposes lower physical pressure on the cell membrane. Two typical methods exist for electronically controlled single-cell manipulation: Dielectrophoresis (DEP) and electroosmosis.

2.2.1. Dielectrophoresis (DEP)

DEP manipulation relies on the use of DEP forces, which are generated by the interaction between the nonuniform electric field and the cells. DEP forces applied to cells depend on the size of the cells, the dielectric properties of the cells and the surrounding solution, the gradient of the electric field, and the frequency of the electric field [67]. As shown in Figure 5a, DEP forces can be categorized as positive or negative [68]. Under positive DEP forces, cells move to strong electric field regions. By contrast, under negative DEP forces, cells move to weak electric field regions. The frequency of the electric field when the DEP force is zero is called cross-over frequency, at which the DEP forces applied to cells is zero. The cell manipulation performance of the DEP chip depends largely on the design of the DEP electrode. DEP can be easily combined with microfluidic systems, is label-free, and has high selectivity in manipulating rare cells. Most DEP cell manipulation systems require a low-conductivity solution. However, physiological solutions, such as blood and urine, are highly conductive. Thus, cell samples require stringent pretreatment. This requirement limits the application of DEP-based approaches and may have prevented the application of DEP-based cell manipulation in the clinical field. Taff et al. first presented a scalable addressable positive-dielectrophoretic single-cell trapping and sorting chip using MEMS technology based on silicon substrate [31]. The chip incorporates a unique "ring-dot" pDEP trap geometry organized in a row/column array format. A passive, scalable architecture for trapping, imaging, and sorting individual microparticles, including cells, using a positive dielectrophoretic (pDEP) trapping array was fabricated. Thomas et al. presented a novel micron-sized particle trap that uses nDEP to trap cells in high conductivity physiological media [13]. The design is scalable and suitable for trapping large numbers of single-cells. Each trap has one electrical connection, and the design can be extended to produce a large array. The trap consists of a metal ring electrode and a surrounding ground plane, which creates a closed electric field cage in the center. The device is operated by trapping the single latex spheres and HeLa cells against a moving fluid. In recent years, Wu. et al., as shown in Figure 5b, reported a design and fabrication of a planar chip for high-throughput cell trapping and pairing by pDEP within only several minutes [20].

The pDEP was generated by applying an alternating current signal on a novel two-pair interdigitated array (TPIDA) electrode. In Figure 5c, Huang et al. reported DEP-based single-cell trap and rotation chip for 3D cell imaging and multiple biophysical property measurements [69]. They firstly trapped a single-cell in constriction and subsequently released it to a rotation chamber formed by four sidewall electrodes and one transparent bottom electrode, which are powered by AC signals.

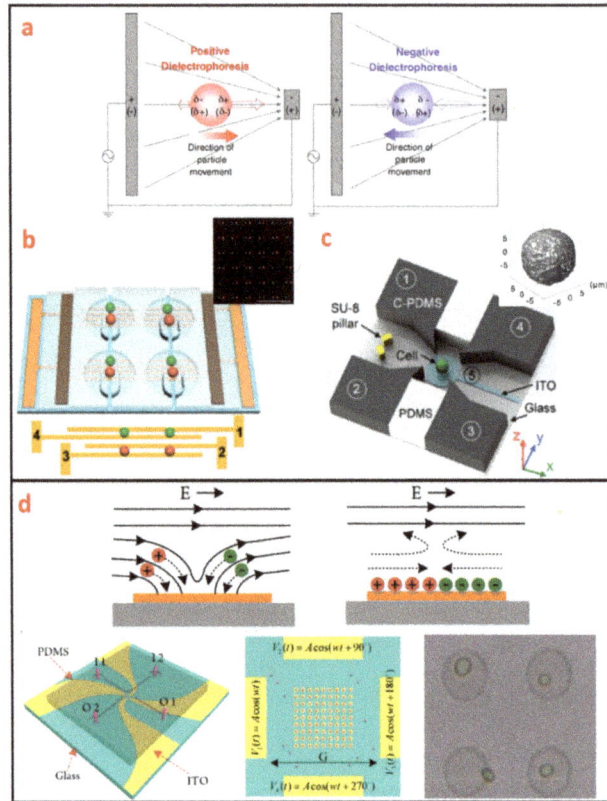

Figure 5. Electrical single-cell manipulation methods and some designs. (**a**) Theory of dielectrophoresis (DEP). Adapted by permission from Reference [68], copyright Elsevier 2005. (**b**) 2D electrode to trap and pair single cells. Adapted by permission from Reference [20], copyright Royal Society of Chemistry 2017. (**c**) 3D electrode to rotate single cell. Adapted by permission from Reference [69], copyright Royal Society of Chemistry 2018. (**d**) Rotating electric field induced-charge Electro-osmosis to trap single cell. Adapted by permission from Reference [70], copyright American Chemical Society 2016.

2.2.2. Electroosmosis

Electroosmotic flow is caused by the Coulomb force that is induced by an electric field on net mobile electric charge in a solution. Two kinds of electroosmosis are usually used for cell manipulation, namely, alternating current electroosmosis (ACEO), and induced charge electroosmosis (ICEO). ACEO is induced by ionic cloud migration in response to a tangentially applied electric field on the electrode surface and only occurs when the applied frequency is far below the charge relaxation frequency of the fluid. ACEO is one of the most promising electrokinetic approaches for developing fully integrated lab-on-a-chip systems because it is a label-free and well-established technique for microelectrode fabrication, as well as low voltage requirement. Gilad Yossifon et al. reported a multifunctional microfluidic platform on-chip electroporation integrated with ACEO-assisted cell trapping [71]. ACEO vortices enable the rapid trapping/alignment of particles at sufficiently low activation frequencies. ICEO is an electrochemical effect that occurs on the surface of an object and manifests as nonlinear fluid flow under electric conditions. As shown in Figure 5d, the induced charge diffusion stimulated by the electric field induces slip-like local fluid flow under the applied electric field. The induced diffusion charges are distributed in the boundary layer on the solid/liquid interface, i.e., a double electrical layer. Electric force can be applied to the fluid molecules to control microfluidic movement given the existence of the double layer [72]. ICEO efficiently enriches cells in specific areas, and high-throughput and noncontact single-cell capture can also be achieved in a given area (Figure 5d) [70]. The ICEO microfluidic chip can enable the small-scale integration of electrode arrays and microchannels. The ICEO chip has a compact structure and is easy to process. Given that the floating electrode does not require an external electrical signal wire connection, the design layout can be flexibly designed in accordance with different application scenarios. However, the performance of the ICEO microfluidic chip is dependent on the induced secondary flow. In the case of continuous sampling, cell manipulation efficiency and precision will greatly decrease with the increase of the flow rate.

2.3. Optical Method

Three types of optical cell manipulation methods currently exist: Optical tweezer, optically induced-dielectrophoresis (ODEP), and opto-thermocapillary.

2.3.1. Optical Tweezer

The use of optical tweezers to move particles was first discovered by the American scientist Arthur Ashkin. He found that a highly focused laser beam could drag an object with a higher refractive index than the medium to the middle of the laser beam. He first studied the "optical tweezer" effect with a single laser beam in 1987 [73]. The tweezer based on monochromatic lasers can manipulate particles within the size range of nanometers to tens of microns. Therefore, the optical tweezer can be used to manipulate biological single-cells. Robotic-assisted optical tweezers have enabled the automated multidimensional manipulation of cells and have been applied in studies on cell mechanics [74], cell transportation [42], and cell migration [75]. Optical tweezers can achieve cell manipulation under static environments and combined with microfluidic chips to realize cell manipulation under continuous flow. The most typical application is optical tweezer-enhanced microfluidic cell sorting [12]. As shown in Figure 6a, the cell sample is first focused on the channel upstream through the sheath flow. Then, the cells are identified through the image processing of the fluorescence characteristics of the cells. The tweezer automatically captures and drags the target single-cells to make them laterally cross the streamline. The target single-cells are thus collected at a specific outlet.

Optical tweezers offer the advantages of high accuracy, non-intrusiveness, and high-throughput single-cell manipulation. However, the manipulative laser force exerted on the cells is typically in the order of pico-Newtons. Thus, cell manipulation under continuous flow requires low fluid velocity; otherwise, the optical tweezers will experience difficulty in deflecting cells. In addition, the massive peripherical optical system required by this technique is difficult to miniaturize and is also very expensive.

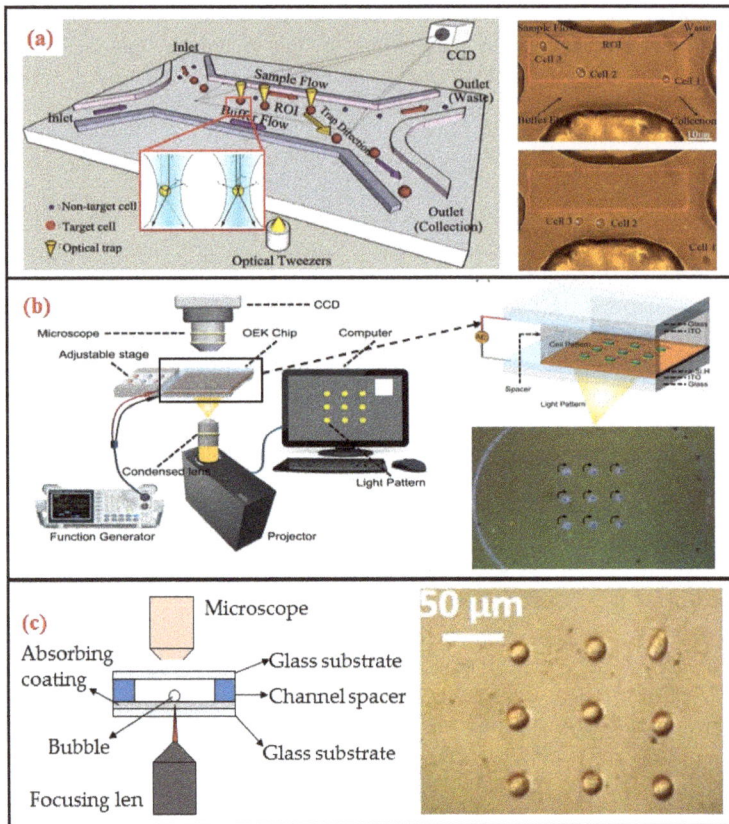

Figure 6. Optical single-cell manipulation methods and some designs. (**a**) Optical tweezer based single-cell sorting. Adapted by permission from Reference [12], copyright Royal Society of Chemistry 2011. (**b**) ODEP based single-cell array rotation. Adapted by permission from Reference [76], copyright Royal Society of Chemistry 2017. (**c**) Opto-thermocapillary based single-cell pattern. Adapted by permission from Reference [77], copyright Royal Society of Chemistry 2013.

2.3.2. Optically Induced Dielectrophoresis (ODEP)

Optically induced dielectrophoresis (ODEP) is a novel particle manipulation technology. The forces used to manipulate particles in ODEP-based chip are the same as those used in traditional DEP technology. That is, a nonuniform electric field is used to polarize cells and generate DEP forces. The difference between the two technologies lies in the technique by which the nonuniform electric field is generated. In contrast to conventional DEP manipulation, ODEP does not need prefabricated electrode patterns. However, digital micromirror device (DMD) projector can be used to project the light pattern onto the chip substrate via a microscope to generate flexible and controllable virtual electrodes [21]. Its working principle is similar to that of photovoltaic power generation. An amorphous silicon substrate material generates photocarriers under light excitation, thereby increasing carrier concentration in the illumination area. Consequently, the electrical conductivity of the illumination area rapidly increases, thereby generating a nonuniform electric field. ODEP can also enable the DEP manipulation of live single-cell. It is contactless and label-free. As shown in Figure 6b, Xie et al. utilized ODEP to effectively trap and transport unicellular swimming algae [76]. They found that the trapped cells started to rotate and demonstrated that functional flagella played a decisive role in the rotation. Furthermore, they also realized homodromous rotation of a live C. reinhardtii cell array in an ODEP trap and the speed of rotation can be controlled by varying the optical intensity.

The ODEP system is considerably simpler than optical tweezer systems and can be miniaturized. Moreover, ODEP can manipulate cells that are not optically transparent, thus exhibiting great flexibility. However, ODEP and its clinical applications are hindered by the same inherent drawbacks as traditional DEP: The manipulation of cells in low-conductivity solutions. Moreover, the substrate of the ODEP chip is opaque because of the deposition of amorphous silicon. The opacity of the substrate precludes the use of an inverted biological microscope for live cell imaging.

2.3.3. Opto-Thermocapillary

Different from the above two optical methods, the opto-thermal method uses light to generate heat for opto-thermophoresis or opto-thermocapillary. Thermophoresis is the thermos-migration or thermos-diffusion of particles subjected to a temperature gradient [78]. The opto-thermophoresis can be used to trap small biological molecules. However, it is difficult for trapping large particle, such as cells. The thermocapillary, also named thermal Marangoni effect, refers to mass transfer along a liquid–gas interface due to a surface tension gradient created by a temperature gradient. Thus, opto-thermocapillary actuation is not dependent on the optical properties of the object. And opto-thermocapillary is not sensitive to the electrical properties of the liquid medium and the object. However, opto-thermocapillary actuation shares the flexibility of optical control, which enables parallel and independent manipulation of multiple micro-objects [79]. Opto-thermocapillary force can be used to actuate microbubbles that enable manipulating single-cells and biomolecules [80]. As shown in Figure 6c, a microbubble can be generated after focusing optical beam on the absorbing coating. This microbubble can be seen as micro-scale actuator to manipulate single cell. Hu et al. used a near-infrared laser focused on indium tin oxide (ITO) glass to generate thermocapillary effect that can trap and transport living single cells with forces of up to 40 pN [77]. Moreover, they also patterned single-cell in two kinds of hydrogels: Polyethylene glycol diacrylate (PEGDA) and agarose. High viability rates were observed in both hydrogels, and single cells patterned in agarose spread and migrated during culture.

2.4. Acoustic Method

Acoustic cell manipulation is based on the complex flow–structure interaction that occurs when acoustic waves enter a microfluidic channel. Acoustic waves can be categorized into body and surface waves. A surface acoustic wave (SAW) is an elastic acoustic wave that can propagate only on the substrate surface. Most of its energy is concentrated on the substrate surface at a depth of several wavelengths. Given their advantages of high frequency, high energy density, good penetrability, and easy integration, SAW chips have been widely used in recent years for cell manipulation on microfluidic chips. Interdigital transducers (IDTs) are generally used to generate SAW. The resonant frequency of the SAW can be controlled by adjusting the interdigital spacing of electrodes. The resonant frequency of SAW devices can reach up to GHz, indicating that they can precisely control micron-sized or even submicron-sized particles. In addition, the distribution of the acoustic field can be regulated by changing the shape of the IDTs, further demonstrating the flexibility of this manipulation method. SAW devices are generally processed using standard microfabrication technology. Thus, SAW devices have excellent reproducibility and consistency. The planar processing method enables the integration of SAW devices into microfluidic chips. When acoustic waves propagate into a fluidic medium, the fluid acquires momentum by absorbing acoustic waves. Bulk flow, in turn, is induced by acoustic wave absorption. This phenomenon is called the acoustic flow effect. The particle moves with the fluid if its size is substantially smaller than the wavelength and its density is small. This phenomenon thus enables the manipulation of particles in the fluid. When two columns of SAWs propagate opposite each other on the same surface, a standing surface acoustic wave (SSAW) is generated. Under the influence of SSAW, particles are subjected to standing wave acoustic radiation force and then accumulate at the antinode or node position depending on the properties of the particles and the surrounding medium, such as density and compressibility. Acoustic radiation is mainly attributed to the effects of particles on sound waves. These effects include reflection, refraction, and absorption and result in the exchange of momentum between sound waves and particles. The magnitude of the acoustic radiation force is related to the physical properties, such as wavelength and amplitude, of the SSAW and the size and density of the particles. In recent years, considerable research effort has been directed toward the establishment of SSAWs on microfluidic chips to realize precise and high-throughput single-cell manipulation. As shown in Figure 7a, Collins et al. introduced multiple high-frequency SSAWs with one cell per acoustic well for the patterning of multiple spatially separated single-cells [23]. They also characterized and demonstrated patterning for a wide range of particle sizes, and patterning of cells, including human lymphocytes and red blood cells infected by the malarial parasite Plasmodium falciparum.

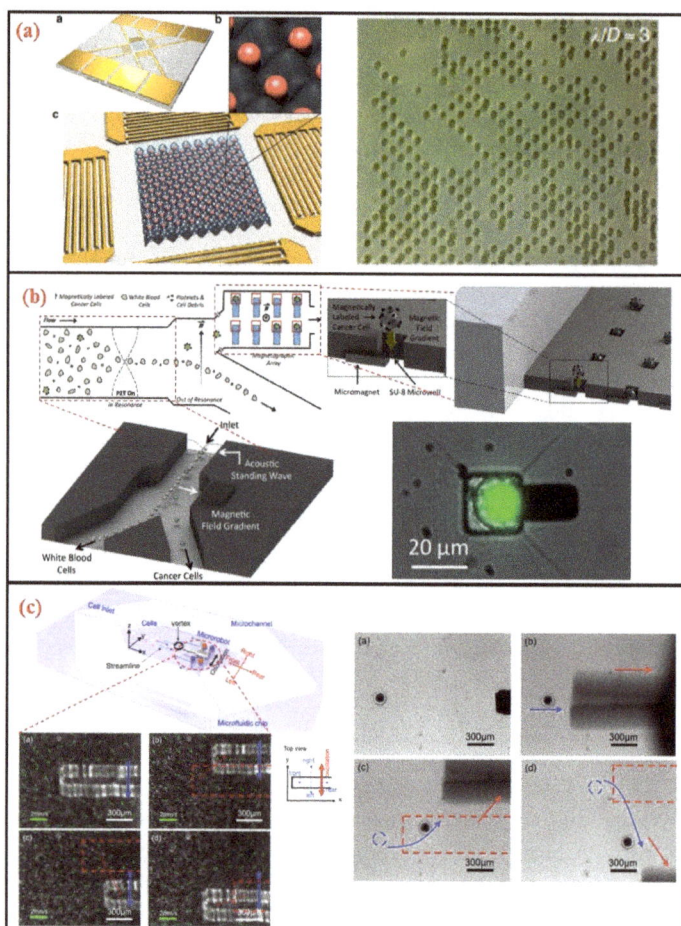

Figure 7. Acoustic, magnetic and microrobot single-cell manipulation methods and some designs. (a) Surface acoustic waves based single-cell pattern. Adapted by permission from Reference [23], copyright Nature Publishing Group 2015. (b) Micromagnet based single-cell trap. Adapted by permission from Reference [81], copyright Royal Society of Chemistry 2016. (c) Noncontact cell transportation by oscillation of microrobot in microfluidic chip. Adapted by permission from Reference [82], copyright AIP Publishing 2017.

Acoustic technology for single-cell manipulation presents the advantages of high frequency, high energy density, good penetrability, easy fabrication, easy integration, and noninvasiveness. However, the application of acoustic control in microfluidic single-cell manipulation remains in its infancy and poses numerous problems that still require resolution. These problems include nonlinear interactions among acoustic waves, fluids, and cells. Moreover, current acoustic chips are mostly based on LiNbO$_3$ substrates, and device substrates based on new piezoelectric materials must be investigated.

2.5. Magnetic Method

Magnetic manipulation refers to the manipulation of cells through using permanent magnets or electromagnets. This method requires the surface label of cells with immunomagnetic beads because cells typically lack paramagnetic or diamagnetic properties. Surface-modified nanomagnetic beads adhere onto the cell surface through specific interactions between the antibody and the antigen. Given this requirement, the magnetic method is a label-based approach. As shown in Figure 7b, Shields IV et al. developed a magnetic microfluidic platform comprised of three modules that offers high throughput separation of cancer cells from blood and on-chip organization of those cells for streamlined analyses [81]. The first module uses an acoustic standing wave to rapidly align cells in a contactless manner. The second module then separates magnetically labeled cells from unlabeled cells, offering purities exceeding 85% for cells and 90% for binary mixtures of synthetic particles. Finally, the third module contains a spatially periodic array of microwells with underlying micromagnets to capture individual cells for on-chip analyses.

Cell sample manipulation using immunomagnetic beads is simple and reliable, and its adoption in many large-scale analytical platforms for clinical applications has matured. However, the application prospects of this method are limited, given that it encounters difficulty in simultaneously separating and purifying multiple types of cells and in separating magnetic beads from sorted cells. In addition, magnetic bead bonding may damage cell membrane proteins and structure. The damaged cells are inconducive for subsequent cell culture.

2.6. Micro-Robot-Assisted Method

Recently, with the improvement of MEMS and NEMS technology, various micro and nanoscale microstructures, which are named microrobots, have been fabricated. In most applications, they can act as microcarrier to delivery drugs or cells in vitro or in vivo [83]. And different driven mechanisms have been developed to control microrobot. Commonly, the propulsion mechanisms of these microrobots can be divided into three categories: Chemical means, physical means, and biological means [84]. In other cases, microrobot can also be used to manipulate single-cell for its micro-scale size. Additionally, microfluidic chip can be combined for it can not only provide a simulated vascular environment or a microscale chamber, but also have high throughput and have high repeatability [83]. Feng et al. fabricated silicon-based microrobot and proposed acoustic levitation driven method to 3-D rotate single oocyte in a microfluidic [39]. The positioning accuracy is less than 1 μm and orientation with an accuracy of one and an average rotation velocity of 3 rad/s were achieved. Moreover, they used a microrobot in microfluidic chip to transport cell in contactless manner, as shown in Figure 7c [82]. A local vortex can be generated after oscillating the microrobot by permanent magnets. Different streamline can be generated when various oscillation amplitudes, frequencies, and relative positions between microrobot and microchannel wall are adapted. The tuning of these parameters changes the viscous fluid dynamics and cells can be transported. Using microrobot on a chip to manipulate single cells is promising owing to the microrobot's high accuracy, high speed, non-contact, enough physical strength, applicability to different sizes cells, and microfluidic chip's sealing property, and also high throughput.

3. Microfluidic Single-Cell Analysis

Various analytical functionalities, including microscopy, microelectrodes array, mass spectrometry, and chromatography, can be integrated with microfluidic components for various qualitative and quantitative single-cell analysis applications [85]. Fluorescence microscopy is the most widely used microfluidic technique for cell analysis [86]. Given the good optical transparency of microfluidic chips, various types of microscopy techniques can be integrated to image the morphology, structure, and migration of cells and specifically labeled subcellular organelles. Optical detectors for absorbance, laser-induce fluorescence (LIF), and chemiluminescence (CL) can also be integrated with microfluidic channels for detecting and quantifying specific biomolecules [87]. Electrical analysis, such as electrochemical impedance spectroscopy (EIS) and patch-clamp, can also be incorporated into microdevices to monitor cell secretion, morphology, and migration [88,89]. Mass spectrometry is a powerful analytical technique that has been coupled with microfluidics for the analysis of cellular contents (DNA, proteins, and glycan) and metabolites [90]. The integration of microfluidic systems with these analytical approaches enables rapid, sensitive, reproducible, and high-throughput single-cell analysis, which promotes the development of basic biological studies and clinical diagnoses and therapies [91]. This section provides review of microfluidics based single-cell analysis by using integrated analytical techniques.

3.1. Cellular Analysis

The study of cellular behavior, such as cell morphology, migration, proliferation, differentiation, and apoptosis, has general scientific and practical value for biology and medicine. However, several important cellular behaviors occur in in vivo environments that cannot be easily implanted with sensors or other types of molecular probes. Thus, an alternative approach is necessary to transfer the cells of interest out of their natural environment to one that is more conducive for the measurement scheme. The disadvantage of this alternative approach is that cell behavior in vitro may be different from that in vivo. In addition, many of the stimuli to which the sample is subjected to in vivo are no longer present in vitro. Microfluidic chips that integrate chemical, mechanical, and electrical functionalities in a lab-on-a-chip system may be powerful tools for mimicking in vivo microenvironments for cell growth. Moreover, many of the materials used to construct the chips are optically transparent, making them ideal for the real-time monitoring of cellular behavior through imaging. Therefore, microfluidics is suitable for cellular behavior analysis, and different microfluidic platforms have been developed for various applications.

3.1.1. Morphology

Cells are surrounded by a myriad of physical and biochemical cues in a cellular microenvironment. Cell morphology is the most intuitionistic parameter that can reflect cellular responses to different stimuli. The quantitative morphological analysis of cells is a key approach for abnormality identification and classification, early cancer detection, and dynamic change analysis under specific environmental stress. Quantitative results guide pathologists in making final diagnostic decisions. By integrating biomimetic cell culture systems with various types of microscopy or electrical techniques, microfluidics offers a robust platform for the real-time monitoring of alterations in cell morphology.

Fluorescence imaging is the most commonly used technique for cell morphology observation. In this approach, cells are cultured in microchannels and labeled with fluorescent dyes or proteins for visualization under fluorescence microscopy. Sung Ke et al. developed simple straight channel arrays as a viable and robust tool for the high-throughput quantitative morphological analysis of single MSCs and the examination of cell–material interactions [92]. Wu et al. developed a novel microfluidic model and studied the influences of interstitial flows on cell morphology. They found that interstitial flows promote amoeboid cell morphology and motility of MDA-MB-231 cells [93]. As shown in Figure 8a, Qin et al. have investigated particular longevity-related changes in cell morphology and characteristics of yeast cells in a microfluidic single-cell analysis chip. They found that cells with the round-budded terminal morphology had longer lifespans than those with the elongated-budded morphology [94]. Electrical techniques can also be incorporated in microfluidic devices for cell morphology analysis. Andreas Hierlemann et al. developed a microfluidic single-cell impedance cytometer that can perform the dielectric characterization of single-cells under frequencies of up to 500 MHz [95]. The increase in working frequency enabled the characterization of subcellular features in addition to the properties that are visible at low frequencies. The capabilities of this electrical cytometer have been demonstrated in the discrimination of a wild-type yeast strain from a mutant strain based on differences in vacuolar size and intracellular fluid distribution. One year later, Andreas Hierlemann et al. reported a microfluidics-based system that can reliably capture single rod-shaped Schizosaccharomyces pombe cells by applying suction through orifices in a channel wall. This system enables the subsequent culturing of immobilized cells in an upright position. Dynamic changes in cell cycle state and morphology are continuously monitored through EIS over a broad frequency range [96]. The obtained results showed that the spatial resolution of the measured cell length is 0.25 μm, which corresponds to a 5 min interval of cell growth under standard conditions. Comprehensive impedance datasets have also been used to determine the occurrence of nuclear division and cytokinesis.

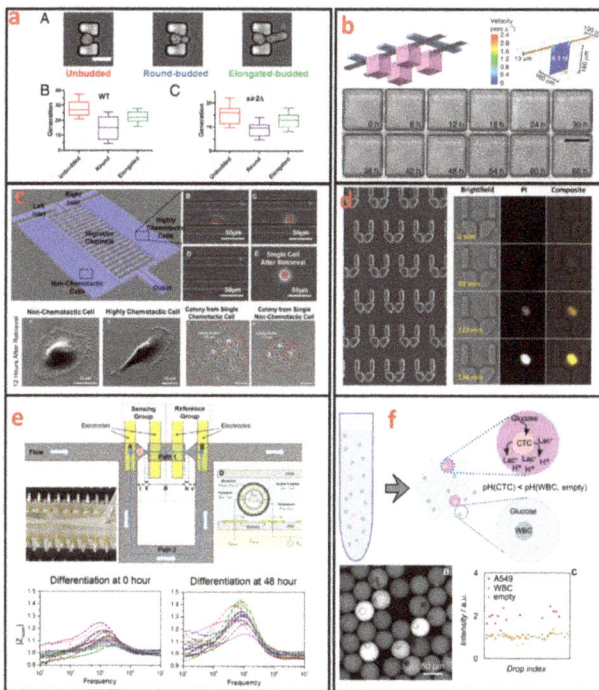

Figure 8. Cellular analysis of single-cells by using microfluidics. (**a**) Characterization of terminal morphology in aging yeast cells. Adapted by permission from Reference [94], copyright National Academy Sciences 2015. (**b**) High-throughput analysis of single hematopoietic stem cell proliferation in microfluidic cell culture arrays. Adapted by permission from Reference [97], copyright Nature Publishing Group 2011. (**c**) Single-cell migration chip for chemotaxis-based microfluidic selection of heterogeneous cell populations. Adapted by permission from Reference [98], under the Creative Commons Attribution License 2015. (**d**) Dynamic analysis of drug-induced cytotoxicity using microfluidic single-cell array. Adapted by permission from Reference [19], copyright American Chemical Society 2009. (**e**) Single-cell studies of mouse embryonic stem cell (mESC) differentiation by electrical impedance measurements in a microfluidic device. Adapted by permission from Reference [99], under the Creative Commons Attribution License 2016. (**f**) Circulating tumor cells (CTCs) detection based on the Warburg effect using single-cell compartmentalization in microdroplets. Adapted by permission from Reference [26], copyright John Wiley and Sons 2016.

3.1.2. Proliferation

Cell proliferation is the process through which the number of cells increases. It is carefully balanced with cell death to maintain a constant number of cells in adult tissues and organs. Cell proliferation analyses are crucial for cell growth and differentiation studies, and they are generally used to evaluate the toxicity of compounds and the inhibition of tumor cell growth during drug development. Microfluidic technology provides precise, controlled, cost-effective, compact, integrated, and high-throughput microsystems that are promising substitutes for conventional biological laboratory methods for the study of single-cell proliferation. Microfluidics allows dynamic cell culture in micro perfusion systems to deliver continuous nutrient supplies for long-term cell culture. In addition, this strategy offers many opportunities for mimicking the cell–cell and cell–extracellular matrix interactions of tissues by creating gradient concentrations of biochemical signals, such as growth factors, chemokines, and hormones. Many applications of cell cultivation in microfluidic systems are aimed toward understanding the proliferation and differentiation of cell populations.

The analyses of clonal cultures established from single-cells are vital for cancer research because heterogeneity plays an important role in tumor formation. Microfluidics-based devices are the most ideal choice for high-throughput single-cell clonal expansion. As shown in Figure 8b, Carl L Hansen et al. presented a simple microfluidic cell-culture design that supports cell growth and replicates standard microcultures [97]. Culture conditions can be precisely controlled on the microfluidic chip, which can also be applied for the in-situ immunostaining and recovery of viable cells. The platform successfully mimics conventional cultures in reproducing the responses of various types of primitive mouse hematopoietic cells, while retaining their functional properties, as demonstrated by the subsequent in vitro and in vivo (transplantation) assays of the recovered cells. Justin Cooper-White et al. reported a two-layered microfluidic device platform for the capture, culture, and clonal expansion of single-cells [35]. Under the manual injection of a cell suspension, hundreds to thousands of single-cells (adherent and nonadherent) are deterministically trapped in a high-throughput manner, and high trapping efficiency is achieved by incorporating a U-shaped hydrodynamic trap in the downstream wall of each microwell. They confirmed that the modified microwells promote the attachment, dispersal, and proliferation of the trapped single-cells for multiple generations over extended periods of time (>7 days) under media perfusion. Chia-Hsien Hsu et al. proposed a microfluidic device with a dual-well (DW) design for high-yielding single-cell loading (~77%) in large microwells (285 and 485 μm in diameter). This device facilitates cell dispersal, proliferation, and differentiation [35]. The architecture of this device allows the size of the "culture" microwells to be flexibly adjusted without affecting single-cell loading efficiency, making it useful for cell culture applications, as demonstrated by their experiments on KT98 mouse neural stem-cell differentiation, A549 and MDA-MB-435 cancer cell proliferation, and A549 single-cell colony formation.

3.1.3. Migration

Cell migration refers to the movement of cells in response to biological signals and environmental cues. This process plays a vital role in key physiological processes, including immune cell recruitment, wound healing, tissue repair, embryonic morphogenesis, and cancer metastasis. The complex processes that govern cell migration must be comprehensively understood to promote the development of novel therapeutic strategies. Cell migration is regulated by several biological, chemical, and physical signals, including mechanotransduction, chemical signaling, and molecular interactions. Given that cell migration is a highly complex biological mechanism, it can only be elucidated through monitoring under defined physiologically conditions. However, in vivo cell migration studies using state-of-the-art imaging methods are hindered by ethical issues associated with animal testing. Moreover, the tracking of cell migration in vivo remains technically challenging. Thus, in vitro migration assays are extensively used by biologists, pharmacologists, medical researchers, and toxicologists for diverse applications. The traditional scratch assay, which is the most convenient and inexpensive method for in vitro cell migration analysis, has several limitations. For example, the 2D scratch assay cannot replicate the 3D environment of cells and the signal gradients that are present in vivo. In addition, this method precludes single-cell analysis and cannot reveal cell heterogeneity, which is a vital factor of cancer metastasis. Microfluidics has emerged as a powerful platform for the study of cancer migration given that it can provide well-defined environmental cues. Microfluidic devices require a low number of cells and are highly suitable for high-throughput single-cell screening. As shown in Figure 8c, Euisik Yoon et al. developed a single-cell migration platform that allows the examination of the migration behavior of individual cells and the sorting of a heterogeneous cell population on the basis of chemotactic phenotype [98]. Highly chemotactic and nonchemotactic cells have been retrieved for the further cellular and molecular analyses of their differences. The migration channel has also been modified to elucidate the movement of certain cancer cells through geometrically confined spaces.

3.1.4. Apoptosis

Programmed cell death, which is known as apoptosis, is a vital component of various processes, including normal cell turnover, proper immune system development and function, hormone-dependent atrophy, embryonic development, and chemical-induced cell death. Inappropriate apoptosis (either inadequate or excessive) is a factor of many human conditions, including neurodegenerative diseases, ischemic damage, autoimmune disorders, and many cancer types. At present, the study of apoptosis is progressing rapidly.

Current methods for evaluating the effects of agents against cell apoptosis are generally expensive, labor-intensive and heterogeneity-ignored because they involve the use of multi-well plates that are operated using cumbersome manual or expensive robotics-based operations to evaluate the average results of a population. Therefore, researchers must urgently develop a technology that can perform such experiments in a cheaper, easier, and higher throughput manner to analyze cell apoptosis at the single-cell level. Microfluidic chips present the advantages of ease of integration and the potential for high-throughput single-cell manipulation, making them attractive platforms for drug metabolism and cell cytotoxicity analyses. As shown in Figure 8d, Wlodkowic et al. used a microfluidic single-cell array chip for the real-time analysis of events leading up to apoptosis in model cell lines [19]. They found that these live-cell, microfluidic microarrays can be readily applied to kinetic analysis of investigational anticancer agents in hematopoietic cancer cells, providing new opportunities for automated microarray cytometry and higher-throughput screening. Through quantifying the anticancer drug induced apoptosis on-chip, they showed that, with small numbers of trapped cells (~300) under careful serial observation, they can achieve results with only slightly greater statistical spread than that can be obtained with single-pass flow cytometer measurements of 15,000–30,000 cells. Kumar et al. used digital microfluidics (DMF) for time-resolved cytotoxicity studies on single non-adherent yeast cells [100]. They achieved real-time monitoring of single yeast cell responses during antifungal treatment in a high-throughput manner, and their DMF platform with microwell arrays is demonstrated as a promising tool for implementing various biological applications concerning single non-adherent cells in a high-throughput manner. Li et al. developed a multifunctional gradients-customizing microfluidic device for high-throughput single-cell multidrug resistance (MDR) analysis [101]. Bithi and Vanapalli reported a microfluidic cell isolation technology for drug testing of single tumor cells and their clusters, and they found that individual tumor cells display diverse uptake profiles of the drug [102]. Experiments with clusters of tumor cells compartmentalized in their microfluidic drops revealed that cells within a cluster have higher viability than their single-cell counterparts when exposed to doxorubicin.

3.1.5. Differentiation

Stem cells can continuously self-renew and have the potential to differentiate into specific tissues. Thus, their roles in tissue engineering, organ regeneration, cell-based therapies, disease models, drug development, and various healthcare applications have been extensively investigated for over 50 years. To date, stem cells have been successfully used to heal damaged tissues and replace nonfunctional organs. The promising applications of stem cells in the biological and therapeutic fields have been hindered by the challenges associated with the maintenance of undifferentiated pluripotency and the reliable direction of stem-cell differentiation. Conventional cell culture methods, such as those based on petri dishes or Transwells, cannot achieve an in vivo-like microenvironment that contains diverse well-controlled stimuli. The emergence and rapid development of microfluidics have presented a possible solution for mimicking an in vivo-like microenvironment. Microfluidic platforms can precisely manipulate the microenvironment to deliver soluble factors to cells, establish well-defined gradients, integrate various biocompatible scaffolds and functional components, and dynamically alter the application of mechanical and electrical signals to cultured cells. The combination of microfluidic technologies with stem-cell analysis could finally provide in-depth insight into stem-cell differentiation mechanisms to enable their application. At present, an increasing number of works have focused on applying microfluidic devices to investigate stem-cell differentiation at single-cell resolution.

Sikorski et al. developed a microfluidic device that supports the robust generation of colonies derived from single human embryonic stem cells (hESCs) [103]. The use of this device to analyze the clonal growth of CAIS hESCs demonstrated its ability to reveal the heterogeneity of differentiation patterns displayed by clonally tracked hESC. In addition to providing controllable microenvironments for directing and observing stem-cell differentiation, microfluidics can also be used to characterize differentiation status. As shown in Figure 8e, Zhou et al. designed and fabricated a microfluidic device that integrates the hydrodynamic trapping of single-cells in predefined locations with the capability to perform electrical impedance measurements [99]. Mouse embryonic stem cells at different states during differentiation (t = 0, 24, and 48 h) were measured and quantitatively analyzed. The magnitude of cell impedance markedly increased. This increase can be attributed to the increase in cell size. The analysis of the measurements suggested that the nucleus-to-cytoplasm ratio decreased during this process. The maximum degree of cell heterogeneity was observed when the cells were in the transition state (24 h).

3.1.6. Metabolism

The intracellular levels and spatial localizations of metabolites reflect the state of a cell and its relationship to its surrounding environment [104]. Microfluidic device is an ideal platform for cellular metabolite profiling both in physiological environment and under drug treatment, owing to the ability of integrating cell culture, stimulation, metabolite enrichment, and detection on a single chip coupled with various analytical instruments [105].

Among diverse analytical techniques, Mass Spectrometry is the most powerful and promising tool for cell metabolite analysis, because of its broad detection range, high sensitivity, high mass resolution, rapid operation, and the ability for multiplexed analysis. Zhang et al. integrated droplet-based microfluidics with mass spectrometry for high-throughput and multiple analysis of single-cells [106]. Specific extraction solvent was used to selectively obtain intracellular components of interest and remove interference of other components. Using this method, matrix-free, selective, and sensitive detection of metabolites in single-cells is easily realized. Optical detecting techniques can also be integrated with microfluidic device for cell metabolite analysis. Wang et al. presented a flexible high-throughput approach that used microfluidics to compartmentalize individual cells for growth and analysis in monodisperse nanoliter aqueous droplets surrounded by an immiscible fluorinated oil phase [107]. The fluorescent assay system was used to measure the concentration of the metabolites (oxidase enzymes), and the assay reaction started when a cell-containing droplet coalesced with an assay droplet. As shown in Figure 8f, Ben et al. proposed a label-free method for exploiting the abnormal metabolic behavior of cancer cells. A single-cell analysis technique is used to measure the secretion of acid from individual living tumor cells compartmentalized in monodisperse, picolitre (pL) droplets. As few as 10 tumor cells can be detected in a background of 200,000 white blood cells and proof-of-concept data was shown on the detection of CTCs in the blood of metastatic patients.

3.2. Genetic Analysis

Genetic analysis is one of the most important and extensively developed field in microfluidic single-cell analysis. Genetic analysis can be categorized into cytogenetic and molecular genetic analysis.

3.2.1. Cytogenetic Analysis

Cytogenetic analysis deals with chromosomes and related abnormalities and is very crucial in the diagnosis of oncologic and hematologic disorders. The methods for cytogenetic analysis usually include Karyotyping and fluorescence in-situ hybridization (FISH). Karyotyping helps detect structural or numerical chromosome abnormalities. Chromosome analyses require cell cultures and involve the harvesting of chromosomes, chromosome banding, microscopic analysis, and the production of karyotypes. FISH involves the determination of the presence, absence, position, and copy number of DNA segments with the help of fluorescence microscopy. The most popular cytogenetic analysis on microfluidic chips is based on the using FISH. Shah et al. described a novel microfluidic FISH preparation device for metaphase FISH slides preparation [108]. The device combines the bioreactor for cell culturing with the splashing device for preparation of the chromosome spreads. As shown in Figure 9a, Sieben et al. have successfully integrated all aspects needed to perform automated FISH on a microfluidic platform [109]. They detected the number of X and Y chromosomes per cell in patient samples; useful for identifying the status of engraftment in patient-donor sex-mismatched transplantation. Zanardi et al. presented a microfluidic-device-based FISH method performed on fresh and fixed hematological samples, which integrated cluster-assembled nanostructured TiO2 (ns-TiO2) as a nanomaterial promoting hematopoietic cell immobilization in conditions of shear stress. By this way, FISH can be performed with at least a 10-fold reduction in probe usage and minimal cell requirements, but had comparable performance to standard FISH, indicating that it is suitable for genetic screenings in research clinical settings.

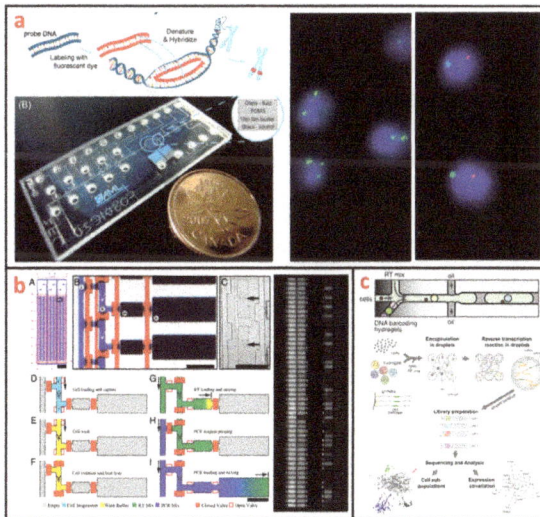

Figure 9. Genetic analysis of single-cells by using microfluidics. (**a**) An integrated microfluidic chip for chromosome enumeration using FISH. Adapted by permission from Reference [109], copyright Royal Society of Chemistry 2008. (**b**) High-throughput microfluidic single-cell RT-qPCR for gene expression analysis. Adapted by permission from Reference [110], copyright National Academy Sciences 2011. (**c**) Capturing single-cells along with a set of uniquely barcoded primers in microfluidic chip generated tiny droplets enables single-cell transcriptomics of many cells in a heterogeneous population. Adapted by permission from Reference [30], copyright Elsevier Inc. 2015.

3.2.2. Molecular Genetic Analysis

Molecular genetic analysis studies the structure and function of genes at a molecular level and thus employs methods of both molecular biology and genetics. Nucleic acid amplification processes play a critical role in sensitive detection and quantification, because the amount of the nucleic acids extracted from cells is small. Polymerase chain reaction (PCR) is the most widely used non-isothermal amplification technique, which performs thermal cycling to amplify a particular DNA sequence to generate thousands to millions of copies. Quantification of RNA can be achieved by performing reverse-transcription PCR (RT-PCR). As for genetic analysis, microfluidic devices have advantages including faster reaction times, low sample consumption, precise temperature distribution and the ease of integrating with separation techniques.

Even genetically identical cells with seemingly identical cell histories and environmental conditions can have significant differences in gene expression levels, due largely to the alteration of mRNA production by random fluctuations or complex molecular switches. Thus, quantitative analysis of gene expression at single-cell level is important for the understanding of basic biological mechanism and disease onset and progression [111]. Currently, several microfluidic-based single-cell RNA-Seq platforms have been developed and applied to study transcriptional heterogeneity of cancer, immune [112], and stem cells [29]. Those microfluidic-based single-cell RNA-Seq platforms are basically based on either active-valve or droplet-based microfluidics. As shown in Figure 9b, White presented a valve-based fully integrated microfluidic device capable of performing high-precision RT-qPCR measurements of gene expression from hundreds of single-cells per run [110]. They applied this technology to 3,300 single-cell measurements of miRNA expression in K562 cells, coregulation of a miRNA and one of its target transcripts during differentiation in embryonic stem cells, and single nucleotide variant detection in primary lobular breast cancer cells. Huang et al. also developed a valve-based strategy for single-cell RNA-Seq that has superior sensitivity and been implemented in a microfluidic platform for single-cell whole-transcriptome analysis [113]. In their approach, single-cells were captured and lysed in a microfluidic device, where mRNAs with poly(A) tails were reverse-transcribed into cDNA. Double-stranded cDNA was then collected and sequenced using a next generation sequencing platform. Droplet microfluidics is among the most promising candidate for capturing and processing thousands of individual cells for whole-transcriptome or genomic analysis in a massively parallel manner with minimal reagent use. As shown in Figure 9c, Klein et al. recently established a method called in Drops, which is based on the using of droplet microfluidics and has the capability to index >15,000 cells in an hour [30]. A suspension of cells was first encapsulated into nanoliter droplets with hydrogel beads (HBs) bearing barcoding DNA primers. Cells were then lysed, and mRNA is barcoded (indexed) by a reverse transcription (RT) reaction.

3.3. Protein Analysis

Proteins are one basic component of cells, which perform and regulate various cellular functions. Owing to the low abundance and high complexity, the development of sensitive and reliable protein analysis techniques is highly desirable. Microfluidics offer rapid, sensitive, reproducible and high-throughput platforms for protein analysis. Various aspects, including protein species, amounts, activity, as well as protein interaction with other biomolecules, can be analyzed using microfluidic devices, with tremendous advantages over conventional methods [114].

Cellular staining assays are commonly used methods that are easy to be applied in microfluidic devices for protein analysis. Proteins in cells are specifically labeled by tags or fluorescent antibodies, and their locations and expressions can be imaged using microscopies. As shown in Figure 10a, Srivastava et al. reported a novel phosphoFlow Chip (pFC) that relies on monolithic microfluidic technology to rapidly conduct signaling studies. The pFC platform integrates cell stimulation and preparation, microscopy, and subsequent flow cytometry [115]. Except for intracellular protein analysis, microfluidics can also be used for the analysis of secreted proteins of single-cells. As shown in Figure 10b, Ma et al. reported a microfluidic platform designed for highly multiplexed (more than ten proteins), reliable, sample- efficient (~1×10^4 cells) and quantitative measurements of secreted proteins from single-cells [116]. They validated the platform by assessment of multiple inflammatory cytokines from lipopolysaccharide-stimulated human macrophages and comparison to standard immune-technologies. Another important microfluidic protein analysis technique is surface-based immunoassay. Proteins are specifically captured by affinity ligands modified on microchannel or microbead surface, and sandwich immunoassays are then performed. Godwin et al. reported an integrated microfluidic approach that enables on-chip immune-isolation and in situ protein analysis of exosomes directly from patient plasma [117]. Specifically, a cascading microfluidic circuit was designed to streamline and expedite the pipeline for proteomic characterization of circulating exosomes, including exosome isolation and enrichment, on-line chemical lysis, protein immunoprecipitation, and sandwich immunoassays assisted by chemi-fluorescence detection. This method enables high level of multiplexing and quantitation, and intracellular, membrane, and secreted proteins can all be analyzed from the same single-cell [114]. Recently, protein immunoblotting assay has been operated on microfluidic devices, and microfluidic single-cell Western blotting (scWestern) has also been developed [44]. Polyacrylamide gels were photo-patterned to form a microwell array, in which single-cells were settled and lysed in situ. Gel electrophoresis was then performed, and separated proteins were immobilized by photoinitiated blotting and detected by antibody probing. This scWestern method enabled multiplexed analysis of 11 protein targets per single-cell with detection thresholds of <30,000 molecules.

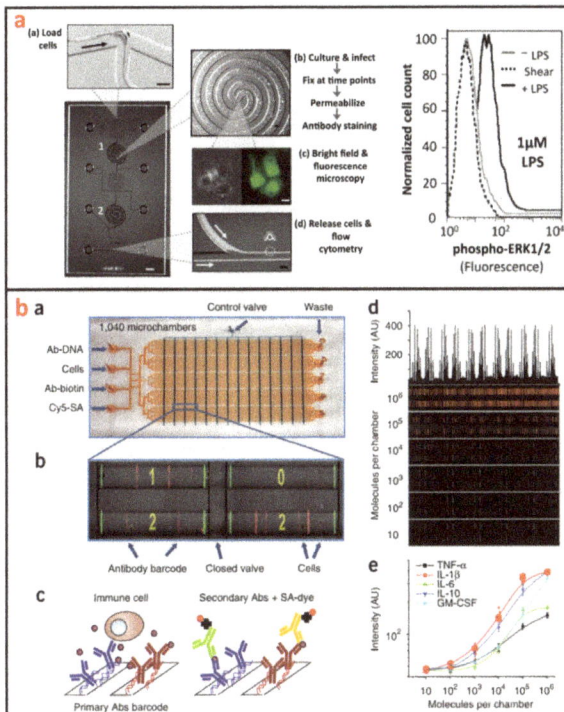

Figure 10. Single-cell protein analysis by using microfluidics. (**a**) A fully integrated microfluidic platform enabling automated phosphoprofiling of macrophage response. Adapted by permission from Reference [115], copyright American Chemical Society 2009. (**b**) A clinical microchip for detecting multiple cytokines of single immune cells reveals high functional heterogeneity in phenotypically similar T cells. Adapted by permission from Reference [116], copyright Nature Publishing Group 2011.

3.4. Biophysical Property Analysis

Cell state is often characterized through measurement of biochemical and biophysical markers. Although biochemical markers have been widely used, intrinsic biophysical markers, such as size, density, and the ability to mechanically deform under a load, are advantageous in that they do not require costly labeling or sample preparation [41]. Great cellular heterogeneity also exists in these biophysical properties, and microfluidics is an ideal technology for analyzing different biophysical properties at the single-cell level.

As shown in Figure 11a, Godin et al. developed a suspended microchannel resonator (SMR) combined with picolitre-scale microfluidic control to measure buoyant mass and determine the 'instantaneous' growth rates of individual cells [118]. The SMR measures mass with femtogram precision, allowing rapid determination of the growth rate in a fraction of a complete cell cycle. They found that for individual cells of Bacillus subtilis, Escherichia coli, Saccharomyces cerevisiae and mouse lymphoblasts, heavier cells grew faster than lighter cells. Not only the mass, but also the size and density of the cells can be measured by using SMR [119–121]. The mechanical property of single-cells can be used to evaluate the status of many diseases including cancer, malaria, and arthritis. Microfluidics is a powerful technology for characterizing the mechanical properties of single-cells at a fast and high-throughput manner. Guo et al. developed a microfluidic chip for measuring the deformability of single-cells using the pressure required to deform such cells through micrometer-scale tapered constrictions [122]. Single-cells are infused into a microfluidic channel, and then deformed through a series of funnel-shaped constrictions. The constriction openings are sized to create a temporary seal with each cell as it passes through the constriction, replicating the interaction with the orifice of a micropipette. They measured the deformability of several types of nucleated cells and determined the optimal range of constriction openings. Hu et al. developed a microfluidic elasticity microcytometer for multiparametric biomechanical phenotypic profiling of live single cancer cells for quantitative, simultaneous characterizations of cell size, and cell deformability/stiffness [123]. The elasticity microcytometer was implemented for measuring and comparing four human cell lines with distinct metastatic potentials. Except for using passive hydrodynamic pressure for cell deformability, active methods such as optical tweezer and DEP can also be combined with microfluidic chip for studying biomechanics of single-cells. For instance, Zhang et al. developed a microfluidic chip for rapid characterization of the biomechanics of drug-treated cells through stretching with dielectrophoresis (DEP) force, and saw a decrease in the stiffness after drug treatment of NB4 cells [124]. Electrical impedance is also an important biophysical marker for label-free identification of different cell types or detecting intracellular changes [125]. Microfluidics has great capability in measuring electrical impedance of single-cells because microelectrode array can be easily integrated into microfluidic chips. Hong et al. proposed a method for differentiating four kinds of cell (HeLa, A549, MCF-7, and MDA-MB-231) using impedance measurements at various voltages and frequencies [126]. According to the impedance measurements, HeLa, A549, and MCF-7 cells and the pathological stages of a given cancer cell line (MCF-7 and MDA-MB-231) can be distinguished. Measuring two or more than two biophysical markers for the same single-cells leads to more comprehensive understanding of the linkage between biological and biophysical things. As shown in Figure 11b, Zhou et al. developed a microfluidic device that can simultaneously characterize the mechanical and electrical properties of individual biological cells in a high-throughput manner (>1000 cells/min) [127]. The combination of mechanical and electrical properties provides better differentiation of cellular phenotypes, which are not easily discernible via single biophysical marker analysis.

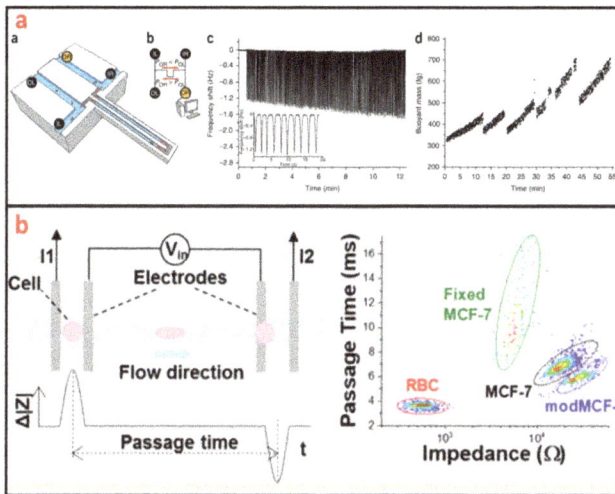

Figure 11. Biophysical property analysis of single-cells by using microfluidics. (**a**) A dynamic fluidic control system that enables the buoyant mass of cells as small as bacteria and as large as mammalian lymphocytes to be repeatedly measured with a suspended microchannel resonator (SMR). Adapted by permission from Reference [119], copyright Nature Publishing Group 2010. (**b**) A microfluidic device that can simultaneously characterize the mechanical and electrical properties of individual biological cells in a high-throughput manner (>1000 cells/min). Adapted by permission from Reference [127], copyright American Chemical Society 2018.

4. Discussion and Conclusions

The use of microfluidic technology for diversified and efficient manipulation and analysis of single biological cells has been a research hotspot in the interdisciplinary field. This review highlighted the microfluidic single-cell manipulation and analysis from the aspects of methods and applications. Specifically, the microfluidic single-cell manipulation can be flexibly realized by using hydrodynamic, electrical, optical, magnetic, and acoustic and micro-robot assisted methods, and microfluidic chips can be combined with various analytical techniques for single-cell analysis ranging from cellular behaviors to secreted proteins. The advantages and disadvantages of some methods are discussed. It is seen that each method has its inherent advantages and disadvantages (Table 2). In general, hydrodynamic method can achieve high throughput manipulation of cell samples, but there are deficiencies in the accuracy and flexibility. Methods such as electrical and optical methods have high accuracy and great flexibility, but they have shortcomings such as low throughput. There is no single method that can fulfill high throughput, high-efficiency, accurate single-cell manipulation and analysis tasks simultaneously. Therefore, to meet the specific requirements of practical applications, multiple methods are integrated. Secondly, existing technologies should be continuously improved for better single-cell manipulation and analysis. For example, the microstructure can be extended from 2-dimensional to 3-dimensional and fabrication crafts with higher precise can also be developed. Thirdly, new mechanisms and technologies must be discovered. The research can be based on theoretical simulations revealing the fundamental theories involved in microfluidics. Fourthly, developing a precise fluidic control system with fast response is very important. For instance, Arai et al. proposed a high-speed local-flow control using dual membrane pumps driven by piezoelectric actuators placed on the outside of microfluidic chip, and their approach can sort single cells at throughput of 23,000 cells/s with a 92.8% success rate, 95.8% purity, and 90.8% cell viability [128]. Finally, microfluidic chip implanted in subcutaneous, blood vessels and other tissues and organs, for achieving precise single-cell manipulation and analysis directly in the human body, is an emerging direction.

Table 2. Comparison of various methods for single-cell manipulation.

Methods		Advantages	Disadvantages	Characteristics		
				Throughput	Efficiency	Accuracy
Hydrody-namic method	Droplet [27]	High-throughput, simple chip structure with great flexibility	Difficult to culture adherent cell, difficult to introduce biochemicals into droplets	250 μL/h	75%	-
	Inertial [55]	High throughput, high cell viability, and simple chip	Only work well under specific flow rates and cell concentrations	3 mL/min	84%	-
	Vortex [58]	Has no strict requirement about the properties of cells and fluid	Require external controller, low single-cell efficiency	-	-	Cell rotation $3.5 \pm 2.1° \text{ s}^{-1}$
	Micro-valve [60]	Reliable and fast for control, suitable for large-scale integration	Require complex and cumbersome external control devices	96 cells/chip	$90.6 \pm 8\%$	-
	Micro-structure [60]	Simple for operation, high throughput	Inflexibility, hard to control specific single-cell	10000 cells/chip	90%	-
Electrical method	Dielectro-phoresis [20]	Contactless, high selectivity, label-free	Require low-conductivity buffer	3264 pair of cells/chip	74.2%	-
	Electro-osmosis [70]	Label-free, easy for integrated fabrication	Low efficiency and accuracy with increased flow rate	81 cells/chip	73%	-
Optical method	Optical tweezer [12]	High accuracy and efficiency	Low throughput, high cost	-	97%	98%
	ODEP [76]	Flexible virtual electrodes, label-free, simple and low-cost	Require low-conductivity solution, opaque substrate	Scalable	-	-
	Opto-thermocapillary [77]	Flexible, can pattern single cells in hydrogel with high viability	High cost for cumbersome peripherical optical system	Low	High	High
Acoustic method [23]		Noninvasive, label-free, good penetrability	Need piezoelectric substrate for chip fabrication	High	-	High
Magnetic method [81]		Reliable and highly efficient	Not label-free	200 μL/min	>85%	>80%
Micro-robot-assisted method [84]		High accuracy, flexible and controllable	Low throughput	Low	High	-

Author Contributions: T.L. and L.F. wrote the manuscript, R.Z. and D.S. revised the manuscript.

Funding: This work was funded by the Research Grants Council of the Hong Kong Special Administrative Region, China (grant numbers N_CityU102/15, and CityU11209917), and Shenzhen Science and Technology Project, China (grant number R-IND13301).

Conflicts of Interest: The authors declare no conflict of interest.

References

1. Altschuler, S.J.; Wu, L.F. Cellular Heterogeneity: Do Differences Make a Difference? *Cell* **2010**, *141*, 559–563. [CrossRef] [PubMed]
2. Shipitsin, M.; Campbell, L.L.; Argani, P.; Weremowicz, S.; Bloushtain-Qimron, N.; Yao, J.; Nikolskaya, T.; Serebryiskaya, T.; Beroukhim, R.; Hu, M.; et al. Molecular Definition of Breast Tumor Heterogeneity. *Cancer Cell* **2007**, *11*, 259–273. [CrossRef]
3. Turner, N.C.; Reis-Filho, J.S. Genetic heterogeneity and cancer drug resistance. *Lancet Oncol.* **2012**, *13*, 178–185. [CrossRef]
4. Bauwens, C.L.; Peerani, R.; Niebruegge, S.; Woodhouse, K.A.; Kumacheva, E.; Husain, M.; Zandstra, P.W. Control of Human Embryonic Stem Cell Colony and Aggregate Size Heterogeneity Influences Differentiation Trajectories. *Stem Cells* **2008**, *26*, 2300–2310. [CrossRef] [PubMed]
5. Gao, D.; Jin, F.; Zhou, M.; Jiang, Y. Recent advances in single cell manipulation and biochemical analysis on microfluidics. *Analyst* **2019**, *144*, 766–781. [CrossRef]
6. Gross, A.; Schoendube, J.; Zimmermann, S.; Steeb, M.; Zengerle, R.; Koltay, P. Technologies for single-cell isolation. *Int. J. Mol. Sci.* **2015**, *16*, 16897–16919. [CrossRef] [PubMed]
7. Hong, J.; Edel, J.B.; deMello, A.J. Micro- and nanofluidic systems for high-throughput biological screening. *Drug Discov. Today* **2009**, *14*, 134–146. [CrossRef] [PubMed]
8. Wyatt Shields Iv, C.; Reyes, C.D.; López, G.P. Microfluidic cell sorting: A review of the advances in the separation of cells from debulking to rare cell isolation. *Lab Chip* **2015**, *15*, 1230–1249. [CrossRef]
9. Rettig, J.R.; Folch, A. Large-scale single-cell trapping and imaging using microwell arrays. *Anal. Chem.* **2005**, *77*, 5628–5634. [CrossRef]
10. Mi, L.; Huang, L.; Li, J.; Xu, G.; Wu, Q.; Wang, W. A fluidic circuit based, high-efficiency and large-scale single cell trap. *Lab Chip* **2016**, *16*, 4507–4511. [CrossRef]
11. Di Carlo, D.; Wu, L.Y.; Lee, L.P. Dynamic single cell culture array. *Lab Chip* **2006**, *6*, 1445–1449. [CrossRef] [PubMed]
12. Wang, X.; Chen, S.; Kong, M.; Wang, Z.; Costa, K.D.; Li, R.A.; Sun, D. Enhanced cell sorting and manipulation with combined optical tweezer and microfluidic chip technologies. *Lab Chip* **2011**, *11*, 3656–3662. [CrossRef] [PubMed]
13. Thomas, R.S.; Morgan, H.; Green, N.G. Negative DEP traps for single cell immobilisation. *Lab Chip* **2009**, *9*, 1534–1540. [CrossRef] [PubMed]
14. Amann, R.; Fuchs, B.M. Single-cell identification in microbial communities by improved fluorescence in situ hybridization techniques. *Nat. Rev. Microbiol.* **2008**, *6*, 339–348. [CrossRef] [PubMed]
15. Huang, W.E.; Stoecker, K.; Griffiths, R.; Newbold, L.; Daims, H.; Whiteley, A.S.; Wagner, M. Raman-FISH: Combining stable-isotope Raman spectroscopy and fluorescence in situ hybridization for the single cell analysis of identity and function. *Environ. Microbiol.* **2007**, *9*, 1878–1889. [CrossRef] [PubMed]
16. Xia, F.; Jin, W.; Yin, X.; Fang, Z. Single-cell analysis by electrochemical detection with a microfluidic device. *J. Chromatogr. A* **2005**, *1063*, 227–233. [CrossRef] [PubMed]
17. Actis, P.; Tokar, S.; Clausmeyer, J.; Babakinejad, B.; Mikhaleva, S.; Takahashi, Y.; Córdoba, A.L.; Novak, P.; Shevchuck, A.I. Electrochemical Nanoprobes for Single-Cell Analysis. *ACS Nano* **2014**, *8*, 875–884. [CrossRef]
18. Banaeiyan, A.A.; Ahmadpour, D.; Adiels, C.B.; Goksör, M. Hydrodynamic cell trapping for high throughput single-cell applications. *Micromachines* **2013**, *4*, 414–430. [CrossRef]
19. Wlodkowic, D.; Faley, S.; Zagnoni, M.; Wikswo, J.P.; Cooper, J.M. Microfluidic single-cell array cytometry for the analysis of tumor apoptosis. *Anal. Chem.* **2009**, *81*, 5517–5523. [CrossRef]
20. Wu, C.; Chen, R.; Liu, Y.; Yu, Z.; Jiang, Y.; Cheng, X. A planar dielectrophoresis-based chip for high-throughput cell pairing. *Lab Chip* **2017**, *17*, 4008–4014. [CrossRef]

21. Chiou, P.Y.; Ohta, A.T.; Wu, M.C. Massively parallel manipulation of single cells and microparticles using optical images. *Nature* **2005**, *436*, 370–372. [CrossRef] [PubMed]
22. Liu, W.; Dechev, N.; Foulds, I.G.; Burke, R.; Parameswaran, A.; Park, E.J. A novel permalloy based magnetic single cell micro array. *Lab Chip* **2009**, *9*, 2381–2390. [CrossRef] [PubMed]
23. Collins, D.J.; Morahan, B.; Garcia-bustos, J.; Doerig, C.; Plebanski, M.; Neild, A. Two-dimensional single-cell patterning with one cell per well driven by surface acoustic waves. *Nat. Commun.* **2015**, *6*, 8686. [CrossRef] [PubMed]
24. Mellors, J.S.; Jorabchi, K.; Smith, L.M.; Ramsey, J.M. Integrated microfluidic device for automated single cell analysis using electrophoretic separation and electrospray ionization mass spectrometry. *Anal. Chem.* **2010**, *82*, 967–973. [CrossRef] [PubMed]
25. Yasukawa, T.; Nagamine, K.; Horiguchi, Y.; Shiku, H.; Koide, M.; Itayama, T.; Shiraishi, F.; Matsue, T. Electrophoretic cell manipulation and electrochemical gene-function analysis based on a yeast two-hybrid system in a microfluidic device. *Anal. Chem.* **2008**, *80*, 3722–3727. [CrossRef] [PubMed]
26. Del Ben, F.; Turetta, M.; Celetti, G.; Piruska, A.; Bulfoni, M.; Cesselli, D.; Huck, W.T.S.; Scoles, G. A Method for Detecting Circulating Tumor Cells Based on the Measurement of Single-Cell Metabolism in Droplet-Based Microfluidics. *Angew. Chem.* **2016**, *128*, 8723–8726. [CrossRef]
27. Zilionis, R.; Nainys, J.; Veres, A.; Savova, V.; Zemmour, D.; Klein, A.M.; Mazutis, L. Single-cell barcoding and sequencing using droplet microfluidics. *Nat. Protoc.* **2017**, *12*, 44–73. [CrossRef]
28. Collins, D.J.; Neild, A.; deMello, A.; Liu, A.Q.; Ai, Y. The Poisson distribution and beyond: Methods for microfluidic droplet production and single cell encapsulation. *Lab Chip* **2015**, *15*, 3439–3459. [CrossRef]
29. Prakadan, S.M.; Shalek, A.K.; Weitz, D.A. Scaling by shrinking: Empowering single-cell "omics" with microfluidic devices. *Nat. Rev. Genet.* **2017**, *18*, 345–361. [CrossRef]
30. Klein, A.M.; Mazutis, L.; Akartuna, I.; Tallapragada, N.; Veres, A.; Li, V.; Peshkin, L.; Weitz, D.A.; Kirschner, M.W. Droplet barcoding for single-cell transcriptomics applied to embryonic stem cells. *Cell* **2015**, *161*, 1187–1201. [CrossRef]
31. Taff, B.M.; Voldman, J. A scalable addressable positive-dielectrophoretic cell-sorting array. *Anal. Chem.* **2005**, *77*, 7976–7983. [CrossRef] [PubMed]
32. Revzin, A.; Sekine, K.; Sin, A.; Tompkins, R.G.; Toner, M. Development of a microfabricated cytometry platform for characterization and sorting of individual leukocytes. *Lab Chip* **2005**, *5*, 30–37. [CrossRef] [PubMed]
33. Wang, M.M.; Tu, E.; Raymond, D.E.; Yang, J.M.; Zhang, H.; Hagen, N.; Dees, B.; Mercer, E.M.; Forster, A.H.; Kariv, I.; et al. Microfluidic sorting of mammalian cells by optical force switching. *Nat. Biotechnol.* **2005**, *23*, 83–87. [CrossRef] [PubMed]
34. Lutz, B.R.; Chen, J.; Schwartz, D.T. Hydrodynamic tweezers: 1. Noncontact trapping of single cells using steady streaming microeddies. *Anal. Chem.* **2006**, *78*, 5429–5435. [CrossRef] [PubMed]
35. Chen, H.; Sun, J.; Wolvetang, E.; Cooper-white, J. High-throughput, deterministic single cell trapping and long-term clonal cell culture in microfluidic devices. *Lab Chip* **2015**, *15*, 1072–1083. [CrossRef] [PubMed]
36. Kobel, S.; Valero, A.; Latt, J.; Renaud, P.; Lutolf, M. Optimization of microfluidic single cell trapping for long-term on-chip culture. *Lab Chip* **2010**, *10*, 857–863. [CrossRef] [PubMed]
37. Zhang, K.; Gao, M.; Chong, Z.; Li, Y.; Han, X.; Chen, R.; Qin, L. Single-cell isolation by a modular single-cell pipette for RNA-sequencing. *Lab Chip* **2016**, *16*, 4742–4748. [CrossRef]
38. Wang, X.; Chen, S.; Sun, D. Automated parallel cell isolation and deposition using microwell array and optical tweezers. In Proceedings of the 2012 IEEE International Conference on Robotics and Automation, Saint Paul, MN, USA, 14–18 May 2012; pp. 4571–4576.
39. Feng, L.; Di, P.; Arai, F. High-precision motion of magnetic microrobot with ultrasonic levitation for 3-D rotation of single oocyte. *Int. J. Robot. Res.* **2016**, *35*, 1445–1458. [CrossRef]
40. Skelley, A.M.; Kirak, O.; Suh, H.; Jaenisch, R.; Voldman, J. Microfluidic control of cell pairing and fusion. *Nat. Methods* **2009**, *6*, 147–152. [CrossRef]
41. Gossett, D.R.; Tse, H.T.K.; Lee, S.A.; Ying, Y.; Lindgren, A.G.; Yang, O.O.; Rao, J.; Clark, A.T.; Di Carlo, D. Hydrodynamic stretching of single cells for large population mechanical phenotyping. *Proc. Natl. Acad. Sci. USA* **2012**, *109*, 7630–7635. [CrossRef]

42. Hu, S.; Sun, D. Transportation of biological cells with robot-tweezer manipulation system. In Proceedings of the 2011 IEEE International Conference on Robotics and Automation, Shanghai, China, 9–13 May 2011; pp. 5997–6002.

43. Gao, J.; Yin, X.-F.; Fang, Z.-L. Integration of single cell injection, cell lysis, separation and detection of intracellular constituents on a microfluidic chip. *Lab Chip* **2004**, *4*, 47–52. [CrossRef] [PubMed]

44. Hughes, A.J.; Spelke, D.P.; Xu, Z.; Kang, C.-C.; Schaffer, D.V.; Herr, A.E. Single-cell western blotting. *Nat. Methods* **2015**, *11*, 749–755. [CrossRef] [PubMed]

45. Braeken, D.; Huys, R.; Jans, D.; Loo, J.; Severi, S.; Vleugels, F.; Borghs, G.; Callewaert, G.; Bartic, C. Local electrical stimulation of single adherent cells using three-dimensional electrode arrays with small interelectrode distances. In Proceedings of the Annual International Conference of the IEEE Engineering in Medicine and Biology Society, EMBC 2009, Minneapolis, MN, USA, 3–6 September 2009; pp. 2756–2759.

46. Huys, R.; Braeken, D.; Jans, D.; Stassen, A.; Collaert, N.; Wouters, J.; Loo, J.; Severi, S.; Vleugels, F.; Callewaert, G.; et al. Single-cell recording and stimulation with a 16k micro-nail electrode array integrated on a 0.18 μm CMOS chip. *Lab Chip* **2012**, *12*, 1274–1280. [CrossRef] [PubMed]

47. Huebner, A.; Sharma, S.; Srisa-Art, M.; Hollfelder, F.; Edel, J.B.; DeMello, A.J. Microdroplets: A sea of applications? *Lab Chip* **2008**, *8*, 1244–1254. [CrossRef] [PubMed]

48. Zhu, P.; Wang, L. Passive and active droplet generation with microfluidics: A review. *Lab Chip* **2017**, *17*, 34–75. [CrossRef] [PubMed]

49. Zhang, Q.; Wang, T.; Zhou, Q.; Zhang, P.; Gong, Y.; Gou, H.; Xu, J.; Ma, B. Development of a facile droplet-based single-cell isolation platform for cultivation and genomic analysis in microorganisms. *Sci. Rep.* **2017**, *7*, 41192. [CrossRef] [PubMed]

50. Adams, L.L.A.; Kodger, T.E.; Kim, S.H.; Shum, H.C.; Franke, T.; Weitz, D.A. Single step emulsification for the generation of multi-component double emulsions. *Soft Matter* **2012**, *8*, 10719–10724. [CrossRef]

51. Thurgood, P.; Baratchi, S.; Szydzik, C.; Zhu, J.Y.; Nahavandi, S.; Mitchell, A.; Khoshmanesh, K. A self-sufficient micro-droplet generation system using highly porous elastomeric sponges: A versatile tool for conducting cellular assays. *Sens. Actuators B Chem.* **2018**, *274*, 645–653. [CrossRef]

52. Di Carlo, D. Inertial microfluidics. *Lab Chip* **2009**, *9*, 3038–3046. [CrossRef]

53. Kuntaegowdanahalli, S.S.; Bhagat, A.A.S.; Kumar, G.; Papautsky, I. Inertial microfluidics for continuous particle separation in spiral microchannels. *Lab Chip* **2009**, *9*, 2973–2980. [CrossRef]

54. Bhagat, A.A.S.; Hou, H.W.; Li, L.D.; Lim, C.T.; Han, J. Pinched flow coupled shear-modulated inertial microfluidics for high-throughput rare blood cell separation. *Lab Chip* **2011**, *11*, 1870–1878. [CrossRef] [PubMed]

55. Nathamgari, S.S.P.; Dong, B.; Zhou, F.; Kang, W.; Giraldo-Vela, J.P.; McGuire, T.; McNaughton, R.L.; Sun, C.; Kessler, J.A.; Espinosa, H.D. Isolating single cells in a neurosphere assay using inertial microfluidics. *Lab Chip* **2015**, *15*, 4591–4597. [CrossRef] [PubMed]

56. SIR JAMES LIGHTHILL Acoustic streaming. *J. Sound Vib.* **1978**, *61*, 391–418. [CrossRef]

57. Yan, B. Oscillatory flow beneath a free surface. *Fluid Dyn. Res.* **1998**, *22*, 1–23. [CrossRef]

58. Hayakawa, T.; Sakuma, S.; Arai, F. On-chip 3D rotation of oocyte based on a vibration-induced local whirling flow. *Microsyst. Nanoeng.* **2015**, *1*, 15001. [CrossRef]

59. Unger, M.A.; Chou, H.P.; Thorsen, T.; Scherer, A.; Quake, S.R. Monolithic Microfabricated Valves and Pumps by Multilayer Soft Lithography. *Science* **2000**, *288*, 113–116. [CrossRef] [PubMed]

60. Shalek, A.K.; Satija, R.; Shuga, J.; Trombetta, J.J.; Gennert, D.; Lu, D.; Chen, P.; Gertner, R.S.; Gaublomme, J.T.; Yosef, N.; et al. Single-cell RNA-seq reveals dynamic paracrine control of cellular variation. *Nature* **2014**, *510*, 363–369. [CrossRef] [PubMed]

61. Di Carlo, D.; Aghdam, N.; Lee, L.P. Single-cell enzyme concentrations, kinetics, and inhibition analysis using high-density hydrodynamic cell isolation arrays. *Anal. Chem.* **2006**, *78*, 4925–4930. [CrossRef]

62. Tran, Q.D.; Kong, T.F.; Hu, D.; Marcos; Lam, R.H.W. Deterministic sequential isolation of floating cancer cells under continuous flow. *Lab Chip* **2016**, *16*, 2813–2819. [CrossRef]

63. Luo, T.; Hou, J.; Chen, S.; Chow, Y.T.; Wang, R.; Ma, D.; Zhu, R.; Sun, D. Microfluidic single-cell array platform enabling week-scale clonal expansion under chemical/electrical stimuli. *Biomicrofluidics* **2017**, *11*, 054103. [CrossRef]

64. Tan, W.H.; Takeuchi, S. A trap-and-release integrated microfluidic system for dynamic microarray applications. *Proc. Natl. Acad. Sci. USA* **2007**, *104*, 1146–1151. [CrossRef] [PubMed]

65. Kim, H.; Lee, S.; Lee, J.H.; Kim, J. Integration of a microfluidic chip with a size-based cell bandpass filter for reliable isolation of single cells. *Lab Chip* **2015**, *15*, 4128–4132. [CrossRef] [PubMed]
66. Lin, C.H.; Hsiao, Y.H.; Chang, H.C.; Yeh, C.F.; He, C.K.; Salm, E.M.; Chen, C.; Chiu, I.M.; Hsu, C.H. A microfluidic dual-well device for high-throughput single-cell capture and culture. *Lab Chip* **2015**, *15*, 2928–2938. [CrossRef] [PubMed]
67. Luo, T.; Fan, L.; Zeng, Y.; Liu, Y.; Chen, S.; Tan, Q.; Lam, R.H.W.; Sun, D. A simplified sheathless cell separation approach using combined gravitational-sedimentationbased prefocusing and dielectrophoretic separation. *Lab Chip* **2018**, *18*, 1521–1532. [CrossRef] [PubMed]
68. Doh, I.; Cho, Y.H. A continuous cell separation chip using hydrodynamic dielectrophoresis (DEP) process. *Sens. Actuators A Phys.* **2005**, *121*, 59–65. [CrossRef]
69. Huang, L.; Zhao, P.; Wang, W. 3D cell electrorotation and imaging for measuring multiple cellular biophysical properties. *Lab Chip* **2018**, *18*, 2359–2368. [CrossRef] [PubMed]
70. Wu, Y.; Ren, Y.; Tao, Y.; Hou, L.; Jiang, H. Large-scale single particle and cell trapping based on rotating electric field induced-charge electroosmosis. *Anal. Chem.* **2016**, *88*, 11791–11798. [CrossRef]
71. Gao, J.; Riahi, R.; Sin, M.L.Y.; Zhang, S.; Wong, P.K. Electrokinetic focusing and separation of mammalian cells in conductive biological fluids. *Analyst* **2012**, *137*, 5215–5221. [CrossRef]
72. Ramos, A.; González, A.; Castellanos, A.; Green, N.G.; Morgan, H. Pumping of liquids with ac voltages applied to asymmetric pairs of microelectrodes. *Phys. Rev. E* **2003**, *67*, 056302. [CrossRef]
73. Ashkin, A.; Dziedzic, J.M.; Yamane, T. Optical trapping and manipulation of single cells using infrared laser beams. *Nature* **1987**, *330*, 769–771. [CrossRef]
74. Wang, K.; Cheng, J.; Han Cheng, S.; Sun, D. Probing cell biophysical behavior based on actin cytoskeleton modeling and stretching manipulation with optical tweezers. *Appl. Phys. Lett.* **2013**, *103*, 083706. [CrossRef]
75. Gou, X.; Yang, H.; Fahmy, T.M.; Wang, Y.; Sun, D. Direct measurement of cell protrusion force utilizing a robot-aided cell manipulation system with optical tweezers for cell migration control. *Int. J. Robot. Res.* **2014**, *33*, 1782–1792. [CrossRef]
76. Xie, S.; Wang, X.; Jiao, N.; Tung, S.; Liu, L. Programmable micrometer-sized motor array based on live cells. *Lab Chip* **2017**, *17*, 2046–2053. [CrossRef] [PubMed]
77. Hu, W.; Fan, Q.; Ohta, A.T. An opto-thermocapillary cell micromanipulator. *Lab Chip* **2013**, *13*, 2285–2291. [CrossRef] [PubMed]
78. Würger, A. Thermal non-equilibrium transport in colloids. *Rep. Prog. Phys.* **2010**, *73*, 126601. [CrossRef]
79. Lee, W.; Tseng, P.; Di, D. *Microtechnology for Cell Manipulation and Sorting*; Springer: Berlin, Germany, 2017.
80. Hu, W.; Ishii, K.S.; Ohta, A.T. Micro-assembly using optically controlled bubbles microrobots. *Appl. Phys. Lett.* **2011**, *99*, 094103. [CrossRef]
81. Shields, C.W.; Wang, J.L.; Ohiri, K.A.; Essoyan, E.D.; Yellen, B.B.; Armstrong, A.J.; López, G.P. Magnetic separation of acoustically focused cancer cells from blood for magnetographic templating and analysis. *Lab Chip* **2016**, *16*, 3833–3844. [CrossRef] [PubMed]
82. Feng, L.; Liang, S.; Zhou, X.; Yang, J.; Jiang, Y.; Zhang, D.; Arai, F. On-chip microfluid induced by oscillation of microrobot for noncontact cell transportation. *Appl. Phys. Lett.* **2017**, *111*, 203703. [CrossRef]
83. Li, J.; Li, X.; Luo, T.; Wang, R.; Liu, C.; Chen, S.; Li, D.; Yue, J.; Cheng, S.H.; Sun, D. Development of a magnetic microrobot for carrying and delivering targeted cells. *Sci. Robot.* **2018**, *3*, eaat8829. [CrossRef]
84. Wang, J.; Gao, W. Nano/microscale motors: Biomedical opportunities and challenges. *ACS Nano* **2012**, *6*, 5745–5751. [CrossRef]
85. Wu, J.; He, Z.; Chen, Q.; Lin, J. Biochemical analysis on microfluidic chips. *Trends Anal. Chem.* **2016**, *80*, 213–231. [CrossRef]
86. Choi, J.; Song, H.; Hwan, J.; Kim, D.; Kim, K. Microfluidic assay-based optical measurement techniques for cell analysis: A review of recent progress. *Biosens. Bioelectron.* **2016**, *77*, 227–236. [CrossRef] [PubMed]
87. Kuswandi, B.; Huskens, J.; Verboom, W. Optical sensing systems for microfluidic devices: A review. *Anal. Chim. Acta* **2007**, *601*, 141–155. [CrossRef] [PubMed]
88. Rackus, D.G.; Shamsi, H.; Wheeler, A.R. Electrochemistry, biosensors and microfluidics: A convergence of fields. *Chem. Soc. Rev.* **2015**, *44*, 5320–5340. [CrossRef]
89. Kiilerich-pedersen, K.; Rozlosnik, N. Cell-Based Biosensors: Electrical Sensing in Microfluidic Devices. *Diagnostics* **2012**, *2*, 83–96. [CrossRef] [PubMed]

90. Gao, D.; Liu, H.; Jiang, Y.; Lin, J.-M. Recent advances in microfluidics combined with mass spectrometry: Technologies and applications. *Lab Chip* **2013**, *13*, 3309–3322. [CrossRef] [PubMed]
91. Khamenehfar, A.; Li, P.C.H. Microfluidic Devices for Circulating Tumor Cells Isolation and Subsequent Analysis. *Curr. Pharm. Biotechnol.* **2016**, *17*, 810–821. [CrossRef]
92. Lam, J.; Marklein, R.A.; Jimenez-torres, J.A.; Beebe, D.J.; Bauer, S.R.; Sung, K.E. Adaptation of a Simple Microfluidic Platform for High-Dimensional Quantitative Morphological Analysis of Human Mesenchymal Stromal Cells on Polystyrene- Based Substrates. *SLAS Technol.* **2017**, *22*, 646–661.
93. Huang, Y.L.; Tung, C.; Zheng, A.; Kim, B.J.; Wu, M. Interstitial flows promote amoeboid over mesenchymal motility of breast cancer cells revealed by a three dimensional microfluidic. *Integr. Biol.* **2015**, *7*, 1402–1411. [CrossRef]
94. Jo, M.C.; Liu, W.; Gu, L.; Dang, W.; Qin, L. High-throughput analysis of yeast replicative aging using a microfluidic system. *Proc. Natl. Acad. Sci. USA* **2015**, *112*, 9364–9369. [CrossRef]
95. Haandbæk, N.; Bürgel, S.C.; Heer, F.; Hierlemann, A. Characterization of subcellular morphology of single yeast cells using high frequency microfluidic impedance cytometer. *Lab Chip* **2014**, *14*, 369–377. [CrossRef] [PubMed]
96. Zhu, Z.; Frey, O.; Haandbaek, N.; Franke, F.; Rudolf, F.; Hierlemann, A. Time-lapse electrical impedance spectroscopy for monitoring the cell cycle of single immobilized S. pombe cells. *Sci. Rep.* **2015**, *5*, 17180. [CrossRef] [PubMed]
97. Lecault, V.; Vaninsberghe, M.; Sekulovic, S.; Knapp, D.J.H.F.; Wohrer, S.; Bowden, W.; Viel, F.; Mclaughlin, T.; Jarandehei, A.; Miller, M.; et al. High-throughput analysis of single hematopoietic stem cell proliferation in microfluidic cell culture arrays. *Nat. Methods* **2011**, *8*, 581–589. [CrossRef] [PubMed]
98. Chen, Y.; Allen, S.G.; Ingram, P.N.; Buckanovich, R.; Merajver, S.D.; Yoon, E. Single-cell Migration Chip for Chemotaxis-based Microfluidic Selection of Heterogeneous Cell Populations. *Sci. Rep.* **2015**, *5*, 9980. [CrossRef] [PubMed]
99. Zhou, Y.; Basu, S.; Laue, E.; Seshia, A.A. Single cell studies of mouse embryonic stem cell (mESC) differentiation by electrical impedance measurements in a microfluidic device. *Biosens. Bioelectron.* **2016**, *81*, 249–258. [CrossRef] [PubMed]
100. Kumar, P.T.; Vriens, K.; Cornaglia, M.; Gijs, M.; Kokalj, T.; Thevissen, K.; Geeraerd, A.; Cammue, B.P.A.; Puers, R.; Lammertyn, J. Digital microfluidics for time-resolved cytotoxicity studies on single non-adherent yeast cells. *Lab Chip* **2015**, *15*, 1852–1860. [CrossRef] [PubMed]
101. Li, Y.; Chen, D.; Zhang, Y.; Liu, C.; Chen, P.; Wang, Y.; Feng, X.; Du, W.; Liu, B.F. High-throughput single cell multidrug resistance analysis with multifunctional gradients-customizing microfluidic device. *Sens. Actuators B Chem.* **2016**, *225*, 563–571. [CrossRef]
102. Bithi, S.S.; Vanapalli, S.A. Microfluidic cell isolation technology for drug testing of single tumor cells and their clusters. *Sci. Rep.* **2017**, *7*, 41707. [CrossRef]
103. Sikorski, D.J.; Caron, N.J.; Vaninsberghe, M.; Zahn, H.; Eaves, C.J.; Piret, J.M.; Hansen, C.L. Clonal analysis of individual human embryonic stem cell differentiation patterns in microfluidic cultures. *Biotechnol. J.* **2015**, *10*, 1546–1554. [CrossRef]
104. Rubakhin, S.S.; Romanova, E.V.; Nemes, P.; Sweedler, J.V. Profiling metabolites and peptides in single cells. *Nat. Methods* **2011**, *8*, S20–S29. [CrossRef]
105. Kraly, J.R.; Holcomb, R.E.; Guan, Q.; Henry, C.S. Review: Microfluidic applications in metabolomics and metabolic profiling. *Anal. Chim. Acta* **2009**, *653*, 23–35. [CrossRef] [PubMed]
106. Zhang, X.C.; Wei, Z.W.; Gong, X.Y.; Si, X.Y.; Zhao, Y.Y.; Yang, C.D.; Zhang, S.C.; Zhang, X.R. Integrated Droplet-Based Microextraction with ESI-MS for Removal of Matrix Interference in Single-Cell Analysis. *Sci. Rep.* **2016**, *6*, 24730. [CrossRef] [PubMed]
107. Wang, B.L.; Ghaderi, A.; Zhou, H.; Agresti, J.; Weitz, D.A.; Fink, G.R.; Stephanopoulos, G. Microfluidic high-throughput culturing of single cells for selection based on extracellular metabolite production or consumption. *Nat. Biotechnol.* **2014**, *32*, 473–478. [CrossRef] [PubMed]
108. Shah, P.; Vedarethinam, I.; Kwasny, D.; Andresen, L.; Skov, S.; Silahtaroglu, A.; Tümer, Z.; Dimaki, M.; Svendsen, W.E. FISHprep: A novel integrated device for metaphase FISH sample preparation. *Micromachines* **2011**, *2*, 116–128. [CrossRef]
109. Sieben, V.J.; Debes-Marun, C.S.; Pilarski, L.M.; Backhouse, C.J. An integrated microfluidic chip for chromosome enumeration using fluorescence in situ hybridization. *Lab Chip* **2008**, *8*, 2151–2156. [CrossRef]

110. White, A.K.; VanInsberghe, M.; Petriv, O.I.; Hamidi, M.; Sikorski, D.; Marra, M.A.; Piret, J.; Aparicio, S.; Hansen, C.L. High-throughput microfluidic single-cell RT-qPCR. *Proc. Natl. Acad. Sci. USA* **2011**, *108*, 13999–14004. [CrossRef]

111. Bennett, M.R.; Hasty, J. Microfluidic devices for measuring gene network dynamics in single cells. *Nat. Rev. Genet.* **2009**, *10*, 628–638. [CrossRef] [PubMed]

112. Kimmerling, R.J.; Lee Szeto, G.; Li, J.W.; Genshaft, A.S.; Kazer, S.W.; Payer, K.R.; De Riba Borrajo, J.; Blainey, P.C.; Irvine, D.J.; Shalek, A.K.; et al. A microfluidic platform enabling single-cell RNA-seq of multigenerational lineages. *Nat. Commun.* **2016**, *7*, 10220. [CrossRef]

113. Streets, A.M.; Zhang, X.; Cao, C.; Pang, Y.; Wu, X.; Xiong, L.; Yang, L.; Fu, Y.; Zhao, L.; Tang, F.; et al. Microfluidic single-cell whole-transcriptome sequencing. *Proc. Natl. Acad. Sci. USA* **2014**, *111*, 7048–7053. [CrossRef]

114. Yu, J.; Zhou, J.; Sutherland, A.; Wei, W.; Shin, Y.S.; Xue, M.; Heath, J.R. Microfluidics-Based Single-Cell Functional Proteomics for Fundamental and Applied Biomedical Applications. *Annu. Rev. Anal. Chem.* **2014**, *7*, 275–295. [CrossRef]

115. Srivastava, N.; Brennan, J.S.; Renzi, R.F.; Wu, M.; Branda, S.S.; Singh, A.K.; Herr, A.E. Fully integrated microfluidic platform enabling automated phosphoprofiling of macrophage response. *Anal. Chem.* **2009**, *81*, 3261–3269. [CrossRef] [PubMed]

116. Ma, C.; Fan, R.; Ahmad, H.; Shi, Q.; Comin-Anduix, B.; Chodon, T.; Koya, R.C.; Liu, C.C.; Kwong, G.A.; Radu, C.G.; et al. A clinical microchip for evaluation of single immune cells reveals high functional heterogeneity in phenotypically similar T cells. *Nat. Med.* **2011**, *17*, 738–743. [CrossRef] [PubMed]

117. He, M.; Crow, J.; Roth, M.; Zeng, Y.; Godwin, A.K. Integrated immunoisolation and protein analysis of circulating exosomes using microfluidic technology. *Lab Chip* **2014**, *14*, 3773–3780. [CrossRef] [PubMed]

118. Godin, M.; Delgado, F.F.; Son, S.; Grover, W.H.; Bryan, A.K.; Tzur, A.; Jorgensen, P.; Payer, K.; Grossman, A.D.; Kirschner, M.W.; et al. Using buoyant mass to measure the growth of single cells. *Nat. Methods* **2010**, *7*, 387–390. [CrossRef] [PubMed]

119. Bryan, A.K.; Hecht, V.C.; Shen, W.; Payer, K.; Grover, W.H.; Manalis, S.R. Measuring single cell mass, volume, and density with dual suspended microchannel resonators. *Lab Chip* **2014**, *14*, 569–576. [CrossRef] [PubMed]

120. Son, S.; Kang, J.H.; Oh, S.; Kirschner, M.W.; Mitchison, T.J.; Manalis, S. Resonant microchannel volume and mass measurements show that suspended cells swell during mitosis. *J. Cell Biol.* **2015**, *211*, 757–763. [CrossRef] [PubMed]

121. Godin, M.; Bryan, A.K.; Burg, T.P.; Babcock, K.; Manalis, S.R. Measuring the mass, density, and size of particles and cells using a suspended microchannel resonator. *Appl. Phys. Lett.* **2007**, *91*, 2–5. [CrossRef]

122. Guo, Q.; Park, S.; Ma, H. Microfluidic micropipette aspiration for measuring the deformability of single cells. *Lab Chip* **2012**, *12*, 2687–2695. [CrossRef]

123. Hu, S.; Liu, G.; Chen, W.; Li, X.; Lu, W.; Lam, R.H.W.; Fu, J. Multiparametric Biomechanical and Biochemical Phenotypic Profiling of Single Cancer Cells Using an Elasticity Microcytometer. *Small* **2016**, *12*, 2300–2311. [CrossRef]

124. Zhang, X.; Chu, H.K.; Zhang, Y.; Bai, G.; Wang, K.; Tan, Q.; Sun, D. Rapid characterization of the biomechanical properties of drug-treated cells in a microfluidic device. *J. Micromech. Microeng.* **2015**, *25*, 105004. [CrossRef]

125. Sun, T.; Morgan, H. Single-cell microfluidic impedance cytometry: A review. *Microfluid. Nanofluid.* **2010**, *8*, 423–443. [CrossRef]

126. Hong, J.L.; Lan, K.C.; Jang, L.S. Electrical characteristics analysis of various cancer cells using a microfluidic device based on single-cell impedance measurement. *Sens. Actuators B Chem.* **2012**, *173*, 927–934. [CrossRef]

127. Zhou, Y.; Yang, D.; Zhou, Y.; Khoo, B.L.; Han, J.; Ai, Y. Characterizing Deformability and Electrical Impedance of Cancer Cells in a Microfluidic Device. *Anal. Chem.* **2018**, *90*, 912–919. [CrossRef] [PubMed]

128. Sakuma, S.; Kasai, Y.; Hayakawa, T.; Arai, F. On-chip cell sorting by high-speed local-flow control using dual membrane pumps. *Lab Chip* **2017**, *17*, 2760–2767. [CrossRef] [PubMed]

micromachines

MDPI

Communication

Bacterial Concentration Detection using a PCB-based Contactless Conductivity Sensor

Xiao-Yan Zhang [1], Zhe-Yu Li [1], Yu Zhang [1], Xiao-Qian Zang [1], Kosei Ueno [2], Hiroaki Misawa [2,3] and Kai Sun [1,*]

[1]   State Key Laboratory of Urban Water Resource and Environment, Harbin Institute of Technology, Harbin 150090, China; xyzhang774985529@163.com (X.-Y.Z.); zhylee@hit.edu.cn (Z.-Y.L.); yuzhang429@126.com (Y.Z.); 1112710105@hit.edu.cn (X.-Q.Z.)
[2]   Research Institute for Electronic Science, Hokkaido University, Sapporo 001-0021, Japan; k-ueno@es.hokudai.ac.jp (K.U.); misawa@es.hokudai.ac.jp (H.M.)
[3]   Department of Applied Chemistry & Institute of Molecular Science, National Chiao Tung University, Hsinchu 30010, Taiwan
*   Correspondence: ksun@hit.edu.cn; Tel.: +86-451-8628-9126

Received: 3 December 2018; Accepted: 10 January 2019; Published: 14 January 2019

Abstract: Capacitively coupled contactless conductivity detection (C^4D) is an improved approach to avoid the problems of labor-intensive, time-consuming and insufficient accuracy of plate count as well as the high-cost apparatus of flow cytometry (FCM) in bacterial counting. This article describes a novel electrode-integrated printed-circuit-board (PCB)-based C^4D device, which supports the simple and safe exchange of capillaries and improves the sensitivity and repeatability of the contactless detection. Furthermore, no syringe pump is needed in the detection, it reduces the system size, and, more importantly, avoids the effect on the bacteria due to high pressure. The recovered bacteria after C^4D detection at excitation of 25 Vpp and 60–120 kHz were analyzed by flow cytometry, and a survival rate higher than 96% was given. It was verified that C^4D detection did not influence the bacterial viability. Moreover, bacteria concentrations from 10^6 cells/mL to 10^8 cells/mL were measured in a linear range, and relative standard deviation (RSD) is below 0.2%. In addition, the effects on bacteria and C^4D from background solutions were discussed. In contrast to common methods used in most laboratories, this method may provide a simple solution to in situ detection of bacterial cultures.

Keywords: bacterial concentration; capacitively coupled contactless conductivity detection (C^4D); capillary; *E. coli*; printed-circuit-board (PCB)

---

1. Introduction

Over the past decades, bacteria have been widely used in many fields, including pharmacy [1], chemistry [2], biology [3], environmental science [4], and fermentation [5]. Bacterial counting, as one of the most important indicators, is used to determine the concentration of bacterial culture, to monitor the water quality, to assess pollution level, and to diagnose patients. Although there are many techniques developed to count bacteria, such as plate count and optical density (OD), these techniques are mostly time-consuming, labor-intensive, and unable to provide sufficient accuracy [6]. Some methods based on fluorescent dyes, for example, fluorescence analysis, and flow cytometry (FCM), are advanced in rapidness, technical simplicity, and efficiency, even in distinguishing different physiological states of bacteria at single cell level [7–9]. Nevertheless, fluorescent quenching and high cost per test [10] are still limiting it from being widely used. Biosensors have been paid more and more attention with the development of micro electro mechanical system (MEMS) [11], which provide a portable and real-time platform [12,13]. In recent years, by making use of MEMS point-of-care testing, technologies have become an effective and rapid method to detect bacteria through analyzing the image [14–18]. Photoelectrochemical and electrochemical biosensors have been widely used for bioanalytical proposes [19], detecting bacterial strains [20], capturing bacteria [21], and monitoring activity of bacterial fermentation [22,23]. Electrochemical detection has a rapid response speed [24] and has the potential to be a useful tool in low-concentration applications [25,26]. However, this method cannot identify those microbes that rely on specific antibodies [27–30]. In addition, electrodes used in electrochemical detection are easily contaminated by the sample, which influences the results. Surface Enhanced Raman Scattering (SERS) avoid these shortcomings, and it can detect microorganisms without specific antibodies [31–33]; however, the weak signal is still a problem to be solved.

Capacitively coupled contactless conductivity detection (C^4D), by analyzing and testing the conductivity change of sample [34], could simply and sensitively detect metal cations, amino acids, and organic ions in beer, wine, milk, potable water, and juice [35–38]. The electrodes of C^4D contain three electrodes, the excitation electrode to which AC signal is applied, the shield electrode, and the pick-up electrode, which measured the detection signal. Two metal needles, with capillary passing through, were used as excitation electrode and pick-up electrode in early C^4D devices; at the same time, to improve the sensitivity of detection, silver conductive adhesive was used to narrow the gap between capillary and needle. To improve the signal-to-noise ratio (S/N) and integration, a printed-circuit-board (PCB) based C^4D with shorter electrical connection of the electrodes to the current/voltage operational amplifier was introduced to connect capillaries and microfluidic chips [39–41]. Silva et al. integrated the circuit and the wire-wrapped electrodes on a 18×18 mm^2 PCB, and optimized amplitude and frequency for each column diameter and electrophoretic buffer [42,43]. Jaanus et al. also integrated electrodes on the PCB and improve detection sensitivity by using an idle capillary for compensation [44].

The application of C^4D was mainly for detecting ions, while there are still some researchers focusing on counting cell by it. Emaminejad et al. [45] was the first to count cells from sheet whole blood based on contactless conductivity. Chen et al. [46] succeeded in counting 9-μm MCF-7 and 15-μm HCM cancer cells on a C^4D integrated chip. They both achieved label-free counting. Although electrical impedance spectroscopy has been used for detection of bacteria counting, there have been few studies focusing on the bacterial counting by C^4D [47]. This is because bacteria are smaller organisms with single cell volume in the range of 0.1–1 μm^3/cell, which is only one tenth of a cell. To detect bacteria by C^4D is much more difficulty comparing to the cells.

In this paper, we developed a type of novel C^4D device with both electrodes and amplifiers integrated on the PCB. Copper via holes of PCB with 400-μm inner diameter were used as C^4D electrodes through which a 360-μm OD capillary passes exactly. The fabrication of the via holes on PCB could provide the inner-diameter (ID) of 300-μm minimum with the tolerance less than 50 μm, which confirms better accuracy and repeatability. The bacterial suspension was dragged into the capillary manually by a syringe, without a syringe pump. The detection is very fast because of the concentration detection mechanisms instead of peak counting. Although the proposed method is not able to distinguish different populations of bacteria, this is a simple, inexpensive, rapid, contactless, and label-free method that enables bacterial counting without damage to the cells.

2. Materials and Methods

2.1. Materials and Reagents

Standard broth was used as a bacterial growth medium, and phosphate-buffered saline (PBS, pH = 7.4, NaCl 8.0 g/L, KCl 0.2 g/L, Na$_2$HPO$_4$ 1.42 g/L, KH$_2$PO$_4$ 0.24 g/L) was prepared in the laboratory. SYBR Green I (Invitrogen, Eugene, OR, USA) was diluted in dimethylsulfoxide (DMSO) to a ratio of 1:100 and then stored at -20 °C prior to use. Propidium iodide (PI) and albumin from bovine serum (BSA) was from Sigma (shanghai, China) and was stored at 4 °C. All samples were incubated in 1 mg/mL PI for 15 min in the dark before flow cytometry (BD, Becton, Dickinson and Company, Franklin Lakes, NJ, USA) measurement. Glutaraldehyde, paraformaldehyde, ethyl alcohol, and isoamyl acetate were from TCI (Tokyo, Japan). All dilutions were carried out in deionized water from a Millipore system.

2.2. Bacterial Culture and Sample Preparation

Escherichia coli (CGMCC 1.2385, China General Microbiological Culture Collection Center, Beijing, China) was cultured in the broth medium at 37 °C for 17 h in an incubator to achieve 10^7 cells/mL. The bacteria were then centrifuged at 4000 rpm for 5 min to remove the broth medium and were re-suspended in PBS followed by centrifuging again. Prior to the experiment, the capillary was flushed with 20 mg/L BSA solution to avoid the non-specific cell adhesion to the surface. To ensure that the bacteria was successfully injected into the capillary, which passed through the detection cell (Figure 1A), a concentration of 1.0×10^7 cells/mL sample was stained by SYBR Green I and was observed under the microscope with a charge-coupled device (CCD) camera (IX-73x microscope and DP80 camera, Olympus, Japan). The polyimide coating of the capillary was peeled off to eliminate fluorescent interference before observation. It can be seen clearly that the stained bacteria with green fluorescence pass through, and there was no aggregation found in the capillary (Figure 1B).

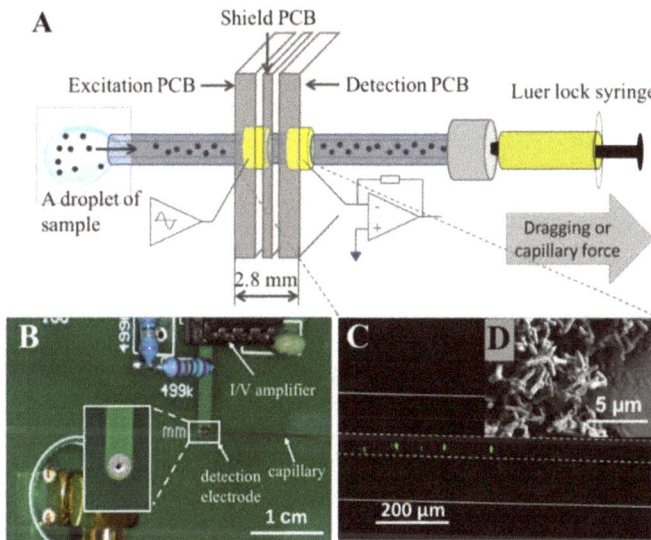

Figure 1. (**A**) Schematic illustration of capacitively coupled contactless conductivity detection (C^{4}D) device on detecting bacterial concentration. (**B**) The printed-circuit-board (PCB) based electrodes. (**C**) Fluorescence-labeled *Escherichia coli* in capillary. (**D**) Scanning electron microscope (SEM) image of *E. coli*, the length of which on average is about 2 μm.

2.3. E.coli Preparation for SEM

Moreover, *E. coli* was characterized by SEM (Figure 1D). It shows that the *E. coli* is approximately 2–3 μm long and 0.5 μm wide. *E. coli* preparation for SEM observation includes the flowing steps. After incubation in nutrient broth for 17 h at 37 °C, *E. coli* was washed by PBS three times to remove the broth and then soaked in 4% glutaraldehyde for 2 h, immersed in 3% paraformaldehyde for 1 h, successively dehydrated in 30%, 50%, 75%, and 80% ethyl alcohol for 10 min and in 95% for 20 min, stored in isoamyl acetate for 30 min, and dried out overnight. The samples should be washed by PBS three times in each step and deposited onto a clean silicon wafer.

2.4. Bacteria Loading

For C^4D detection, different concentrations of the samples were loaded into 200-µL tubes and were then injected into a 150-µm-ID and 10-cm-long capillary (Yongnian Optic Fiber Plant, Yongnian, Hebei, China) for 5 s by pulling back a syringe manually instead of a syringe pump. The outlet diameter of untreated polyimide coated fused silicon capillary is 360 µm. The output voltage signal was then recorded after stopping the injection. This rapid injection method could avoid the contamination from the syringe and the Luer lock. Furthermore, pump-free injection could avoid the destruction of the sample and conductivity changing due to high pressure in the capillary from the pump's pushing.

3. Results and Discussion

3.1. Design Strategy and Instrumentation

The experimental setup comprised a home-made C^4D and a data acquisition system. A block diagram of the C^4D circuitry, which contains an excitation PCB, a detection PCB, and a shielding PCB, is given in Figure 1A. Two solder pads, working as an excitation electrode and pick-up electrode, were located in the corresponding PCBs. The ID of the electrodes was 400 µm, which was decided by the 360-µm OD of the capillary used in the following experiments. The width of both the excitation and detection electrodes was 1.0 mm which was decided by the thickness of the excitation and detection PCBs. The thickness of the shielding PCB is 0.8 mm, which served as the gap between the two electrodes. Hence, the detection cell size was 2.8 mm (shown in Figure 1A). The capillary was washed with deionized (DI) water, PBS, and BSA in turn prior to use, then passed the excitation, shielding, and detection PCBs in sequence. The three PCBs were assembled with quite accurate alignment holes. The shielding PCB could effectively lower the noise and could control the gap between two electrodes, which could avoid the shortcut problem. Hence, the above design supported a smooth, safe, and simple exchange of capillaries and improved the sensitivity and repeatability of the contactless detection. A field programmable gate array (FPGA) was used to generate a sine-wave excitation signal with the designed frequency. The sinusoidal signal with 100 kHz frequency and 25 Vpp (peak-to-peak) amplitude was applied to the excitation electrode after optimization. The pick-up amplifier consisted of two cheap (unction Field-Effect Transistor) JFET-input operational amplifiers (LF-357, Texas Instruments, Austin, TX, USA) as the transimpedance amplifier and voltage follower. The signal was then rectified with a lock-in synchronous amplifier (AD630, Analog Devices, Norwood, MA, USA), two-order Chebyshev low-pass filtered, and programmable gain amplified (PGA205, Burr-Brown, Austin, TX, USA). The output was finally fed via a coaxial cable to a 16-bit data acquisition system (NI 6229, National Instruments, Austin, TX, USA). The system was controlled under LabVIEW software. Figure 1B shows the photograph of the prototype C^4D system. The signal-to-noise (S/N) of this system was detected by the different concentrations of the potassium chloride solution from 1 to 10 µM (Figure S1). The detection at the level of 1 µM is illustrated, and a linear relationship ($R^2 = 0.9885$) was obtained at this low concentration. Figure S2 shows the detection limit of 0.1 µM potassium salt solution, the S/N of which is higher than 9.

3.2. Electrical Effect of C^4D on Bacteria Viability

Bacteria can be killed when placed to voltage pulses of high strength for sufficient time, thus, there is a risk of bacteria damage or viability affection by the electrical detection. Especially for some rare uncultured bacteria, which are difficult to culture because of their complex growing conditions and long growing period, the risk may lead to unexpected loss. Hence, it is very important to discuss the bacterial viability before and after detection [48]. C^4D is an electrical detection method that measures conductivity corresponding to bacteria concentrations; thus, the influence of voltage amplitude and frequency of the wave on bacteria viability will be discussed. Wang et al. reported that the cell membrane would be disintegrated by the stimulation over 1000 V/cm of the electric field strength, which resulted in nearly 100% cell death; however, low field strength less than 300 V/cm did not affect the cell viability in the exposure period of 30–40 s [49,50]. In addition, the membrane potential induced by an external field is another factor need to be considered in detection. The membrane potential of bacteria is the difference in and out of cell. The critical valve is 1.1 Vm for bacteria to be in stationary growth phase [51]. The permeabiltization of cell membrane would increase when the external field was applied, which would lead to bacterial lysis and death. In our C^{4}D system, the measured current through the media in the capillary was less than 1 µA at the excitation of 25 Vpp, though the voltage was larger than the one used in electric double layer [52–54]. The maximum conductivity of the media was 0.286 mS/cm measured by METTLER TOLEDOLE 740 (METTLER TOLEDO, Shanghai, China), that is, the resistivity is 416.7 Ω·cm. The length of the C^{4}D cell is 2.8 mm (shown in Figure 1A); and the diameter of C^{4}D cell is 150 µm, which is the ID of capillary. Hence, the electric field strength in the capillary was about 2.36 V/cm, which is far below 300 V/cm. According to the theoretical equation [51], the membrane potential of 0.288 mV for *E. coli* cell at the excitation of 25 Vpp is revealed. It is far below the critical membrane potential of 1.1 Vm, which is only 0.1% of the Vm.

Frequency is another electrical parameter that will influence the bacterial viability. Shawki [55] recently reported that the current with low frequency less than 100 Hz (about 10 V/cm, 130 s exposure) can be used as a physical method to kill bacteria, with a death rate over 40%, however, bacteria exposure to higher frequencies showed insignificant lethal effect; the death rate at 100 kHz was less than 1%. 100 kHz is the working frequency of C^{4}D. In our assessment experiment, the *E. coli* samples with the concentration of 10^7 cells/mL in PBS buffer were exposed in the capillary with external sinusoidal excitation of 25 Vpp and 60–120 kHz. The viability of bacteria before and after conductivity detection was analyzed using FCM after being fluorescent-labeled by Propidium Iodide (PI). This essay is a commonly used method to analyze the cell viability [56]. PI is unable to pass through the membrane of live bacteria, thus, the damaged bacteria will be labeled by PI. Figure 2 shows that the survival rate after exposure at the frequencies of 60 kHz, 80 kHz, 100 kHz, and 120 kHz were all above 96%, with very small error bars. It can be concluded that the working frequency in range from 60–120 kHz of C^{4}D may not influence the viability of the bacteria obviously. Additionally, numerous studies have provided evidence that alternating current (AC) frequency of 5 kHz–10 MHz [57] can been used in the application of dielectrophoretic cell separation. Cells could be successfully cultured after dielectrophoretic separation.

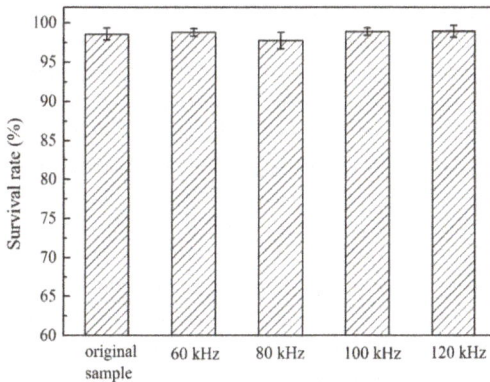

Figure 2. Influence on bacteria viability due to C^{4}D detection by analyzing the ratio of live to dead cells. The recovered bacteria were labeled with Propidium Iodide (PI) and analyzed by flow cytometry (FCM). Error bars are standard deviations of four measurements.

3.3. Optimization of PBS Concentration for Bacteria Counting

The conductivity of the membrane of live bacteria is only ~10^{-3} µS/cm, thus, bacterial membrane is highly insulated [58]. The conductivity of bacteria's interior can be as high as ~10 µS/cm, which is much higher than 0.055 µS/cm of DI water. Thus, when live bacteria are suspended in DI water, ions will release from the *E. coli* and generate osmotic shock. At the same time, water will enter into the bacteria, and cause death of most bacteria due to the bacterial membrane broken [47]. Therefore, PBS buffer is widely used for substance solution and cell container rinsing, which is able to balance pH for better bacterial viability and provide same ion concentration. Electrical impedance spectroscopy makes use of this characteristic to detect bacteria; however, the electrodes are easily contaminated and corroded due to direct sample contacting [47].

The osmolality and ion concentrations of PBS solutions are equal to those of bacteria. However, because the conductivity of PBS buffer is also roughly the same as the bacterial interior, it is hard to detect the conductivity change after suspending the bacteria with the concentration from 10^4 cells/mL to 10^8 cells/mL in the 10 mM PBS buffer (see Figure 3). Hence, it is essential to reduce the conductivity of background solution. It was reported that bacteria could survive in the solution, in which the concentration of sodium chloride is higher than the critical value of 1.7 mM [59]. To understand the effect of salt concentration on the bacteria, the bacteria survival rate in the original PBS buffer and different diluted PBS buffer were measured by FCM. Figure 4A shows that the survival rate of bacteria in 10 mM PBS buffer for 2 h was 96%, and decreased with the descent of the PBS concentration. Variation between 2 h and 24 h detection for 0.2 mM and 0.1 mM PBS dilution was less than 10%.

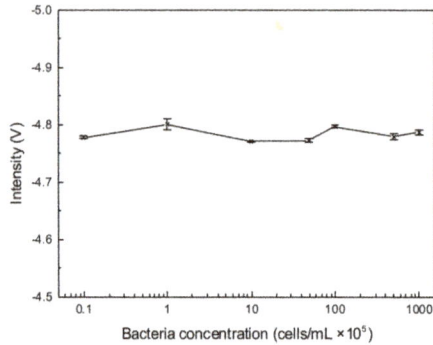

Figure 3. The C^4D response to the bacteria suspended in original phosphate-buffered saline (PBS) buffer with concentrations from 10^4 to 10^8 cells/mL. Error bars are standard deviations of four measurements.

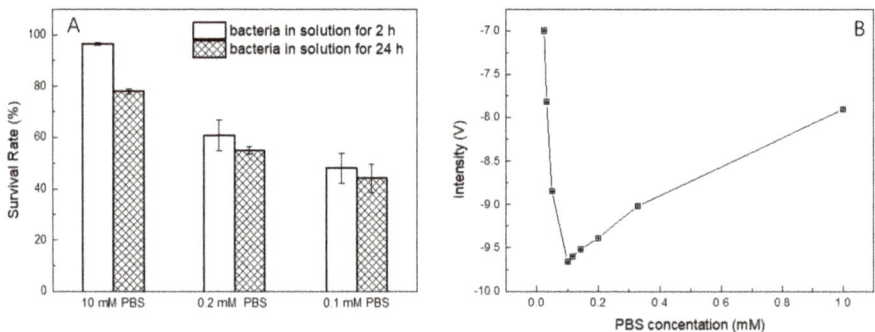

Figure 4. (**A**) Bacteria survival rate in 10 mM PBS solution, 0.2 mM PBS dilution, 0.1 mM PBS dilution and water after 2 h and 24 h were detected by FCM. Bacteria were dyed by PI for 10 min. Error bars are standard deviations of 3 measurements. (**B**) The relationship between PBS dilution and C4D intensity. Error bars are standard deviations of 4 measurements.

The 10 mM PBS buffer was then diluted to different concentrations to measure corresponding C4D intensities (shown in Figure 4B). It is noted that the C4D intensities indicate a negative correlation in case of the PBS concentration higher than 0.1 mM dilution, however, a strong positive correlation from 0.1 mM to 0.025 mM. It could be explained that in the case of low conductivity detection the C4D impedance is essentially a function of the capillary inner solution conductivity, while at high conductivities, impedance is mainly determined by capacitances of the capillary wall [60]. Low concentrations, less than 0.1 mM PBS solution, exhibit a high quality of linearity, whereas the ion concentration of sodium chloride is much lower than the critical value. Hence, we finally selected 0.2 mM and 0.1 mM PBS dilution as C4D background for discussion.

Figure 5 shows the relationship between the C^4D intensity and different bacteria concentrations from 10^4 to 10^8 cells/mL in 0.2 mM and 0.1 mM PBS dilutions. In both PBS dilutions, C^4D intensities increase with the increase of bacteria concentration, and give higher sensitivities when detecting bacteria in the range of 10^6–10^8 cells/mL in contrast with the concentration less than 10^6 cells/mL. It seems that the detection in 0.2 mM PBS indicates a little better response than that in 0.1 mM PBS in the case of low bacterial concentration. The S/N of 10^4 and 10^5 cells/mL bacteria in 0.2 mM PBS is 55.7 and 124.3, respectively.

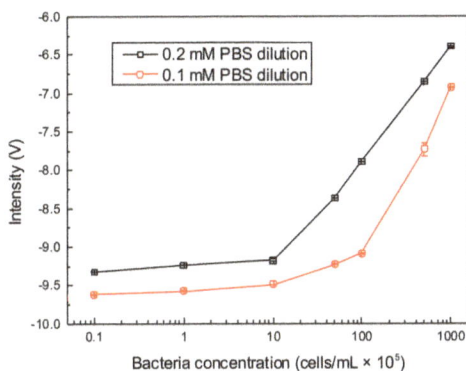

Figure 5. The C^4D response to the bacteria suspended in 0.2 mM PBS dilution and 0.1 mM PBS dilution with the different concentrations of 10^4–10^8 cells/mL. Error bars are standard deviations of four measurements.

4. Conclusions

In summary, we designed and fabricated a novel C^4D device to count *E. coli*. The electrodes and the circuits were integrated on three PCBs. Thus, the design supports a simple and safe change of capillaries to improve the sensitivity and repeatability comparing to traditional C^4D methods. Firstly, we discussed the electric field strength and the frequency effect on the bacteria after the C^4D detection; results indicate that the survival rate is above 96% at the excitation of 25 Vpp and 100 kHz. Secondly, the background solutions for C^4D detection of bacteria were discussed. For the in situ detection of bacteria culture, we suggest 0.2 mM PBS solution as the C^4D background, which adapts to detecting the bacteria concentration in the range of 10^6–10^8 cells/mL. In addition, the consumption solution is nanoliter scale, and no syringe pump is needed, which avoids the high pressure effect on the bacteria. Hence, this method may provide a good technical solution for many applications and the comparison of developed devices and this work is in Table A1. Reasonable direction for further research is to improve the detection range by increasing the ID of capillary, which can enlarge the volume of detection cells. In addition, narrow-ID-capillary array or holey fibers can be used to improve the detection limit by lower the background from the PBS buffer.

Supplementary Materials: The following are available online at http://www.mdpi.com/2072-666X/10/1/55/s1, Figure S1: The C^4D detection of different concentrations of the potassium chloride solution from 1 to 10 μM (R^2 = 0.9885), Figure S2: Response to the injection of 0.1 μM KCl solution. The flow velocity is 3 μL/min of DI water, and the injection volume of sample is 1 nL.

Author Contributions: Conceptualization, K.S. and X.-Y.Z.; methodology, X.-Y.Z.; software, K.S.; validation, Z.-Y.L., and X.-Q.Z.; formal analysis, X.-Y.Z. and Z.-Y.L.; investigation, X.-Y.Z.; resources, K.S. and H.M.; data curation, X.-Y.Z.; writing—original draft preparation, X.-Y.Z.; writing—review and editing, K.U. and K.S.; visualization, Y.Z.; supervision, K.S.; project administration, K.S.; funding acquisition, K.S. and H.M.

Funding: This research was funded by the National Natural Science Foundation of China (grant number 61674046); State Key Laboratory of Urban Water Resource and Environment (Harbin Institute of Technology), grant number 2017TS04.

Acknowledgments: The authors acknowledge the Research Institute for Electronic Science (RIES) International Exchange Program of "Dynamic Alliance for Open Innovation Bridging Human, Environment and Materials" from the Ministry of Education, Culture, Sports, Science and Technology of Japan (MEXT), Hokkaido University Program for Invitation of International Faculty.

Conflicts of Interest: The authors declare no conflict of interest.

Appendix A

Table A1. The comparison of developed devices and this work.

Method	Operation	Recording	Label	Accuracy	Reference
Optical density (OD)	simple	automatic monitoring	no	providing the growth trend	[6]
Plate count	simple	manual reading	no	depending on labor	[6]
Flow cytometry	skilled operator needed	automatic monitoring	fluorescence	good	[7–9]
This work	simple	automatic monitoring	no	Better than OD in range of 10^6–10^8 cells/ml	

References

1. Bergeron, S.; Boopathy, R.; Nathaniel, R.; Corbin, A.; LaFleur, G. Presence of antibiotic resistant bacteria and antibiotic resistance genes in raw source water and treated drinking water. *Int. Biodeterior. Biodegrad.* **2015**, *102*, 370–374. [CrossRef]
2. Chen, C.W.; Hsu, C.Y.; Lai, S.M.; Syu, W.J.; Wang, T.Y.; Lai, P.S. Metal nanobullets for multidrug resistant bacteria and biofilms. *Adv. Drug Deliv. Rev.* **2014**, *78*, 88–104. [CrossRef] [PubMed]
3. Huang, X.; Zou, X.; Zhao, J.; Shi, J.; Zhang, X.; Li, Z.; Shen, L. Sensing the quality parameters of Chinese traditional Yao-meat by using a colorimetric sensor combined with genetic algorithm partial least squares regression. *Meat Sci.* **2014**, *98*, 203–210. [CrossRef] [PubMed]
4. Hammes, F.; Berney, M.; Wang, Y.Y.; Vital, M.; Koster, O.; Egil, T. Flow-cytometric total bacterial cell counts as a descriptive microbiological parameter for drinking water treatment processes. *Water Res.* **2008**, *42*, 269–277. [CrossRef] [PubMed]
5. Li, Z.H.; Hu, X.T.; Shi, J.Y.; Zou, X.B.; Huang, X.W.; Zhou, X.C.; Haroon, E.T.; Mel, H.; Malcolm, P. Bacteria counting method based on polyaniline/bacteria thin film. *Biosens. Bioelectron.* **2016**, *81*, 75–79. [CrossRef] [PubMed]
6. Song, Y.X.; Zhang, H.P.; Chon, C.H.; Chen, S.; Pan, X.X.; LI, D.Q. Counting bacteria on a microfluidic chip. *Anal. Chim. Acta* **2010**, *681*, 82–86. [CrossRef] [PubMed]
7. Liu, G.; Van der Mark, E.J.; Verberk, J.Q.; Van Dijk, J.C. Flow cytometry total cell counts: A field study assessing microbiological water quality and growth in unchlorinated drinking water distribution systems. *BioMed Res. Int.* **2013**, *2013*. [CrossRef]
8. Czechowska, K.; Johnson, D.R.; van der Meer, J.R. Use of flow cytometric methods for single-cell analysis in environmental microbiology. *Curr. Opin. Microbiol.* **2008**, *11*, 205–212. [CrossRef]
9. Lepowsky, E.; Amin, R.; Tasoglu, S. Assessing the reusability of 3D-printed photopolymer microfluidic chips for urine processing. *Mciromachines* **2018**, *9*, 520. [CrossRef]
10. Chen, Y.C.; Nawaz, A.A.; Zhao, Y.H.; Huang, P.H.; McCoy, J.P.; Levine, S.J.; Wang, L.; Huang, T.J. Standing surface acoustic wave (SSAW)-based microfluidic cytometer. *Lab Chip* **2014**, *10*, 916–923. [CrossRef]
11. Steil, D.; Pohlentz, G.; Legros, N.; Mormann, M.; Mellmann, A.; Karch, H.; Müthing, J. Combining Mass Spectrometry, Surface Acoustic Wave Interaction Analysis, and Cell Viability Assays for Characterization of Shiga Toxin Subtypes of Pathogenic Escherichia coli Bacteria. *Anal. Chem.* **2018**, *90*, 8989–8997. [CrossRef] [PubMed]

12. Duarte, C.; Costa, T.; Carneiro, C.; Soares, R.; Jitariu, A.; Cardoso, S.; Piedade, M.; Bexiga, R.; Freitas, P. Semi-quantitative method for streptococci magnetic detection in raw milk. *Biosensors* **2016**, *6*, 19. [CrossRef] [PubMed]

13. Ghali, H.; Chibli, H.; Nadeau, J.; Bianucci, P.; Peter, Y.A. Real-time detection of staphylococcus aureus using whispering gallery mode optical microdisks. *Biosensors* **2016**, *6*, 20. [CrossRef] [PubMed]

14. Hu, L.; Ge, A.; Wang, X.; Wang, S.; Yue, X.; Wang, J.; Feng, X.; Du, W.; Liu, B.F. Real-time monitoring of immune responses under pathogen invasion and drug interference by integrated microfluidic device coupled with worm-based biosensor. *Biosens. Bioelectron.* **2018**, *110*, 233–238. [CrossRef] [PubMed]

15. Khan, S.M.; Misra, K.S.; Schwartz-Duval, S.A.; Daza, E.; Ostadhossein, F.; Bowman, M.; Jain, A.; Taylor, G.; McDonagh, D.; Labriola, T.L.; et al. Real-time monitoring of post-surgical and post-traumatic eye injuries using multilayered electrical biosensor chip. *Appl. Mater. Interfaces* **2017**, *9*, 8609–8622. [CrossRef] [PubMed]

16. Liu, L.; Xu, Y.; Cui, F.; Xia, Y.; Chen, L.; Mou, X.; Lv, J. Monitoring of bacteria biofilms forming process by in-situ impedimetric biosensor chip. *Biosens. Bioelectron.* **2018**, *112*, 86–92. [CrossRef]

17. Xu, D.; Huang, X.; Guo, J.; Ma, X. Automatic smartphone-based microfluidic biosensor system at the point of care. *Biosens. Bioelectron.* **2018**, *110*, 78–88. [CrossRef]

18. Saucedo, M.N.; Gao, Y.; Pham, T.; Mulchandani, A. Lectin- and saccharide-functionalized nano-chemiresistor arrays for detection and identification of pathogenic bacteria infection. *Biosensors* **2018**, *8*, 63. [CrossRef]

19. Zhang, N.; Ruan, Y.F.; Zhang, L.B.; Zhao, W.W.; Xu, J.J.; Chen, H.Y. Nanochannels photoelectrochemical biosensor. *Anal. Chem.* **2017**, *90*, 2341–2347. [CrossRef]

20. Khan, S.M.; Misra, S.K.; Dighe, K.; Wang, Z.; Schwartz-Duval, A.S.; Sar, D.; Pan, D. Electrically-receptive and thermally-responsive paper-based sensor chip for rapid detection of bacterial cells. *Biosens. Bioelectron.* **2018**, *110*, 132–140. [CrossRef]

21. Liebana, S.; Spricigo, A.D.; Cortes, M.P.; Barbe, J.; Llagostera, M.; Alegret, S.; Pividori, M.I. Phagomagnetci separation and electrochemical magneto-genosensing of pathogenic bacteria. *Anal. Chem.* **2013**, *85*, 3079–3086. [CrossRef] [PubMed]

22. Chistyako, V.A.; Prazdnova, E.V.; Mazanko, M.S.; Bren, A.B. The use of biosensors to explore the potential of probiotic strains to reduce the SOS response and mutagenesis in bacteria. *Biosensors* **2018**, *8*, 25. [CrossRef] [PubMed]

23. Le, D.Q.; Takai, M.; Suekuni, S.; Tokonami, S.; Nishino, T.; Shiigi, H.; Nagaoka, T. Development of an observation platform for bacterial activity using polypyrrole films doped with bacteria. *Anal. Chem.* **2015**, *87*, 4047–4052. [CrossRef] [PubMed]

24. Wang, R.; Lum, J.; Callaway, Z.; Lin, J.; Bottje, W.; Li, Y. A label-free impedance immunosensor using screen-printed interdigitated electrodes and magnetic nanobeads for the detection of *E. coli* O157:H7. *Biosensors* **2015**, *5*, 791–803. [CrossRef] [PubMed]

25. Ward, A.C.; Hannah, A.J.; Kendrick, S.L.; Tucker, N.P.; MacGregor, G.; Connolly, P. Identification and characterisation of *Staphylococcus aureus* on low cost screen printed carbon electrodes using impedance spectroscopy. *Biosens. Bioelectron.* **2018**, *110*, 65–70. [CrossRef] [PubMed]

26. Ma, F.; Rehman, A.; Sims, M.; Zeng, X. Antimicrobial susceptibility assays based on the quantification of bacterial lipopolysaccharides via a label free lectin biosensor. *Anal. Chem.* **2015**, *87*, 4385–4393. [CrossRef] [PubMed]

27. Zhang, D.; Bi, H.; Liu, B.; Qiao, L. Detection of pathogenic microorganisms by microfluidics based analytical methods. *Anal. Chem.* **2018**, *90*, 5512–5520. [CrossRef]

28. Thakur, B.; Zhou, G.; Chang, J.; Pu, H.; Jin, B.; Sui, X.; Yuan, X.; Yang, C.H.; Magruder, M.; Chen, J. Rapid detection of single *E. coli* bacteria using a graphene-based field-effect transistor device. *Biosens. Bioelectron.* **2018**, *110*, 16–22. [CrossRef]

29. Braiek, M.; Rokbani, K.B.; Chrouda, A.; Mrabet, B.; Bakhrouf, A.; Maaref, A.; Jaffrezic-Renault, N. An electrochemical immunosensor for detection of staphylococcus aureus bacteria based on immobilization of antibodies on self-assembled monolayers-functionalized gold electrode. *Biosensors* **2012**, *2*, 417–426. [CrossRef]

30. Miyajima, K.; Koshida, T.; Arakawa, T.; Kudo, H.; Saito, H.; Yano, K.; Mitsubayashi, K. Fiber-optic fluoroimmunoassay system with a flow-through cell for rapid on-site determination of *Escherichia coli* O157:H7 by monitoring fluorescence dynamics. *Biosensors* **2013**, *3*, 120–131. [CrossRef]

31. Xiao, N.; Wang, C.; Yu, C. A self-referencing detection of microorganisms using surface enhanced raman scattering nanoprobes in a test-in-a-tube platform. *Biosensors* **2013**, *3*, 312–326. [CrossRef] [PubMed]
32. Mosier-Boss, A.P. Review on SERS of bacteria. *Biosensors* **2017**, *7*, 51. [CrossRef] [PubMed]
33. Wang, Y.; Knoll, W.; Dostalek, J. Bacterial pathogen surface plasmon resonance biosensor advanced by long range surface plasmons and magnetic nanoparticle assays. *Anal. Chem.* **2012**, *84*, 8345–8350. [CrossRef] [PubMed]
34. Lyu, Y.C.; Ji, H.F.; Yang, S.J.; Huang, Z.Y.; Wang, B.L.; Li, H.Q. New C^4D sensor with a simulated inductor. *Sensors* **2018**, *16*, 165. [CrossRef] [PubMed]
35. Gao, L.D.; Li, X.H.; Fan, L.; Zheng, L.; Wu, M.M.; Zhang, S.X.; Huang, Q.L. Determination of inorganic cations and anions in chitooligosaccharides by ion chromatography with conductivity detection. *Mar. Drugs* **2017**, *15*, 51–59.
36. Gubartallah, E.A.; Makahleh, A.; Quirino, J.P.; Saad, B. Determination of biogenic amines in seawater using capillary electrophoresis with capacitively coupled contactless conductivity cetection. *Molecules* **2018**, *23*, 1112. [CrossRef] [PubMed]
37. Kaml, I.; Vcelakova, K.; Kenndler, E. Characterisation and identification of proteinaceous binding media (animal glues) from their amino acid profile by capillary zone electrophoresis. *J. Sep. Sci.* **2004**, *27*, 161–166. [CrossRef] [PubMed]
38. Kuban, P.; Hauser, P.C. Application of an external contactless conductivity detector for the analysis of beverages by microchip capillary electrophoresis. *Electrophoresis* **2005**, *26*, 3169–3178. [CrossRef]
39. Guijit, M.R.; Armstrong, P.J.; Candish, E.; Lefleur, V.; Percey, J.W.; Shabala, S.; Hauser, C.P.; Breadmore, C.M. Microfluidic chips for capillary electrophoresis with integrated electrodes for capacitively coupled conductivity detection based on printed circuit board technology. *Sens. Actuator B Chem.* **2011**, *159*, 307–313. [CrossRef]
40. Coltro, W.K.T.; Da Silva, J.A.F.; Carrilho, E. Rapid prototyping of polymeric electrophoresis microchips with integrated copper electrodes for contactless conductivity detection. *Anal. Methods* **2011**, *3*, 168–172. [CrossRef]
41. Vazquez, M.; Frankenfeld, C.; Coltro, W.K.T.; Carrilho, E.; Diamond, D.; Lunte, S.M. Dual contactless conductivity and amperometric detection on hybrid PDMS/glass electrophoresis microchips. *Analyst* **2010**, *1*, 96–103. [CrossRef]
42. Da Silva, J.A.F.; Guzman, N.; do Lago, L.C. Contactless conductivity detection for capillary electrophoresis Hardware improvements and optimization of the input-signal amplitude and frequency. *J. Chromatogr. A* **2002**, *942*, 249–258. [CrossRef]
43. Francisco, K.J.M.; do Lago, C.L. A compact and high-resolution version of a capacitively coupled contactless conductivity detector. *Electrophoresis* **2009**, *30*, 3458–3464. [CrossRef] [PubMed]
44. Jaanus, M.; Udal, A.; Kukk, V.; Umbleja, K.; Gorbatsova, J.; Molder, L. Improved C5D electronic realization of conductivity detector for capillary electrophoresis. *Electron. Elektrotech.* **2016**, *22*, 29–32. [CrossRef]
45. Emaminejad, S.; Javanmard, M.; Dutton, R.W.; Davis, R.W. Microfluidic diagnostic tool for the developing world: Contactless impedance flow cytometry. *Lab Chip* **2012**, *12*, 4499–4507. [CrossRef]
46. Sun, D.P.; Lu, J.; Chen, Z.G. Microfluidic contactless conductivity cytometer for electrical cell sensing and counting. *RSC Adv.* **2015**, *5*, 59306–59313. [CrossRef]
47. Yang, L. Electrical impedance spectroscopy for detection of bacterial cells in suspensions using interdigitated microelectrodes. *Talanta* **2004**, *74*, 1621–1629. [CrossRef]
48. Coates, J.D.; Ellis, D.J.; Blunt-Harris, E.L.; Gaw, G.V.; Roden, E.E.; Lovley, D.R. Recovery of humic-reducing bacteria from a diversity of environments. *Appl. Environ. Microbiol.* **1998**, *61*, 1504–1509.
49. Wang, H.Y.; Bhunia, A.K.; Lu, C. A microfluidic flow-through device for high throughput electrical lysis of bacterial cells based on continuous dc voltage. *Biosens. Bioelectron.* **2006**, *22*, 582–588. [CrossRef]
50. Islam, M.S.; Shahid, A.; Kuryllo, K.; Li, Y.F.; Deen, M.J.; Selvaganapathy, P.R. Electrophoretic concentration and electrical lysis of bacteria in a microfluidic device using a nanoporous membrane. *Micromachines* **2017**, *8*, 45. [CrossRef]
51. Hulsheger, H.; Potel, J.; Niemann, E.G. Electric field effects on bacteria and yeast cells. *Radiat. Environ. Biophys.* **1983**, *22*, 149–162. [CrossRef] [PubMed]

52. Khan, M.S.; Dosoky, N.S.; Mustafa, G.; Patel, D.; Berdiev, B.; Williams, J.D. Electrophysiology of epithelial sodium channel (ENaC) embedded in supported lipid bilayer using a single nanopore chip. *Langmuir* **2017**, *33*, 13680–13688. [CrossRef] [PubMed]

53. Zhu, Z.W.; Wang, Y.; Zhang, X.; Sun, C.F.; Li, M.G.; Yan, J.W.; Mao, B.W. Electrochemical impedance spectroscopy and atomic force microscopic studies of electrical and mechanical properties of nano-black lipid membranes and size dependence. *Langmuir* **2012**, *28*, 14739–14746. [CrossRef] [PubMed]

54. Khan, M.S.; Dosoky, N.S.; Patel, D.; Weimer, J.; Williams, J.D. Lipid bilayer membrane in a silicon based micron sized cavity accessed by atomic force microscopy and electrochemical impedance spectroscopy. *Biosensors* **2017**, *7*, 26. [CrossRef] [PubMed]

55. Shawki, M.M.; Gaballah, A. The effect of low AC electric field on bacterial cell death. *Rom. J. Biophys.* **2015**, *25*, 163–172.

56. Syed, M.S.; Rafeie, M.; Vandamme, D.; Asadnia, M.; Henderson, R.; Taylor, R.A.; Warkiani, M.E. Selective separation of microalgae cells using inertial microfluidics. *Bioresour. Technol.* **2018**, *252*, 91–99. [CrossRef]

57. Liu, W.; Wang, C.; Ding, H.; Shao, J.; Ding, Y. AC electric field induced dielectrophoretic assembly behavior of gold nanoparticles in a wide frequency range. *Appl. Surf. Sci.* **2016**, *370*, 184–192. [CrossRef]

58. Lapizco-Encinas, B.H.; Simmons, B.A.; Cummings, E.B.; Fintschenko, Y. Dielectrophoretic concentration and separation of live and dead bacteria in an array of insulators. *Anal. Chem.* **2004**, *76*, 1571–1579. [CrossRef]

59. Rojas, E.; Theriot, A.J.; Huang, K.C. Response of *Escherichia coli* growth rate to osmotic shock. *Proc. Natl. Acad. Sci. USA* **2014**, *111*, 7807–7812. [CrossRef]

60. Brito-Neto, J.G.A.; da Silva, J.A.F.; do Lago, L.; Blanes, C.L. Understanding Capacitively Coupled Contactless Conductivity Detection in Capillary and Microchip Electrophoresis. Part 1. Fundamentals. *Electroanalysis* **2005**, *17*, 1198–1206. [CrossRef]

micromachines

MDPI

Article

Microfluidic Device for Screening for Target Cell-Specific Binding Molecules by Using Adherent Cells

Maho Kaminaga [1,*], Tadashi Ishida [1], Tetsuya Kadonosono [2], Shinae Kizaka-Kondoh [2] and Toru Omata [1]

[1] Department of Mechanical Engineering, School of Engineering, Tokyo Institute of Technology, Kanagawa 226-8503, Japan; ishida.t.ai@m.titech.ac.jp (T.I.); omata.t.aa@m.titech.ac.jp (T.O.)
[2] Department of Life Science and Technology, School of Life Science and Technology, Tokyo Institute of Technology, Kanagawa 226-8503, Japan; tetsuyak@bio.titech.ac.jp (T.K.); skondoh@bio.titech.ac.jp (S.K.-K.)
* Correspondence: kaminaga.m.ab@m.titech.ac.jp; Tel.: +81-45-924-5468

Received: 13 December 2018; Accepted: 4 January 2019; Published: 9 January 2019

Abstract: This paper proposes a microfluidic device for screening molecules such as aptamers, antibodies, proteins, etc. for target cell-specific binding molecules. The discovery of cancer cell-specific binding molecules was the goal of this study. Its functions include filtering non-target cell-binding molecules, trapping molecules on the surface of target cells, washing away unbound molecules, and collecting target cell-specific binding molecules from target cells. These functions were effectively implemented by using our previously developed micro pillar arrays for cell homogeneous dispersion and pneumatic microvalves for tall microchannels. The device was also equipped with serially connected filter chambers in which non-target cells were cultured to reduce the molecules binding to non-target cells as much as possible. We evaluated the performance of the device using cancer cell lines (N87 cells as target cells and HeLa cells as non-target cells) and two fluorescent dye-labeled antibodies: Anti-human epidermal growth factor receptor 2 (anti-HER2) antibody that binds to target cells and anti-integrin antibody that binds to non-target cells. The results showed that the device could reduce anti-integrin antibodies to the detection limit of fluorescent measurement and collect anti-HER2 antibodies from the target cells.

Keywords: microfluidic device; target cell-specific binding molecules; screening; adherent cells; pneumatic microvalve; cell homogenous dispersion structure

---

1. Introduction

Anti-cancer drugs are widely used for cancer treatment. Conventional anti-cancer drugs not only damage cancer cells but also harm normal cells [1]. One approach for suppressing the damage caused to normal cells is to deliver combinations of cancer cell-specific binding molecules and anti-cancer drugs that act only on cancer cells [2]. Some cancer cell-specific binding molecules have been identified. For example, humanized anti-HER2 antibody (trastuzumab) [3] is used in clinical applications. Cancer cell-specific binding molecules such as cyclic arginine-glycine-aspartic acid tripeptide that specifically binds to malignant brain tumor cells in glioma [4], an aptamer that specifically binds to ovarian cancer cells [5], and a protein that binds to the protein disulfide isomerase, which is highly expressed on the surface of tumor cells [6] have been reported. However, few combinations of cancer cell-specific binding molecules and targets are known. The search for combinations is necessary and is performed by the screening of molecular libraries. To screen for cancer cell-specific binding molecules using cancer cells, conventional screening procedures involve the following steps: (Step 1) normal cells and the molecular library are mixed to filter out molecules that bind to normal cells; (Step 2) unbound molecules and target cancer cells are mixed to capture target cancer cell-specific binding molecules; (Step 3) washing of the target cancer cells; (Step 4) collecting bound molecules; and (Step 5) amplifying collected molecules. These amplified molecules are used in the next round of screening. Cancer cell-specific binding molecules are condensed by repeating steps 1 to 5. In addition to these complicated steps, this screening procedure requires precise manual operations, which are laborious and time-consuming. To conduct screening without human errors and decrease the screening time, automation with precise manipulation is required.

Microfluidic technology is suitable for automation owing to the following advantages: (1) Easy manipulation of liquid and cells owing to the dimensions as small as the size of the cells, (2) multiple processing capability on a single chip, (3) low sample and reagent consumption. Because of these advantages, several microfluidic screening devices have recently been developed. The microfluidic phage selection (MiPS) device can perform the screening using adherent cells [7]. Because this device does not filter the peptides that bind to normal cells, it collects not only cancer cell-specific binding peptides but also non-specific binding peptides. The cell-systematic evolution of ligands by exponential enrichment (Cell-SELEX) chip can isolate target cell-binding single strand deoxyribonucleic acids (ssDNAs) from a combinatorial ssDNA library [8]. The device uses magnetic microbeads attached to the cell surface and traps the beads, and consequently the attached cells, with a magnet. The chambers of normal cells and target cells are serially connected to search for target cell-specific binding ssDNA using a small quantity of ssDNA binding to normal cells in a single chip. Screening devices [9–11] also attach magnetic beads on cells to manipulate them under floating conditions. However, the molecules present on the surface of such cells may differ from those present under adherent conditions. Microfluidic devices developed in previous studies do not satisfy the following points: (1) Adherent culture conditions for adherent cells rather than floating condition and (2) filtering out the molecules that bind to non-target cells, which cause side effects.

Therefore, the purpose of this paper was to develop a microfluidic device for removing non-target cell-binding molecules to select target cell-specific binding molecules by using adherent cells in an adhered state. Our microfluidic device can perform steps 1 to 4 of the screening procedure in one chip. Another possible application of the microfluidic device is to detect changes in the expression of molecules depending on the malignancy of cancer cells. For example, the microfluidic device can introduce normal cancer cells as non-target cells and cancer stem cells as target cells. Additionally, if the microfluidic device introduces primary cells obtained from patients, cancer cell-specific binding molecules can be searched for every patient, leading to custom treatment.

For effective filtering, the cultured cells should be uniformly distributed. The cell chambers integrate our previously developed micro pillar arrays (MPAs) [12]. The MPA generates random flow caused by a repetitive cell clog-and-release process at the gaps between the micro pillars, resulting in a uniform distribution of cells. To perform steps 1 to 4 in the microfluidic device, microvalves that open and close the microchannels are necessary. The height of the microchannels is 50 μm, which is relatively tall and enables cells to pass through while minimizing physical interactions in the microchannels. A microvalve fabricated by a reflow process can open and close a microchannel of 50 μm in height. However, a reflow process is not compatible with the fabrication of the MPA, as it makes the shape of the micro pillars round. Therefore, the microfluidic device implements our previously developed pneumatic valves fabricated by using inclined lithography [13]. We evaluated the performances of the developed microfluidic device using known combinations of cells and antibodies, which was necessary to achieve the goal of this study of screening by using molecular libraries to discover cancer cell-specific binding molecules.

This paper is organized as follows: Section 2 describes the fabrication process, design, and operation of the microfluidic device, Section 3 describes evaluation of the performance of the filtering anti-integrin antibody that binds to non-target cells and the performance of collecting HER2 antibodies bound to target cells. Section 4 concludes this paper.

2. Materials and Methods

2.1. Microfluidic Device for Screening for Target Cell-Specific Binding Molecules

2.1.1. Fabrication Process of the Microfluidic Device

The microfluidic device consisted of three layers: A layer for liquid channels (liquid layer for short), a thin membrane, and a layer for pneumatic channels (pneumatic layer for short). The pneumatic and liquid channels were crossed at the position of the microvalves. All the layers of the microfluidic device were fabricated by soft lithography, and the fabricated layers were bonded to each other (Figure 1). The microchannels in the liquid layer had a parallelogram-shaped cross section, which could be obtained by inclined photolithography.

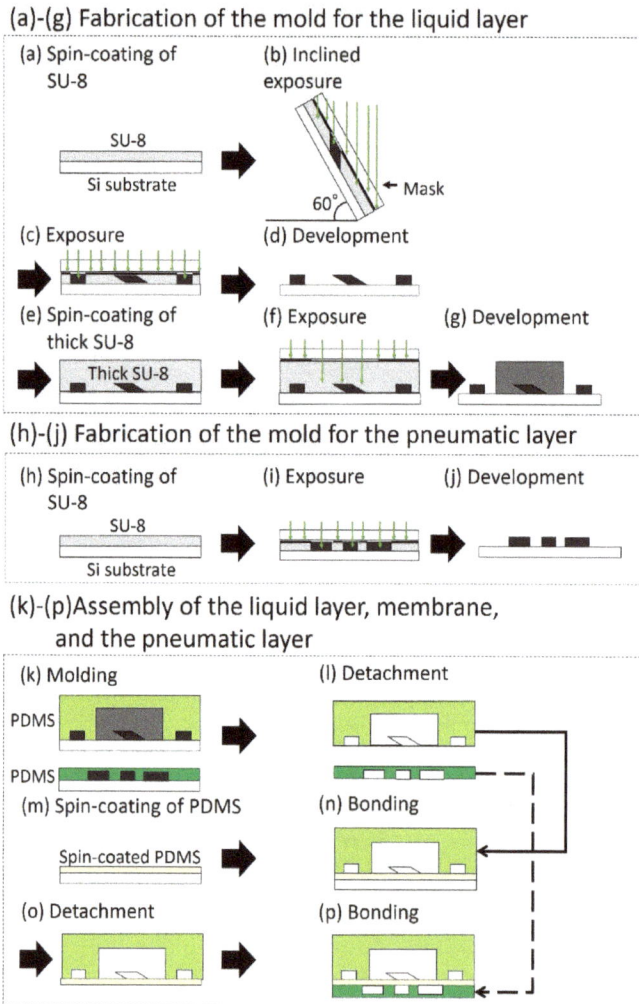

Figure 1. Fabrication process of the microfluidic device. (**a–g**) Fabrication of the mold for the liquid layer. (**h–j**) Fabrication of the mold for the pneumatic layer. (**k–p**) Assembly of the liquid layer, membrane, and the pneumatic layer.

The complete fabrication process is as follows: (a) SU-8 (SU-8 3025, Microchem, Westborough, MA, USA) was spin-coated on a Si substrate. (b) A photomask of the liquid channels was aligned on the Si substrate. The substrate with the photomask was inclined at 60 degrees and exposed to ultra violet (UV) light [13]. (c) A photomask of the MPAs was aligned on the substrate and exposed to UV light without inclining the substrate [12]. (d) The unexposed SU-8 was etched away by the development process, resulting in the mold of the liquid microchannel. (e) Thick SU-8 (SU-8 2150, Microchem) was spin-coated on the developed substrate. (f) A photomask of the microchambers was aligned and exposed to UV light without inclining the substrate. (g) After the development, the patterned thick SU-8 was obtained. (h–j) The mold for the pneumatic microchannels was fabricated by photolithography without inclining the substrate. (k) polydimethylsiloxane (PDMS; Silpot 184 W/C, Dow Corning Toray, Tokyo, Japan. Base polymer to curing agent ratio was 10:1 by weight.) was casted on the molds of the liquid and pneumatic layers. (l) The PDMS structures for the liquid and pneumatic layers were detached from their molds. (m) A thin membrane between liquid and pneumatic layers was fabricated by spin-coating PDMS on a flat Si substrate. (n) The PDMS structure for the liquid layer was bonded to the thin membrane on the flat Si substrate by the surface activation of vacuum UV irradiation. (o) The bonded PDMS structure was detached from the Si substrate. (p) The bonded PDMS structure was again bonded to the pneumatic layer. Figure 2 shows the fabricated microfluidic device. The liquid and pneumatic microchannels were filled with red and blue dyed water, respectively, to improve visibility.

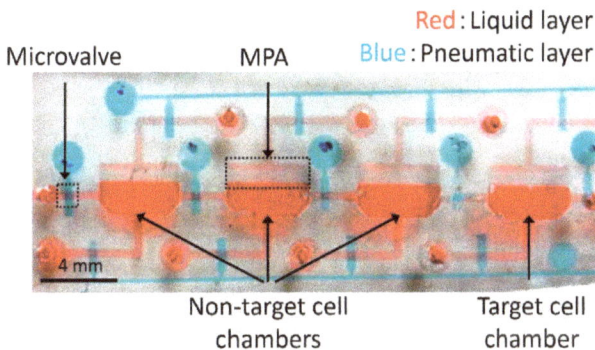

Figure 2. Photograph of the fabricated microfluidic device. The liquid layer was filled with red dye and the pneumatic layer blue dye.

2.1.2. Design of the Microfluidic Device

A schematic illustration of the microfluidic device is shown in Figure 3. It has chambers capable of adhesively culturing both non-target and target cells. The three chambers upstream that contain non-target cells capture non-target cell-binding molecules and then reduce contamination of non-target cell-binding molecules into the target-cell chamber. The reason for introducing three non-target cell chambers is described in Section 2.1.3.

For effective filtering, the cultured cells should be uniformly distributed. For this purpose, MPAs are equipped between the cell inlet and the cell chamber [12]. The chambers are gently connected by the microchannels between them to prevent molecules from remaining at the corners of the chamber. Pneumatic microvalves developed in [13] were used to open and close all the microchannels. Each pneumatic microvalve can be independently driven by applying compressed air separately.

Figure 4a shows the dimensions of the liquid channel in the microfluidic device. The length and width of the chamber excluding MPA were 5 mm and 2 mm, respectively. The height of the chambers was set to 200 μm to facilitate cell culture. To prevent the introduction of bubbles entering the liquid channels, the heights of the inlet and outlet were also set to 200 μm, whereas the liquid channels connected to the chamber was 500 μm in width and 50 μm in height for easy passage of the cells and closing of the microvalves [13]. The diameter and interval of the micro pillars were 100 μm and 5 μm, respectively. The thickness of the thin membrane was 40 μm. Figure 4b shows the dimensions of the pneumatic channel to control the microvalves.

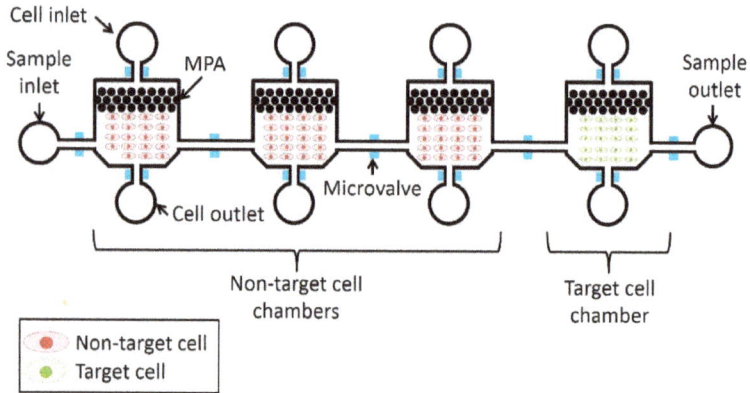

Figure 3. Schematic illustration of the microfluidic device.

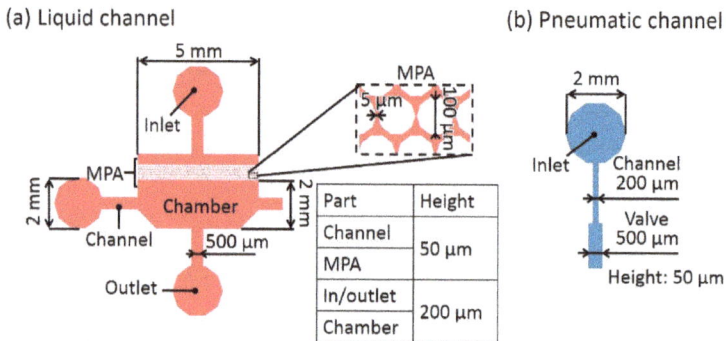

Figure 4. Dimensions of the microfluidic device. (**a**) Liquid channel. (**b**) Pneumatic channel.

2.1.3. Operation of the Microfluidic Device

Figure 5 shows the operation of the microfluidic device that performed the screening procedure from steps 1 to 4. Open and closed microvalves are shown in blue and red, respectively. (a) Cell Introduction: Non-target and target cells were introduced into the chambers from the cell inlets, (Figure 5a). In this step, microvalves at the cell inlets—V_{u1}, V_{u2}, V_{u3}, and V_{u4}—and microvalves at the cell outlets—V_{l1}, V_{l2}, V_{l3}, and V_{l4}—were open, and the microvalves between the cell chambers—V_{m1}, V_{m2}, V_{m3}, V_{m4}, and V_{m5}—were closed to prevent cross-contamination of cells. The chambers cultured the introduced cells, maintaining the microvalve condition. (b) Sample introduction: A molecular sample (or library in the case of screening) was introduced from the molecular sample inlet into the non-target cell chamber (Figure 5b). The microvalve condition becomes opposite; microvalves V_{u1}, V_{u2}, V_{u3}, V_{u4}, V_{l1}, V_{l2}, V_{l3}, and V_{l4} were closed to prevent the leakage of the molecular sample, and microvalves V_{m1}, V_{m2}, V_{m3}, V_{m4}, and V_{m5} were kept open. (c) Reaction and transportation of the molecular sample to the next chamber: Non-target cell-binding molecules were captured in each of the non-target cell chambers in a certain reaction time. During the reaction, microvalves V_{m1}, V_{m2}, V_{m3}, V_{m4}, and V_{m5} were closed. Introducing oil from the sample inlet transfers the sample to the next chamber. During the transportation, microvalves V_{m1}, V_{m2}, V_{m3}, V_{m4}, and V_{m5} were open and other microvalves were closed (Figure 5c). This step was repeated until the molecular sample arrived at the target cancer cell chamber. (d) Washing: The unbound molecules were washed by introducing a washing buffer from the cell inlet of the target cell chamber (Figure 5d). In this step, microvalves V_{u4}, V_{l4}, V_{m1}, V_{m2}, V_{m3}, V_{m4}, and V_{m5} were open and the others were closed. (e) Collection of target cell-specific binding molecules: The specific molecules that bound to target cells were collected by introducing a collection buffer (Figure 5e). Only microvalves V_{u4} and V_{l4} were open and the others were closed to prevent contamination of molecules from one chamber to another.

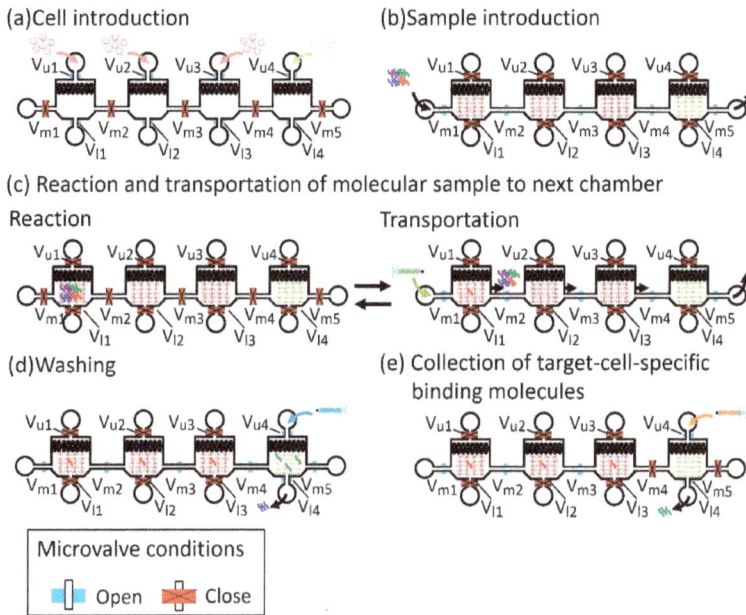

Figure 5. Operation of the microfluidic device to perform the screening procedure from step 1 to step 4. (**a**) Cell introduction. (**b**) Sample introduction. (**c**) Reaction and transportation of molecular sample to the next chamber. (**d**) Washing. (**e**) Collection of target-cell-specific binding molecules.

Note: While introducing a sample solution, it pushes out the culture medium filling the chamber. If the chamber volume is larger than the sample volume, the remaining culture medium in the chamber dilutes the sample solution. Therefore, the chamber volume must be the same as the sample volume. The number of non-target cell chambers (not its volume) is the factor that can be altered to improve the filtering performance. In this study, the number of non-target cell chambers was three. This was because a microfluidic device with only one non-target cell chamber did not achieve adequate performance of filtering in our preliminary experiments. Furthermore, when the number of the chambers exceeded four, the reaction time also exceeded by 10 h, which was too long for one cycle of screening.

2.1.4. Cells

The human gastric carcinoma cell line N87 and human cervical cancer cell line HeLa were purchased from the American Type Culture Collection (Manassas, VA, USA). N87 cells were used as target cells and HeLa cells were used as non-target cells (Table 1).

Table 1. Cell lines used in the experiments.

	Name of Cell Lines	Expressing Surface Molecules	
Target cells	N87 cell lines	HER2 (Target molecules)	Integrin $\alpha_v\beta_5$
Non-target cells	HeLa cell lines	-	Integrin $\alpha_v\beta_5$

N87 cells express HER2 (target molecules) and HeLa cells do not, while both the cell lines express integrin. Both the cell lines were cultured in Dulbecco's Modified Eagle's Medium (DMEM; D-MEM (High Glucose) with L-Glutamine, Phenol Red and Sodium Pyruvate, Wako, Osaka, Japan) supplemented with 10% fetal bovine serum (FBS; Fetal Bovine Serum regular, Wako) and 100 UI/mL penicillin–streptomycin (Penicillin-Streptomycin solution ($\times$100), Wako). They were incubated at 37 °C under 5% CO_2 condition. Cell suspensions were made by detaching cells from the cell culture dish, and mixing with DMEM. Both N87 and HeLa cells were detached using trypsin solution (0.25 w/v% Trypsin Solution with Phenol Red, wako) at room temperature for 10 min or 2 min, respectively.

Figure 6 shows the HeLa and N87 cells cultured inside the chambers. The N87 cell suspension (1×10^4 cells/μL) was introduced into the target cell chamber at 2 μL/min for 1 min, and incubated for 6 h at 37 °C under 5% CO_2 condition. The HeLa cell suspension (1×10^4 cells/μL) was introduced into the non-target cell chambers at 2 μL/min for 1 min, and incubated for 6 h at 37 °C under 5% CO_2 conditions. Owing to the MPA, both HeLa and N87 cells spread to the full width of the chambers. Besides, cell morphology in the chambers indicated that there was no contamination of the cells due to the pneumatic microvalve between the chambers. Video S1 shows that cells did not flow into the adjacent chamber by closing the valve. After washing the chambers with phosphate buffer salts (PBS, Wako) at 10 μL/min for 10 min, the cells were fixed by using the 4% paraformaldehyde solution (4%-Paraformaldehyde Phosphate Buffer Solution, Nacalai Tesque, Kyoto, Japan) (20 °C for 15 min). In this study, fixation was necessary because the antigen-antibody binding took a long time, which made the cells in the chambers inactive or mortal. After the fixation, blocking was performed using 5% fetal bovine serum -PBS (20 °C for 60 min).

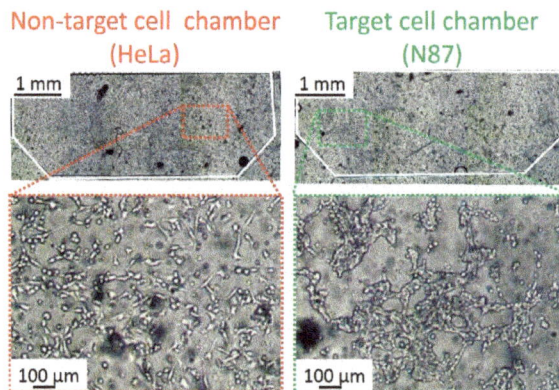

Figure 6. Micrographs of the HeLa and N87 cells cultured in the cell chambers. The HeLa and N87 cells were separately cultured in non-target and target cells without cross contamination.

2.1.5. Reagent

Fluorescent dye-labeled antibodies were used as binding molecules. Anti-HER2 antibody conjugated to AF 488(Alexa Fluor® 488 anti-human CD340 (erbB2/HER2) Antibodies, BioLegend, San Diego, CA, USA) was used as an antibody that specifically binds to target cells (abbreviated as target-specific Ab). AF 488 dye has 490 nm in excitation wavelength and 525 nm in emission wavelength

(Green fluorescence). Anti-integrin antibodies conjugated to AF 555 (Anti-integrin $\alpha_v \beta_5$ Antibody, Alexa Fluor® 555 Conjugated, Bioss, Woburn, MA, USA) was used as an antibody that binds to non-target cells (abbreviated as non-specific Ab). AF 555 dye has 555 nm in excitation wavelength and 580 nm in fluorescent wavelength (Red fluorescence) (Table 2). The concentration of the non-specific antibody (Ab) solution was 10 µg/mL, and that of target-specific Ab was 4 µg/mL. The mixture showed similar fluorescence intensity. The reaction time for both Abs was 2 h at 37 °C. Canola oil (cooking canola oil, Ajinomoto, Tokyo, Japan) was used for sample transportation. It contained 64.0% of oleic acid, 20.6% of linoleic acid, 9.5% of alpha-linolenic acid, 4.1% of palmitic acid, and 1.8% of stearic acid. The mixture of 1% Sodium Dodecyl Sulfate (SDS; Sodium Dodecyl Sulfate, Wako), 0.1% Tween 20 (Polyoxyethylene Sorbitan Monolaurate, TCI, Tokyo, Japan), and PBS was used as a collection buffer, which was similar to the collection buffer described in the T7 phage screening protocol [14].

Table 2. Antibodies used in the experiments.

	Name of Molecules	Binding Cells		Fluorescent Dye
Target cell-specific	Anti-HER2 antibody	N87 cells	-	AF 488
binding molecules	(Target-specific Ab)	(Target cells)		(Green fluorescence)
Non-target cell-binding	Anti-inegrin antibody	N87 cells	HeLa cells	AF 555
molecules	(Non-specific Ab)	(Target cells)	(Non-target cells)	(Red fluorescence)

2.2. Experimental Procedure

2.2.1. Experimental Setup

The experiments in this paper were conducted under a fluorescence microscope (IX-83, Olympus, Tokyo, Japan). The fluorescence intensities of the cells in the chambers were measured using a fluorescence microscope. Filters U-FGWA (Olympus) and U-FBNA (Olympus) were used for red and green fluorescence imaging, respectively. The fluorescence intensity of a solution was measured using a flourometer (Infinite F500 microplate reader, Tecan, Männedorf, Switzerland) with Ex/Em filters (485 ± 20 nm/ 535 ± 25 nm for AF488 or 535 ± 25 nm/ 590 ± 20 nm for AF555). A syringe pump (KDS-210, KD scientific, Holliston, MA, USA) was connected to the inlets of the microfluidic device to introduce cells, fluorescent dye-labeled antibodies, culture medium, and PBS. A pneumatic pressure source (OFP-07005, Iwata, Kanagawa, Japan) was connected to the inlets of pneumatic channels of the microfluidic device via a solenoid valve array (SY114-5LZ, SMC, Tokyo, Japan) and a regulator (IR1020-01BG-A, SMC) to switch the microvalves.

2.2.2. Filtering Non-Specific antibodies (Abs)

The performance of the filtering, which removes non-specific Abs, was examined. We prepared four microfluidic devices (devices (a), (b), (c) and (d)) of three different types as shown in Figure 7: (type A) Three blank chambers and one target cell chamber; (type B) two blank chambers, one non-target cell chamber and one target cell chamber; (type C) three non-target cell chambers and one target cell chamber. The chambers of each microfluidic device were numbered 1, 2, 3, and 4 on the upstream side.

The performance of the filtering can be assessed by the amount of non-specific Abs bound to the target cancer cells. The mixture of the fluorescent dye-labeled target-specific Ab and non-specific Ab solutions were introduced to devices (a), (b) and (c) at 2 µL/min for 1 min in the same operations. As the number of non-target cell chambers increased, it was expected that they filtered more non-specific Abs and red fluorescence intensity decreased in the target cell chamber. The ratio of red to green fluorescence intensities per unit area from target cells was used to evaluate the performance of the filtering. For the autofluorescence measurement, only target-specific Ab solution was introduced to type A device (d) at 2 µL/min for 1 min. The temperature of the chambers was maintained at 37 °C for

2 h. Introducing canola oil from the inlet at 2 μL/min for 1 min transported the solution to the next chamber. This operation was repeated until the mixture or solution reached chamber 4. The mixture or solution was kept in chamber 4 for 2 h. Then, chamber 4 was washed off using PBS at 100 μL/min for 10 min. After washing chamber 4, the fluorescence image of chamber 4 was observed using the fluorescence microscope.

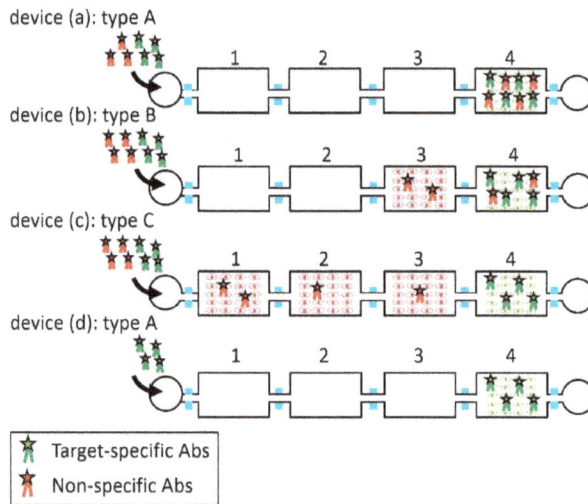

Figure 7. Four microfluidic devices with three types for the experiment of filtering the non-specific antibodies (Abs). The mixture of the fluorescent dye-labeled target-specific Ab and non-specific Ab solutions was introduced to the devices. (**a**) Type A: Three blank and one target cell chambers. (**b**) Type B: Two blank, one non-target cell and one target cell chambers. (**c**) Type C: Three non-target and one target cell chambers. (**d**) Type A: Only target-specific Ab solution was introduced for autofluorescence measurement.

2.2.3. Collecting Target-Specific Antibodies (Abs)

The target-specific Abs on the surface of the target cells need to be collected for amplification or identification for screening. To detach the target-specific Abs from the target cells, a collection buffer was introduced into the target cell chamber. To evaluate the collection method, the fluorescence intensities of the target cells and the collected solution were measured. The detachment of the target-specific Abs on the target cells was measured by comparing the fluorescence intensity before the detachment with that after the detachment. It was expected that the fluorescence intensity of the target cells decreased and that of the collected solution increased.

We prepared four microfluidic devices for this experiment and used only the target cell chamber of each device (chamber 4). Three of them were used to find the sufficient reaction time and one was used as a control. A target-specific Ab solution was introduced into the target cell chamber (4 μg/mL) at 2 μL/min for 1 min and incubated for 2 h at 37 °C. Subsequently, unbound antibodies were washed by PBS at 100 μL/min for 10 min. The target cell chamber was filled with the collection buffer (PBS for the control device) at 2 μL/min for 1 min. The reaction times were 15 min, 30 min, and 60 min, respectively, at 20 °C to find the sufficient reaction time. After the reaction, the buffer in the target cell chamber was collected from the cell outlet by PBS at 100 μL/min for 5 min.

3. Results and Discussion

3.1. Results of Filtering Non-Specific Antibodies (Abs)

Figure 8 shows the results of the experiment of filtering non-specific Abs described in Section 2.2. Figure 8a shows the bright-field and fluorescence images of the target cells in the target cell chambers of devices (a), (b), (c) and (d) and Figure 8b shows the ratio of red and green fluorescence intensities of the target cells in intensity per unit area. The result of the autofluorescence measured in device (d) was a control.

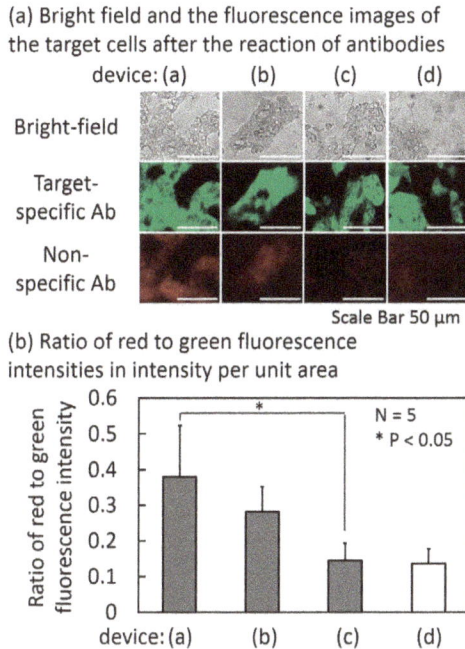

Figure 8. Bright field and the fluorescence images of the target cells (N87 cells) after the reaction of antibodies (**a**) and the ratio of red to green fluorescence intensities in intensity per unit area (**b**).

In device (a), which had no non-target-cell chambers, the average red to green ratio ($N = 5$) was 0.38. The ratio decreased as the number of the non-target cell chambers increased. In the cases of one non-target cell chamber and three non-target cell chambers (devices (b) and (c), respectively), the average red to green ratios were 0.28 and 0.14, respectively. The reduction ratio in the case of three non-target cell chambers reached 63% in comparison with no non-target cell chamber. The red to green ratio was comparable to that of the autofluorescence measurement in device (d). The result suggests that the device was able to reduce the non-specific Abs to the detection limit of the fluorescent measurement (smaller than 2 µg/mL as shown in Figure S1) when the number of the non-target cell chambers was three. However, the non-specific Abs may not have been completely removed because fluorescence weaker than autofluorescence cannot be measured. Because the cells were fixed with 4% paraformaldehyde solution, non-target cells did not float and drift into the next chamber during sample transportation as shown in Video S2.

The MPAs will influence the binding procedure of molecules in the future study of screening. Because the oil pushed the sample solution from the sample inlet, the sample around the micro pillars remained between the cell inlet and cell culture area in the chamber without flowing into the next chamber. Partial loss of the sample molecules did not affect the results of this paper because known antibodies were used in the experiments. However, this sample transportation method should be improved for screening using a molecular library. By introducing oil from the cell inlet, the remaining sample around the micro pillars can be pushed out.

3.2. Results of Collecting Target-Specific Antibodies (Abs)

The fluorescence images of the target cells before and after the reaction with the collection buffer are shown in Figure 9a. The reaction time was 60 min. The average fluorescence intensity over the cell region was calculated for each fluorescence image and the ratio of that after the reaction to that before the reaction (denoted as R) was calculated. Figure 9b shows the results. When only PBS was introduced instead of the collection buffer, R was 1.06. When the collection buffer filled the target cell chamber for 15 min, 30 min and 60 min for reaction, R was 0.35, 0.26, and 0.18, respectively. According to a student's test, R significantly decreased in all reaction times between the target cells and the collection buffer, compared with the case where only PBS was introduced.

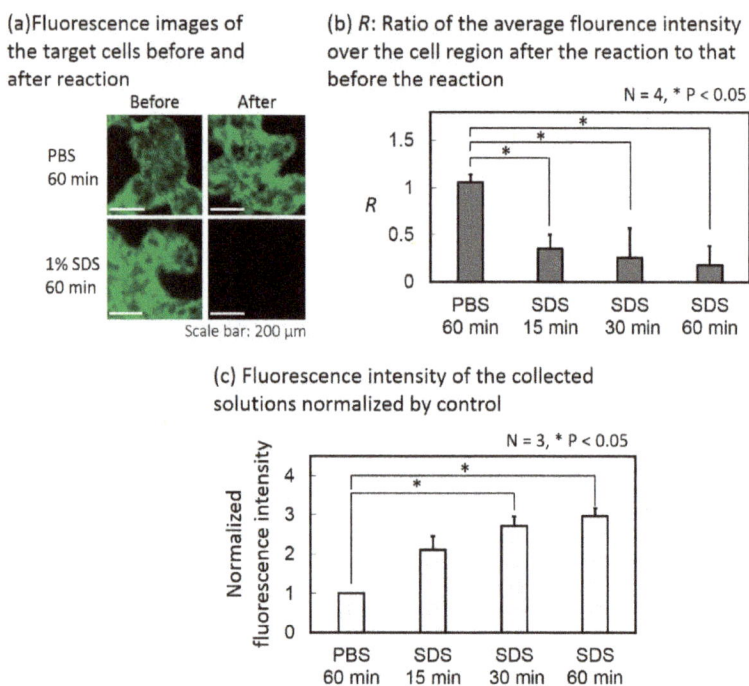

(a) Fluorescence images of the target cells before and after reaction

(b) R: Ratio of the average flourence intensity over the cell region after the reaction to that before the reaction

(c) Fluorescence intensity of the collected solutions normalized by control

Figure 9. Collection of the target-specific antibodies (Abs) attached on the target cells. (**a**) Fluorescence images of target cells before and after the reaction. (**b**) Ratio of the average fluorescence intensity over the cell region after the reaction to that before the reaction. (**c**) Fluorescence intensities of the collected solutions normalized by control.

The fluorescence intensity of the solution collected from each device was measured and normalized by that of the control device. Figure 9c shows the results. When the collection buffer filled the target cell chamber for 15 min, 30 min and 60 min for reaction, respectively, the average fluorescence intensities of the collected solutions ($N = 3$) were 2.1, 2.7, and 3.0 times higher than the case where only PBS was used instead of the collection buffer. According to a Student's T-test, the fluorescence intensity increased significantly when the reaction times were 30 min and 60 min compared with the case where only PBS was introduced.

From Figure 9b,c, the fluorescence intensity measured from the collected solution increased when the fluorescence intensity of the target cells decreased, as the reaction time increased. Therefore, the target-specific Abs bound on the surface of the target cells were successfully detached and collected.

3.3. Future Prospects on the Development of the Device

The microfluidic device proposed here carries out molecular selection using two-dimensionally cultured cells. However, differences between in vitro and in vivo environments are a general problem in the screening of target cell-specific binding molecules. For clinical applications, the screening results must be very precise. Investigation of cell culture methods is necessary for improving precision.

In recent years, many techniques for three-dimensional cell culture have been developed. It is said that such cell models are closer to the in vivo environment. More recently, a mesh cell culture method was proposed [15], which used a mesh for a scaffold. According to [15], it can reduce cell-substrate adhesion and promote cell-cell adhesion, thus the cell model is closer to the in vivo environment. This device can easily incorporate the mesh cell culture method by modifying the substrates of the chambers.

4. Conclusions

We developed a microfluidic device to screen for molecules that specifically bind to target cells by filtering nonspecifically binding molecules. By using the pneumatic microvalves, the microfluidic device was able to perform the following functions: Filtering of non-target cell-binding molecules, trapping of molecules on the surface of target cells, washing of unbinding molecules, and collecting target cell-specific binding molecules from the target cells. We evaluated the performance of the microfluidic device using cancer cell lines and fluorescent dye-labeled antibodies. The results showed that the microfluidic device could reduce the antibodies that bound to non-target cells to the detection limit of fluorescent measurement and could collect antibodies that specifically bound to target cells. To perform the screening in the microfluidic device, conditions such as the blocking conditions, reaction time, and collection buffer must be optimized. In future studies, we will conduct screening experiments using molecular libraries to identify effective cancer cell-specific binding molecules.

Supplementary Materials: The following are available online at http://www.mdpi.com/2072-666X/10/1/41/s1, Figure S1: Relationship between concentration of non-specific antibodies and mean fluorescence intensity of the cell area normalized by autofluorescence of the target cells. Video S1: Non-target cell chamber in the cell introduction. Cell suspension was introduced into the chamber from the cell inlet. Because the valves between the chambers were closed, the cells introduced from the cell inlet did not flow into the next chamber, but rather into the cell outlet. Video S2: Non-target cell chambers during sample transportation. Because the cells were fixed with 4% paraformaldehyde solution, non-target cells did not float and drift into the next chamber. The video is at 4 times speed.

Author Contributions: M.K., T.I., T.K. conceived, designed, performed the experiments, analyzed the data, M.K., T.I, T.O. wrote the paper, S.K.-K., T.O. supervised the research.

Funding: This research was funded by the Japan Society for the Promotion of Science (grant number KAKENHI 16H02320).

Conflicts of Interest: The authors declare no conflict of interest.

References

1. DeVita, V.T., Jr.; Chu, E. A History of Cancer Chemotherapy. *Cancer Res.* **2008**, *68*, 8643–8653. [CrossRef] [PubMed]
2. Allen, T.M.; Cullis, P.R. Drug Delivery Systems: Entering the Mainstream. *Science* **2004**, *303*, 1818–1822. [CrossRef] [PubMed]
3. Boekhout, A.H.; Beijnen, J.H.; Schellens, J.H. Trastuzumab. *Oncologist* **2011**, *16*, 800–810. [CrossRef] [PubMed]
4. Danhier, F.; Breton, A.L.; Préat, V. RGD-Based Strategies to Target Alpha(v) Beta (3) Integrin in Cancer Therapy and Diagnosis. *Mol. Pharmaceutics* **2012**, *9*, 2916–2973. [CrossRef] [PubMed]

5. Henri, J.L.; Macdonald, J.; Strom, M.; Duan, W.; Shigdar, S. Aptamers as potential therapeutic agents for ovarian cancer. *Biochimie* **2018**, *145*, 34–44. [CrossRef] [PubMed]

6. Ishima, Y.; Chen, D.; Fang, J.; Maeda, H.; Minomo, A.; Kragh-Hansen, U.; Kai, T.; Maruyama, T.; Otagiri, M. S-Nitrosated Human Serum Albumin Dimer is not only a Novel Anti-Tumor Drug but also a Potentiator for Anti-Tumor Drugs with Augmented EPR Effects. *Bioconjugate Chem.* **2012**, *23*, 264–271. [CrossRef] [PubMed]

7. Wanga, J.; Liu, Y.; Teesalu, T.; Sugahara, K.N.; Kotamrajua, V.R.; Adams, J.D.; Ferguson, B.S.; Gong, Q.; Oh, S.S.; Csordas, A.T.; et al. Selection of phage-displayed peptides on live adherent cells in microfluidic channels. *Proc. Natl. Acad. Sci.* **2011**, *108*, 6909–6914. [CrossRef] [PubMed]

8. Hung, L.-Y.; Wang, C.-H.; Hsu, K.-F.; Chou, C.-Y.; Lee, G.-B. An on-chip Cell-SELEX process for automatic selection of high-affinity aptamers specific to different histologically classified ovarian cancer cells. *Lab Chip* **2014**, *14*, 4017–4028. [CrossRef] [PubMed]

9. Che, Y.-J.; Wu, H.-W.; Hung, L.-Y.; Liu, C.-A.; Chang, H.-Y.; Wang, K.; Lee, G.-B. An integrated microfluidic system for screening of phage-displayed peptides specific to colon cancer cells and colon cancer stem cells. *Biomicrofluidics* **2015**, *9*, 054121. [CrossRef] [PubMed]

10. Weng, C.-H.; Hsieh, I.-S.; Hung, L.-Y.; Lin, H.-I.; Shiesh, S.-C.; Chen, Y.-L.; Lee, G.-B. An automatic microfluidic system for rapid screening of cancer stem-like cell-specific aptamers. *Microfluid. Nanofluid.* **2013**, *14*, 753–765. [CrossRef]

11. Wang, C.-H.; Weng, C.-H.; Che, Y.-J.; Wang, K.; Lee, G.-B. Cancer Cell-Specific Oligopeptides Selected by an Integrated Microfluidic System from a Phage Display Library for Ovarian Cancer Diagnosis. *Theranostics* **2015**, *5*, 431–442. [CrossRef] [PubMed]

12. Kaminaga, M.; Ishida, T.; Kadonosono, T.; Kizaka-Kondoh, S.; Omata, T. Uniform Cell Distribution Achieved by Using Cell Deformation in a Micropillar Array. *Micromachines* **2015**, *6*, 409–422. [CrossRef]

13. Kaminaga, M.; Ishida, T.; Omata, T. Fabrication of Pneumatic Microvalve for Tall Microchannel Using Inclined Lithography. *Micromachines* **2016**, *7*, 224. [CrossRef] [PubMed]

14. Nowak, J.E.; Chatterjee, M.; Mohapatra, S.; Dryde, S.C.; Tainsky, M.A. Direct production and purification of T7 phage display cloned proteins selected and analyzed on microarrays. *BioTechniques* **2006**, *40*, 220–227. [CrossRef] [PubMed]

15. Hori, T.; Kurosawa, O. A Three-dimensional Cell Culture Method with a Micromesh Sheet and Its Application to Hepatic Cells. *Tissue Eng. Part C: Methods* **2018**, *24*. [CrossRef]

micromachines

MDPI

Article

In Situ Analysis of Interactions between Fibroblast and Tumor Cells for Drug Assays with Microfluidic Non-Contact Co-Culture

Hongmei Chen [1,*], Wenting Liu [2,3], Bin Wang [1,*] and Zhifeng Zhang [4,*]

[1] School of Mathematics and Physics of Science and Engineering, Anhui University of Technology, Maanshan 243002, China
[2] Division of Nanobionic Research, Suzhou Institute of Nano-Tech and Nano-Bionics, Chinese Academy of Sciences (CAS), Suzhou 215123, China; wentingliu@sinano.ac.cn
[3] University of the Chinese Academy of Sciences, Beijing 100049, China
[4] Department of Engineering Science and Mechanics, The Pennsylvania State University, State College, PA 16802, USA
* Correspondence: hongmeichen@semi.ac.cn (H.C.); bin2013@ahut.edu.cn (B.W.); alex.zf.zhang@gmail.com (Z.Z.)

Received: 5 November 2018; Accepted: 11 December 2018; Published: 17 December 2018

Abstract: Fibroblasts have significant involvement in cancer progression and are an important therapeutic target for cancer. Here, we present a microfluidic non-contact co-culture device to analyze interactions between tumor cells and fibroblasts. Further, we investigate myofibroblast behaviors induced by lung tumor cells as responses to gallic acid and baicalein. Human lung fibroblast (HLF) and lung cancer cell line (A549) cells were introduced into neighboring, separated regions by well-controlled laminar flows. The phenotypic behavior and secretion activity of the tumor cells indicate that fibroblasts could become activated through paracrine signaling to create a supportive microenvironment for cancer cells when HLF is co-cultured with A549. Furthermore, both gallic acid (GA) and baicalein (BAE) could inhibit the activation of fibroblasts. In situ analysis of various cell communications via the paracrine pathway could be realizable in this contactless co-culture single device. This device facilitates a better understanding of interactions between heterotypic cells, thus exploring the mechanism of cancer, and performs anti-invasion drug assays in a relatively complex microenvironment.

Keywords: laminar flows; paracrine signaling; co-culture

1. Introduction

Cancer progression has an affinity relationship with the tumor microenvironment, including the extracellular matrix (ECM), fibroblasts, immune cells, and endothelial cells, and the blood vessels and proteins produced [1,2]. Fibroblasts synthesize components of the extracellular matrix (ECM), thus forming the structural framework of the tumor microenvironment. Fibroblasts are normally quiescent. Under some conditions, fibroblasts could become activated and acquire an activated phenotype. Activated fibroblasts located adjacent to cancer cells as carcinoma-associated fibroblasts (CAFs) could support tumor epithelial growth and invasion [3]. On one side, normal fibroblasts have mostly been thought to play a more passive role in cancer. On the other side, some researchers compared the effect of normal fibroblasts with that of CAFs on tumor cells and demonstrated that the former inhibit cancer progression [4,5]. Other researchers found that normal fibroblasts could induce tumor growth [6–8]; further, quiescent fibroblasts can become activated and might be key regulators of paracrine signaling during cancer progression [9]. Therefore, it is necessary to understand the mechanism between normal fibroblasts and tumors for personalized, targeted anticancer therapies.

The traditional co-culture platforms for studying cell–cell communications are limited to culture dishes and Transwell assays, which cannot avoid the drawbacks of complicated manual operations and large consumption of reagents. Microfluidic chips can integrate various experimental operations and mimic an in vivo microenvironment [10]. Over the past decade, considerable progress has been made in culturing cells with microfluidic chips [11,12]. Xie et al. reported a microchip that achieved co-culture and made a "wound" by removing a narrow barrier after cell seeding [13]. This device could co-culture different types of cells for cell migration assay. However, physical removal of a cell monolayer could damage cells, thus affecting the experimental result of cell migration. Businaro et al. realized an on-chip model consisting of two center end-closed channels to investigate the interactions between cancer and the immune system [14]. This model elucidated that the reciprocal interactions were heterogeneous; however, the model was difficult to process. Besides this, typical cell–cell communication can be classified into direct and indirect contact modes [15]. To date, much research has been performed on the direct mode, but focus on the indirect mode has been rare. It is not clear whether cells affect each other through cell–cell contacts or paracrine signals in the direct mode, but cells could only communicate through diffusible signals secreted in the indirect.

To overcome these limitations, we developed a microfluidic device with a novel technique, well-controlled laminar flow, allowing two types of cells to be in non-contact and only experience paracrine interactions with limited amounts of reagent consumption. A similar design has been presented previously [16,17]. Cellular responses can be observed immediately without physical damage and distinguished after fluorescent labeling. With this device, we are able to investigate interactions between tumor cells and normal fibroblasts and analyze the function of antitumor drugs with different concentrations in tumor-induced fibroblasts. This rarely reported co-culture mode provides a cost-effective approach to accomplish multiple functions, including cell loading with passive pumping, heterogeneous cell compartmentalization, and an accurate and reliable cellular assay directly monitored in real time. The precise and well-controlled laminar flow makes it a perfect non-contact co-culture platform. Moreover, multiple types of cells can be contactlessly co-cultured by designing several entrance channels, which enables us to mimic complicated microenvironments in vivo. Thus, it provides a compartmentalized co-culture model to elucidate reciprocal interactions between heterogeneous cell types within tumors, posing a relevant impact on antitumor therapeutic strategies.

2. Materials and Methods

2.1. Design and Fabrication of the Microfluidic Co-Culture Device

The microfluidic device was composed of a polydimethylsiloxane (PDMS, Silgard 184, Dow Corning, Midland, MI, USA) layer and a glass substrate. The upper layer, in Figure 1a, which contained three inlet channels (5 mm × 300 μm × 100 μm) that converged into a main channel (5 mm × 900 μm × 100 μm), was fabricated using PDMS following a well-established replica molding process. A mold with micrometer-sized structures was prepared by standard soft lithography methods. A mixture of PDMS followed the manufacturer's instruction was poured onto the mold at 85 °C for 45 min. After cooling, the cured PDMS layer with the desired structures was gently peeled off from the mold and punched to form inlets and an outlet. Then, PDMS was bonded to the glass slide after oxygen plasma treatment for 50 s.

Before use, the device was sterilized with 70% (*v/v*) ethanol for 1 h and exposed to UV light for 30 min.

2.2. Formation of Laminar Flow in Microchannels

As shown in Figure 1b, samples containing different kinds of cells were placed into the three inlets. Continuous flow was obtained due to gravitational forces created by pressure differences from a fluid level discrepancy between the inlets and outlet. In order to confirm laminar flow formation inside the main microchannel, rhodamine B (RB), PBS solution, and blue ink (BI) were injected into the different inlets. A paper tip was placed at the outlet to remove the waste and continuous laminar flow of all solutions was maintained toward the outlet. The laminar flow was observed using an inverted microscope (Nikon Ti-Eclipse, Tokyo, Japan).

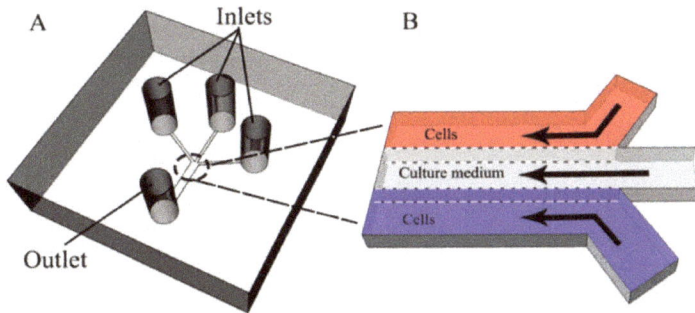

Figure 1. Schematic of the microfluidic platform employed for cell co-culture. (**A**) Diagram of the microfluidic device with three branching microchannels and one main microchannel. (**B**) Three parallel flows underway on the microfluidic platform.

2.3. Cell Culture

A549 cells were purchased from ATCC, and human lung fibroblast (HLF) cells were kindly offered by Dr. Yimin Zhu (Suzhou Institute of Nano-Tech and Nano-Bionics, CAS, Beijing, China). Cells were cultured in RPMI-1640 (Hyclone Corp., San Angelo, TX, USA) supplemented with 10% (*v/v*) fetal bovine serum (FBS; Hyclone Corp., San Angelo, TX, USA), 100 U/mL penicillin, and 100 U/mL streptomycin. A culture incubator with a humidified atmosphere of 5% CO_2 and at 37 °C was used for cell culture.

2.4. Co-Culture of A549 and HLF on the Microfluidic Device

Cells were harvested using 0.25% trypsin/EDTA (Hyclone Corp., San Angelo, TX, USA) when they reached 70%–80% confluence and were gently re-suspended in cell culture medium at a density of 5×10^6 cells/mL. In order to investigate interactions between tumor cells and normal fibroblasts, cells were divided into experimental and control groups. In the experimental group, 3 μL HLF and 3 μL A549 were separately and simultaneously injected into inlets on both outer sides. After a while, when both cell types had settled down, 6 μl of cell culture medium without cells was loaded from the middle. HLF and A549 went straight along the channel, without any mixture, and finally formed a non-contact co-culture model. In the control groups, homogeneous cell models of HLF–HLF and A549–A549 were cultured in the microchannels. Cells were grown on the surface of the glass substrate inside the microchannels, and cell morphologies were monitored using an inverted microscope (Nikon Ti-Eclipse, Tokyo, Japan).

2.5. Cell Viability Assay

Cells with a density of 5×10^6 cells/mL were loaded into the device and driven into the disjunctive side areas of the main microfluidic channel. Then, the device was placed into a cell culture incubator. After 48 h, a cell viability assay was performed with two specific fluorescent probes: Hoechst 33342 (Molecular Probes, Solarbio Corp., Beijing, China) and propidium iodide (PI, Molecular Probes, Solarbio Corp., Beijing, China). Briefly, we introduced 5 μg/mL of each reagent into the main channel and incubated the device at 4 °C for 20 min. Images were acquired using an inverted fluorescent microscope (Nikon Ti-Eclipse, Tokyo, Japan). Cell viability was statistically quantified by the proportion of living cells to total cells from several optional fields.

2.6. Cell Migration on the Microfluidic Co-Culture Device

In order to investigate how fibroblasts and tumor cells interact, the boundary perimeter of the regions containing migrated cells was captured to record cell positions using an inverted microscope (Olympus CKX41-A32PH, Tokyo, Japan) every 4 hours after HLF and A549 were compartmentalized into opposite sides of the main passage. The migration distance is defined as average length of all cells migrating into the blank space between the two compartments and can represent the migration ability of cells. Migration distances can be quantified at different times.

2.7. Immunocytochemistry and Fluorescence Imaging

To identify whether HLF was activated, expression of α-SMA in HLF was recognized by an immunofluorescence technique. After 2 days of co-culture, cells were fixed in 4% paraformaldehyde for 20 min, rinsed twice in PBS, and permeabilized in 0.25% Triton X-100 for 15 min. After three washes in PBS, nonspecific hybridization was blocked in 5% Bovine Serum Albumin (BSA) for 40 min at 37 °C. Cells were immunostained with primary antibody (mouse anti-α-SMA, Boster Corp., Wuhan, China) at 1:200 overnight. Then, they were incubated with secondary antibody (FITC-labeled goat antimouse IgG, Boster Corp., Wuhan, China) at 1:46 dilution for 50 min. After washing in PBS, they were incubated with 5 µg/mL Hoechst 33342 staining nucleus for 20 min. Images were taken using an inverted fluorescent microscope (Nikon Ti-Eclipse, Tokyo, Japan).

2.8. Enzyme-Linked Immunosorbent Assay

Cells were cultured in the chip, and then the supernatant was collected at the outlet after 24 h and centrifuged at 5000 rpm for 10 min. To measure the level of TGF-β1 released in the medium, an enzyme-linked immunosorbent assay (ELISA) was performed using commercially available kits for cytokine detection (Human TGF-β1 ELISA kit Boster Corp., Wuhan, China). The optical density was determined within 30 min at 450 nm on an ELISA microplate reader (Multilabel Plate reader PerkinElmer, Waltham, MA, USA).

2.9. Anticancer Drug Assays on the Microfluidic Co-Culture Device

It was elucidated that gallic acid (GA) or baicalein (BAE) could interfere with the interactions between HLF and A549 cells. After cells were co-cultured for 8 h, drug media with different concentrations of GA or BAE varying from 0 to 80 µg/mL were introduced into the co-culture device. After 24 h of incubation, apoptosis of HLF cells was analyzed after treatment with GA or BAE. The migration ability and expression of α-SMA in HLF cells were analyzed.

2.10. Statistical Analysis

All these experiments were replicated three or more independent times, and data are presented as the mean ± standard deviation (SD). Distances of fibroblast migration and comparisons with controls were evaluated statistically using Student's *t*-test. A value $p < 0.05$ was considered statistically significant.

3. Results and Discussion

3.1. HLF and A549 Indirect Co-Culture on the Microfluidic Device

In this work, a non-contact co-culture microfluidic platform was developed for the study of interactions between tumor cells and fibroblasts (Figure 2). Three branches converging to a single main microchannel were subjected to laminar fluid flow, thus providing a flexible approach to the co-culture of two different types of cells. A wound could be formed automatically. Before cell seeding, different color indicators (rhodamine B, PBS solution, and blue ink) were loaded to confirm laminar flow formation inside the main microchannel. As shown in Figure 3a, three streams with clear linear boundaries indicated that laminar flow was stably linearly formed. Based on this well-controlled laminar flow, HLF and A549 cells went straightly and stably along both sides of the main channel automatically, without any mixing (Figure 3b), as expected. A blank region with clear edges gently appeared between heterotypic cells to generate a non-contact co-culture model, facilitating the observation of cell behavior in real time. It could also be used as a wound area.

Viability testing of HLF and A549 was performed on this contactless platform after 48 h. Cells maintained high viability in the microchannels after 48 h (93.6% alive for HLF and 90% alive for A549) (Figure 4), demonstrating the compatibility and feasibility of this device for further cellular assays.

With this device, different types of cells are able to be co-cultured in indirect contact for further tests such as cellular events. Passive loading facilitated rapid incubation. The size of the bare area for cell migration assay can be accurately controlled through adjusting the amounts of reagents. The edges of different compartments were neat without any cell debris or substances which would produce an unfavorable influence on cell migration.

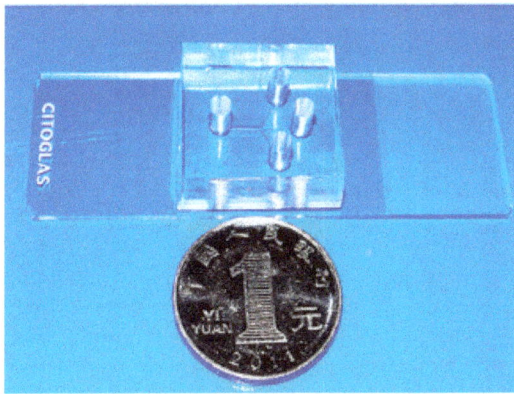

Figure 2. Photograph of the tiny microfluidic device.

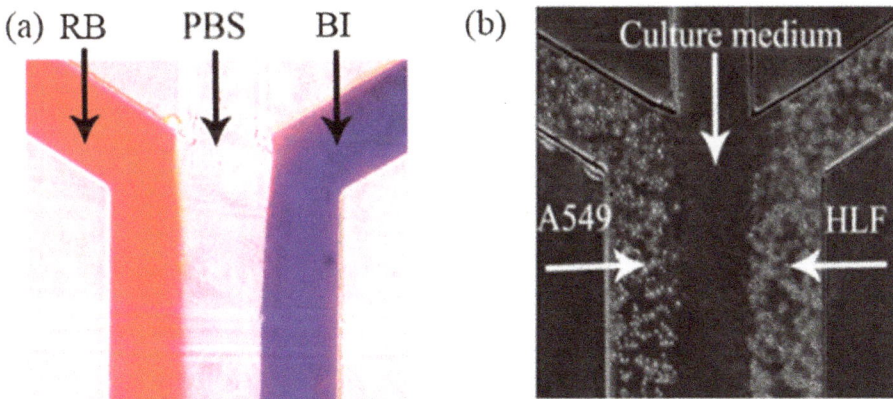

Figure 3. (**a**) Parallel flow testing. Red flow is rhodamine B; blue flow is blue ink; bright flow is PBS. (**b**) Different cells adhered on opposite sides of the main microchannel.

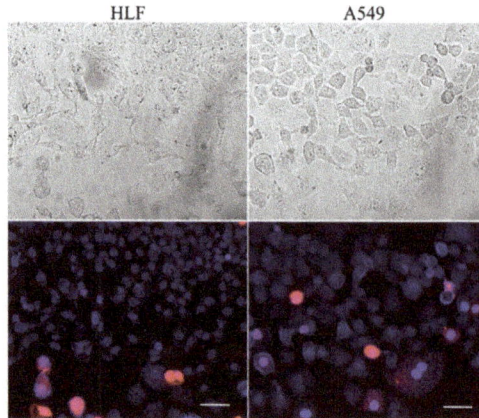

Figure 4. Viability of cells cultured on the co-culture device after 48 h. The top pictures are bright field and the bottom ones are merged fluorescence images. Fluorescent staining showed mostly living cells (blue) mixed with a few apoptotic cells (red). Scale bar = 50 μm.

3.2. Activation of HLF Indirectly Co-Cultured with A549

In order to demonstrate whether A549 can activate fibroblasts, a series of tests of cellular behaviors, such as cell morphology and cell migration, including immunofluorescence and enzyme assays were carried out on HLF.

After HLF and A549 were successfully seeded into the co-culture device and non-contact co-cultured for 24 h, the morphological characterization of cells was assessed (Figure 3). Compared to the control groups (HLF–HLF), HLF co-cultured with A549 became nonuniform, with more protrusions stretching out extensively along the direction of A549. These protrusions, termed pseudopods, are specialized cellular structures containing an array of different proteins like matrix metalloproteinase and fibrous actins [8]. These extension structures can regulate cell membrane and cytoskeleton remodeling and are the prerequisite for the maintenance of cell motility [18]. The morphological change in HLF here was conducive to the progression and metastasis of tumor cells, implying that HLF might become activated and acquire a stronger capability for invasion in the presence of A549.

Total cell migration including cell proliferation of HLF and A549 was analyzed by tracking cells migrating from the boundary perimeter of each compartment to the middle vacant space every 4 h (Figure 5). Obviously, the migration distances of both HLF and A549 cells corresponding to cell propagation and invasion increased greatly with prolonged time under the co-culture condition compared with the other two groups. In contrast, cells moved steady in the HLF–HLF control group. However, in the HLF–A549 co-culture model, HLF started to migrate significantly at the point of 16 h, and the average migration distance of 74 μm was evidently larger than the 53.5 μm of the control ($p < 0.05$) (Figure 6a), implying that HLF located beside A549 could become activated to increase its invasion ability. Similarly, the migration of A549 was remarkably enhanced after 12 h to 62 μm at 16 h and further increased highly to 81.4 μm after 24 h when co-cultured with HLF ($p < 0.05$) (Figure 6b). This suggested that normal fibroblasts might play a vital role in tumor progression through paracrine cytokines.

The transdifferentiation of HLF cells was further characterized after 48 h of culture. As illustrated in Figure 7, α-SMA was highly up-regulated in the HLF–A549 co-cultured group. However, only mild α-SMA expression was detected in the culture of HLF alone (HLF–HLF). This secretion distinction further suggested that tumor cells were able to activate fibroblasts into myofibroblasts.

There was no cell communication relying on direct contact between these separated cells, indicating that chemical signals coordinated the communication between HLF and A549. In order to further validate the effect of signaling pathways, the TGF-β1 content was investigated using an ELISA method. As shown in Figure 8, the results verified that TGF-β1 would be expressed by both A549 and HLF. However, the concentration of TGF-β1 secreted in the supernatant medium of HLF–A549 of 180.5 pg/mL was remarkably higher than those of the control groups: 110.5 pg/mL for A549 and 50 pg/mL for HLF ($p < 0.05$).

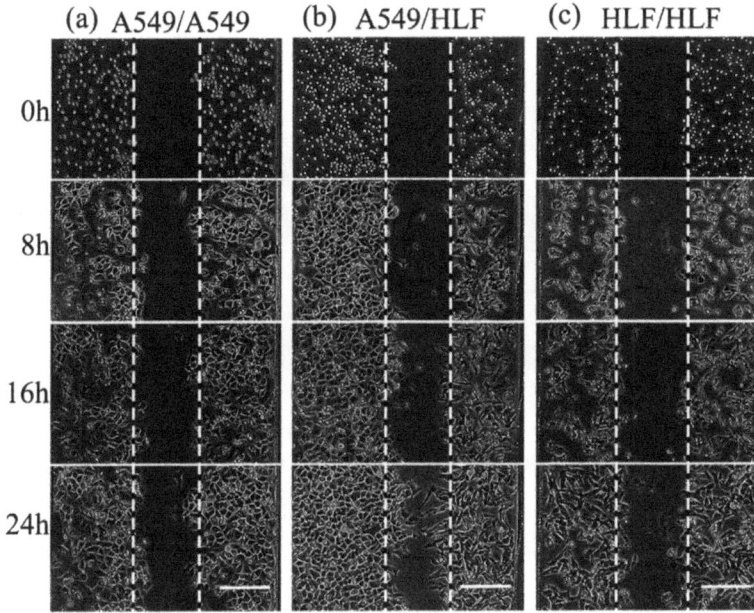

Figure 5. Contrast images of cells migrating on the micro-fluidic device at different times for different groups. (**a**) A549–A549. (**b**) HLF–A549. (**c**) HLF–HLF. Scale bar = 200 μm.

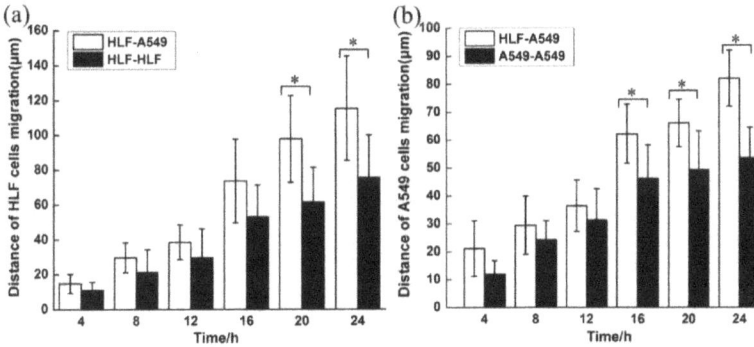

Figure 6. Histograms of average migration distances of cells for co-culture group compared with control groups. (**a**) Average migration distance of HLF at corresponding different times for HLF–A549 and HLF–HLF groups. (**b**) Average migration distance of A549 at corresponding different times for HLF–A549 and A549–A549 groups. (* $p < 0.05$)

Therefore, we deduced that under the condition of co-culture, HLF cells were possibly continuously influenced by high levels of TGF-β1 secreted by A549 and converted to activated fibroblasts. A549 promoted transdifferentiation of fibroblasts to acquire a favorable microenvironment for tumor growth characterized by more α-SMA. Once activated, fibroblasts characterized by a migratory spindle-shaped phenotype might generate increased TGF-β1 secretion of tumor cells with stronger invasion ability and higher expression of α-SMA.

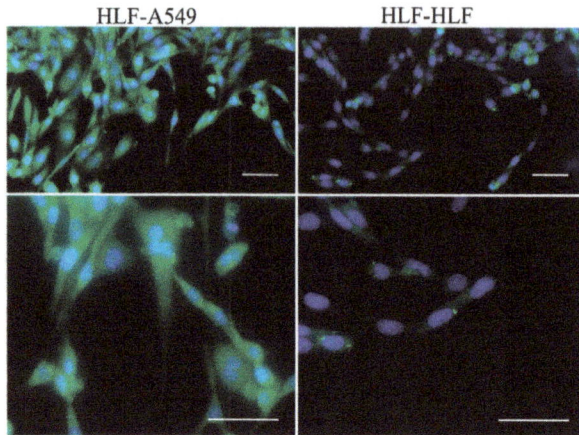

Figure 7. Transdifferentiation of HLF. α-SMA (green fluorescence) was used to identify activation of normal fibroblasts and the nucleus was stained blue by Hoechst 33342 (blue fluorescence). HLF could express more α-SMA after 48 h co-culture with A549 (green fluorescence). Scale bar = 50 μm.

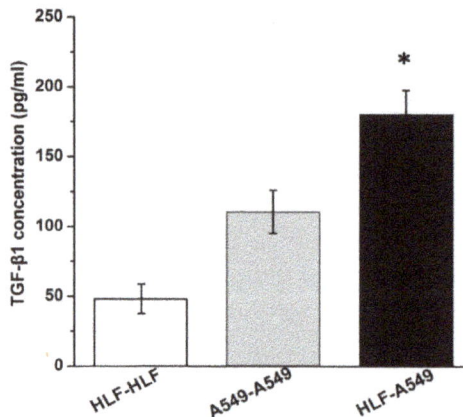

Figure 8. Concentrations of TGF-β1 secreted in the supernatant medium. TGF-β1 expressed by both A549 and HLF. The TGF-β1 level in HLF–A549 was remarkably higher than those in A549–A549 and HLF–HLF (* $p < 0.05$).

The transdifferentiation of fibroblasts into myofibroblasts via indirect contact might be modulated by tumor-cell-derived cytokines and further shows that tumor cells can indirectly stimulate fibroblasts and change their function to create a more survivable microenvironment.

3.3. Cellular Assay on HLF Cells after Being Treated by GA and BAE

The evidence presented above elucidated that normal fibroblasts are an efficacious therapeutic target for cancer therapy. Therefore, GA and BAE were introduced to interfere with the interactions between them.

In order to test the efficacy of GA and BAE, the total cell migration combined with the cell proliferation of HLF was noted. As shown in Fig. 9, compared with the non-drug-treated group (HLF–A549 control) and fibroblasts-only group (HLF–HLF), the fibroblasts showed a dramatic restriction in migration at different concentrations in a dose-dependent manner ($p < 0.05$) (Figure 9a). When the concentration of GA reached 20 µg/mL, the migratory ability of HLF co-cultured with A549 was controlled at a low level, as was that for HLF cultured alone. Similarly, BAE could inhibit the activation of fibroblasts ($p < 0.05$) (Figure 9b), and the best concentration of BAE for this was 10 µg/mL.

Furthermore, immunofluorescence tests were carried out on HLF cells. As shown in Figure 10, the intensity of the green fluorescence of α-SMA displayed in HLF cells treated with GA or BAE was obviously weak compared with that in the non-treated tumor-induced HLF cells, which means that GA and BAE can decrease the expression of α-SMA in fibroblasts greatly. These results suggest that both GA and BAE could suppress the effect of A549 on HLF. The effective and optimum dosages of these two drugs are 20 µg/mL and 10 µg/mL, respectively.

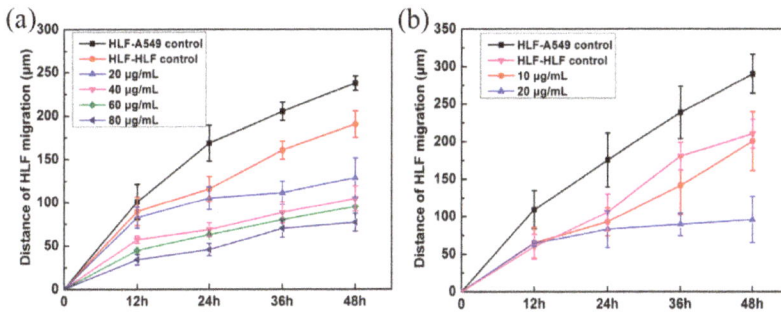

Figure 9. Average migration distances of tumor-induced HLF treated by gallic acid (GA) or baicalein (BAE) (* $p < 0.05$).

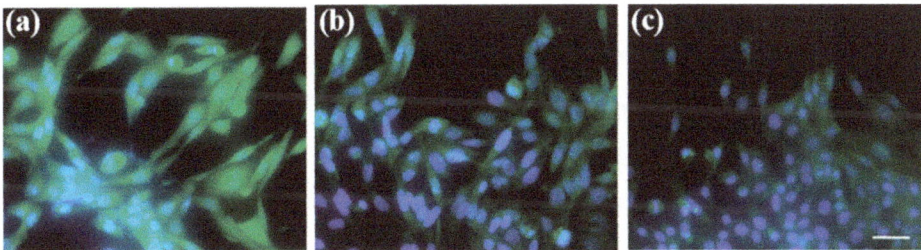

Figure 10. Transdifferentiation of tumor-induced HLF cells treated by GA or BAE. (**a**) Fluorescent images of HLF cells without any drugs. (**b**) Morphology of HLF cells under GA. (**c**) Cellular shapes of HLF treated with BAE.

4. Conclusions

Under well-controlled laminar flow, two heterogeneous cell types interacted under non-contact within a microfluidic co-culture device. Cell co-culture and migration were examined, and protein in situ detection and cytokine detection were thoroughly performed. To elucidate the mechanism of cancer cell progression for improved cancer therapy, we accomplished in situ and label-free analysis of the interactions between tumor cells and normal fibroblasts. Our investigation revealed several valuable facts. Firstly, molecular cross-talk between tumor cells and fibroblasts was demonstrated in a contactless co-culture mode. Secondly, cytokines from tumor cells effectively transformed the co-cultured fibroblasts into myofibroblasts through indirect contact, creating a favorable microenvironment intimately associated with tumor growth and metastasis. Thus, anti-invasion cancer therapy strategies could be established through protecting fibroblasts against the influence of tumor cells. Thirdly, GA and BAE could inhibit the tumor-induced activation of fibroblasts, making them a potential source of antitumor drugs with low toxicity. Therefore, this non-contact co-culture device enables various biological assays to be performed, such as the analysis of cellular events between fibroblasts and tumor cells. Moreover, a complicated, close-to-real tumor microenvironment could be mimicked through the compartmentalization of multiple heterotypic cells within the microfluidic device for analyzing cancer invasion mechanisms and targeting anti-invasion therapeutics.

Author Contributions: H.C. designed the experiment and revised the manuscript and got involved in performing the experiment. W.L. performed the experiment. B.W. supported the experiment and revised. Z.Z. contributed some ideas and revised the manuscript.

Acknowledgments: This research work was supported by National Natural Science Foundation of China (2011CB933102, 21271182) and Anhui Natural Science Foundation of China (1708085QF145).

Conflicts of Interest: The authors declare no conflict of interest.

References

1. Karlou, M.; Tzelepi, V.; Efstathiou, E. Therapeutic targeting of the prostate cancer microenvironment. *Nat. Rev. Urol.* **2010**, *7*, 494–509. [CrossRef] [PubMed]
2. Quail, D.F.; Joyce, J.A. Microenvironmental regulation of tumor progression and metastasis. *Nat. Med.* **2013**, *19*, 1423–1437. [CrossRef] [PubMed]
3. Bhowmick, N.A.; Neilson, E.G.; Moses, H.L. Stromal fibroblasts in cancer initiation and progression. *Nature* **2004**, *432*, 332–337. [CrossRef] [PubMed]
4. Silzle, T.; Randolph, G.J.; Kreutz, M.; Kunz-Schughart, L.A. The fibroblast: Sentinel cell and local immune modulator in tumor tissue. *Int. J. Cancer* **2004**, *108*, 173–180. [CrossRef] [PubMed]
5. Kuperwasser, C.; Chavarria, T.; Wu, M.; Magrane, G.; Gray, J.W.; Carey, L.; Richardson, A.; Weinberg, R.A. Reconstruction of functionally normal and malignant human breast tissues in mice. *Proc. Natl. Acad. Sci. USA* **2004**, *101*, 4966–4971. [CrossRef] [PubMed]
6. Hao, Y.; Zhang, L.; He, J.; Guo, Z.; Ying, L.; Xu, Z.; Zhang, J.; Lu, J.; Wang, Q. Functional investigation of NCI-H460-inducible myofibroblasts on the chemoresistance to VP-16 with a microfluidic 3D co-culture device. *PLOS ONE* **2013**, *8*, e61754.
7. Tripathi, M.; Billet, S.; Bhowmick, N.A. Understanding the role of stromal fibroblasts in cancer progression. *Cell Adhes. Migr.* **2012**, *6*, 231–235. [CrossRef] [PubMed]
8. Ma, H.; Liu, T.; Qin, J.; Lin, B. Characterization of the interaction between fibroblasts and tumor cells on a microfluidic co-culture device. *Electrophoresis* **2010**, *31*, 1599–1605. [CrossRef] [PubMed]
9. Lorusso, G.; Rüegg, C. The tumor microenvironment and its contribution to tumor evolution toward metastasis. *Histochem. Cell Biol.* **2008**, *130*, 1091–1103. [CrossRef] [PubMed]
10. Mehling, M.; Tay, S. Microfluidic cell culture. *Curr. Opin. Biotechnol.* **2014**, *25*, 95–102. [CrossRef]
11. Nie, F.Q.; Yamada, M.; Kobayashi, J.; Yamato, M.; Kikuchi, A.; Okano, T. On-chip cell migration assay using microfluidic channels. *Biomaterials* **2007**, *28*, 4017–4022. [CrossRef]
12. Bischel, L.L.; Casavant, B.P.; Young, P.A.; Eliceiri, K.W.; Basu, H.S.; Beebe, D.J. A combined microfluidic coculture and multiphoton FAD analysis enables insight into the influence of the bone microenvironment on prostate cancer cells. *Integr. Biol.* **2014**, *6*, 627–635. [CrossRef] [PubMed]
13. Xie, Y.; Zhang, W.; Wang, L.; Sun, K.; Sun, Y.; Jiang, X. A microchip-based model wound with multiple types of cells. *Lab Chip* **2011**, *11*, 2819–2822. [CrossRef] [PubMed]
14. Businaro, L.; De, N.A.; Schiavoni, G.; Lucarini, V.; Ciasca, G.; Gerardino, A.; Belardelli, F.; Gabriele, L.; Mattei, F. Cross talk between cancer and immune cells: Exploring complex dynamics in a microfluidic environment. *Lab Chip* **2012**, *13*, 229–239. [CrossRef] [PubMed]
15. Orimo, A.; Gupta, P.B.; Sgroi, D.C.; Arenzana-Seisdedos, F.; Delaunay, T.; Naeem, R.; Carey, V.J.; Richardson, A.L.; Weinberg, R.A. Stromal fibroblasts present in invasive human breast carcinomas promote tumor growth and angiogenesis through elevated SDF-1/CXCL 12 secretion. *Cell* **2005**, *121*, 335–348. [CrossRef] [PubMed]

16. Tourovskaia, A.; Fiqueroa-Masot, X.; Folch, A. Differentiation-on-a-chip: A microfluidic platform for long-term cell culture studies. *Lab Chip* **2005**, *5*, 14–19. [CrossRef] [PubMed]

17. Jahn, A.; Vreeland, W.N.; Gaitan, M.; Locascio, L.E. Controlled vesicle self-assembly in microfluidic chanelswith hydrodynamic focusing. *J. Am. Chem. Soc.* **2004**, *126*, 2674–2675. [CrossRef] [PubMed]

18. Friedl, P.; Wolf, K. Proteolytic and non-proteolytic migration of tumor cells and leucocytes. *Biochem. Soc. Symp.* **2003**, *70*, 277–285. [CrossRef]

micromachines

MDPI

Article

Separation and Characterization of Prostate Cancer Cell Subtype according to Their Motility Using a Multi-Layer CiGiP Culture

Lin-Xiang Wang [1], Ying Zhou [1], Jing-Jing Fu [1], Zhisong Lu [1] and Ling Yu [1,2,*]

[1] Key Laboratory of Luminescent and Real-Time Analytical Chemistry (Southwest University),
Ministry of Education, Institute for Clean Energy and Advanced Materials, Faculty of Materials and Energy,
Southwest University, Chongqing 400715, China; wlx3952@email.swu.edu.cn (L.-X.W.);
zy0200@email.swu.edu.cn (Y.Z.); jingjing1991@email.swu.edu.cn (J.-J.F.); zslu@swu.edu.cn (Z.L.)
[2] Guangan Changming Research Institute for Advanced Industrial Technology, Guangan 638500, China
* Correspondence: lingyu12@swu.edu.cn; Tel.: +86-23-6825-4842

Received: 26 November 2018; Accepted: 13 December 2018; Published: 14 December 2018

Abstract: Cancer cell metastasis has been recognized as one hallmark of malignant tumor progression; thus, measuring the motility of cells, especially tumor cell migration, is important for evaluating the therapeutic effects of anti-tumor drugs. Here, we used a paper-based cell migration platform to separate and isolate cells according to their distinct motility. A multi-layer cells-in-gels-in-paper (CiGiP) stack was assembled. Only a small portion of DU 145 prostate cancer cells seeded in the middle layer could successfully migrate into the top and bottom layers of the stack, showing heterogeneous motility. The cells with distinct migration were isolated for further analysis. Quantitative PCR assay results demonstrated that cells with higher migration potential had increased expression of the ALDH1A1, SRY (sex-determining region Y)-box 2, NANOG, and octamer-binding transcription 4. Increased doxorubicin tolerance was also observed in cells that migrated through the CiGiP layers. In summary, the separation and characterization of prostate cancer cell subtype can be achieved by using the multi-layer CiGiP cell migration platform.

Keywords: cells-in-gels-in-paper; cancer metastasis; cell motility; cancer stem cell; drug resistance

1. Introduction

Cell migration is a fundamental cellular function implicated in many biological and pathological processes, such as embryonic morphogenesis, wound repair, and cancer invasion [1–3]. Cancer cell metastasis has been recognized as one hallmark of malignant tumor progression [4]. Metastatic relapse or distant progression is one of the most frequent causes of death from cancer, which clearly emphasizes the urgent need to develop strategies for the prevention of metastasis [2,5,6]. Measuring the motility of cells, especially tumor cell migration, is one approach for evaluating the therapeutic effects of anti-tumor drugs [7]. The Boyden chamber and scratch/would healing assay are routinely conducted in biological sectors to evaluate the migration abilities of tumor cells [8,9]. However, it is not practical to use those methods to harvest and separate cells with different motilities. Recently, Cui et al. developed a microfluidic device containing a biocompatible porous membrane and an array of independently controlled microchambers to isolate and collect migrating cancer cell [10]. By using the microfluidic platform, migrating speed and persistence of breast cells, as well as the morphology and cytoskeletal structures of migrating cells were investigated. However, the gene profile which can underline the migration capability have not been directly studied. Characterizing the phenotype or gene profile of subpopulations that show higher motility will help further elucidate the relationship between the molecular characteristics of tumor cells and their functions [11].

In 2009, the novel cells-in-gels-in-paper (CiGiP) method was developed by Dr. Whitesides [12]. The merits of this system include its ease of use and the ability to mimic the three-dimensional (3D) biological, chemical, and mechanical properties of native tissues [13,14]. Since its development, the CiGiP method has been modified to investigate cell invasion, proliferation, and differentiation [15–18]. For instance, a stacked CiGiP was developed to study cancer cell migration under a gradient of oxygen tension [17]. The monotonically decreasing gradient of oxygen through the CiGiP stack can be easily controlled with the aid of a special holder [16]. Moreover, the impact of the oxygen gradient on the sensitivity of lung cancer cells to ionizing radiation has been evaluated with the CiGiP platform [19]. The flexibility in stacking and destacking the paper [20,21] allows the evaluation of the metabolism of cells at different layers. The CiGiP method provides a versatile tool for exploring the areas of fundamental cell biology and the development of novel therapeutics [12,22,23].

Similar to running competitions, we hypothesized that a 3D CiGiP platform could be established to model a cell race (migration), in which the migratory response of cells could be measured by peeling apart the stacked layers. To study the characteristics of cells with different motilities, we established CiGiPs for the model cell, DU 145 prostate cancer cell line. The cells that migrated into different paper layers were isolated and putative biomarkers for prostate cancer stem cells (CSCs) were determined by quantitative PCR (qPCR). To the best of our knowledge, this is the first study to separate and isolate cancer cell subtypes according to their motility, and further compare the CSC-related genes expression of cells with different motility.

2. Materials and Methods

2.1. Materials and Reagents

Prostate cancer cells DU 145 were obtained from (Chinese Academy of Sciences Cell Bank, Shanghai, China). The cells were maintained in Dulbecco's TM®Modified Eagle Medium (DMEM, Gibco, Gaithersburg, MD, USA) containing 10% fetal bovine serum (Gibco), penicillin (100 U/mL) and streptomycin (100 μg/mL) at 37 °C in a 5% CO_2 atmosphere. Whatman®105 lens paper and Parafilm®M were purchased from Sigma Aldrich (St. Louis, MO, USA). Matrigel was purchased from Corning (Corning, NY, USA). All other chemicals were acquired from Sigma-Aldrich unless otherwise indicated. All solution was prepared with deionized water produced by PURELAB flex system (ELGA, High Wycombe, UK).

2.2. Preparation of the Multi-Layer CiGiP Platform

Preparation of the paper sheet: Lens paper (Whatman®105, GE Healthcare, Buckinghamshire, UK) was used to prepare the paper scaffold. According to the product information, the thickness of the lens paper is 35 to 40 μm. To form hydrophobic regions on lens paper, commercial wax film Parafilm®(Bemis company. Inc., Neenah, USA) was used. The pattern containing eight holes (diameter of 6 mm) was designed and crafted by a desktop paper crafter. Then a lens paper–Parafilm®-lens paper sandwich structure was assembled and fed into a hot lamination machine (110 °C) to form a single paper stack layer (Figure 1A). The paper stack was sterilized by ethanol washing and ultraviolet light irradiation for 1 h.

CiGiP-based migration assays: Cultures of DU145 cells were maintained in tissue culture flasks. Prior to preparing the migration assay, Trypsin-EDTA was used to detach the cells from the tissue culture flask. The Matrigel was thawed at 4 °C. A pre-cooled culture plate and paper sheets were used. All culture plate, patterned lens paper and cell suspension were plated on an ice box during the cell loading process. First, cells were mixed with Matrigel to a final concentration of 12,500 cells/μL. 8 μL cell/Matrigel mixture was spotted into each zone of the seeding layer. The zones in the migration layers were spotted with Matrigel only. Then the paper sheets were incubated at 37 °C for 12 h prior to stacking. The cell-seeded paper sheet was sandwiched between gel-embedded only paper sheets. The multi-layer paper sheets were fastening into a chip holder to ensure the conformal contact of the sheets (Figure 1B). Each multi-layer culture consisted of five layers. The cell-seeded layer was referred to as layer 0. The paper sheets above layer 0 (L0) were layer+1 (L+1) and layer+2 (L+2); the paper sheets below layer 0 were layer-1 (L-1) and layer-2 (L-1). The multi-layer paper sheets were placed in a homemade poly(methyl methacrylate) (PMMA) holder and fastened with screws (Figure 1B). The assembled multi-layer migration stack was placed in petri dishes containing dulbecco's modified eagle medium (DMEM) and incubated at 37 °C for seven days.

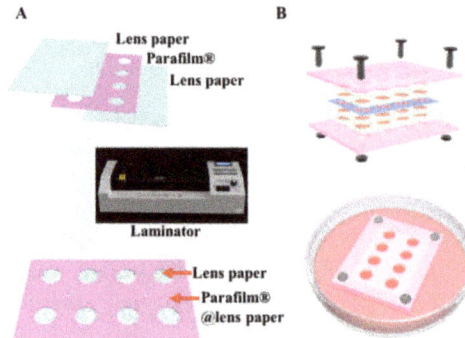

Figure 1. Fabrication of cell-in-gel-in-paper (CiGiP) cell invasion platform. (**A**) Preparing of paper sheet for cell invasion: lens paper–Parafilm®film-lens paper sandwich was sealed with a hot lamination machine to produce single layer of paper sheet; and (**B**) multi-layer paper sheets were assembled and fasten by a chip holder.

2.3. Characterization of Cell Invasion on Multi-Layer Cultures

To quantify cell migration in CiGiP platforms, multi-layer cultures were de-stacked by disassembling the holder and separating the individual sheets by tweezers. Each sheet was washed with culture medium and phosphate-buffered saline (PBS). To characterize and quantify cell migration, the following experiment was conducted.

Fluorescent staining of cells: The live/dead cell double staining kit, calcein-acetoxymethyl ester (Calcein-AM) and propidium iodide (PI) kit (KeyGEN BioTECH, Nanjing, China), was used for measuring viable and dead cells. In brief, 100 μL Calcein-AM/PI solution was pipetted onto the paper sheets and followed by a 35 min incubation. Then the paper sheets were washed with PBS to remove unstained fluorescence dye. Fluorescence images of the paper-based scaffolds were obtained on a fluorescent microscope (TS100-F, Nikon, Tokyo, Japan). The excitation and emission wavelength for Calcein-AM (staining the living cells) is 495 nm and 515 nm, respectively. While the excitation and emission wavelength for PI (staining of the dead cells) is 535 nm, 617 nm, respectively.

Total RNA quantification: Cells on paper sheets were harvested by Accutase™ cell dissociation reagent (Gibco, Gaithersburg, MD, USA). In brief, 1 mL Accutase™ cell dissociation reagent was added to the paper sheets and incubated at 37 °C for 15 min. Then the paper sheets were washed three times with 0.01 M PBS (pH 7.4). The collected solution was centrifuged at 1000 rpm for 5 min after which the cell pellet was collected. Total RNA was isolated with the TaKaRa MiniBEST Universal RNA Extraction Kit (TaKaRa, Tokyo, Japan) and quantified by DeNovix®DS-11+ spectrophotometry (Gene, Hong Kong, China) at 260 nm.

2.4. qPCR to Evaluate Gene Expression in Cells that Invade Different Paper Layers

Cells on each paper sheet were harvested for real time quantitative PCR (RT-qPCR) experiments. Total RNA was reverse transcribed to cDNA using the PrimeScript RT Reagent Kit with gDNA Eraser (Perfect Real Time). PCR primers for OCT4, SRY (sex determining region Y)-box 2 (SOX2), SONG, aldehyde dehydrogenase 1 (ALDH1A1), hypoxia-inducible factor 1-alpha (HIF-1α), and hypoxia-inducible factor 2-alpha (HIF-2α) genes were designed and validated in accordance with the guidelines recommended by the Minimum Information for Publication of Quantitative Real-Time PCR experiments (MIQE) (Table 1). Then qPCR was conducted on the CFX96 Real-Time PCR Detection System (Bio-Rad, Hercules, CA, USA) with SYBR®Premix Ex Taq™ II (Tli RNaseH Plus, Beyotime, Shanghai, China). Glyceraldehyde-3-phosphate dehydrogenase (GAPDH) was used as an internal control. For all reactions, cycling conditions were 95 °C for 2 min followed by 40 cycles of 95 for 15 s and 60 °C for 30 s with a temperature ramp rate of 1.6 °C/s. Amplification profiles were analyzed with QuantStudio 6 and 7 Flex Real-Time PCR System software (MyGo, Gene, Hong Kong, China). The relative expression levels of each gene in cells were normalized to the GAPDH gene using the 2-ΔΔCt method. Three independent experiments were performed.

Table 1. Primers for RT-qPCR.

Gene	Primer	Sequence (5′—3′)
OCT4	F	AACCGAGTGAGAGGCAACCT
	R	ACAGAACCACACTCGGACCA
SOX 2	F	TGACCAGCTCGCAGACCTAC
	R	TCGGACTTGACCACCGAAC
NANOG	F	CGCGTTGTGATCTCCTTCTG
	R	GTCTGGTTGCTCCAGGTTGA
ALDH1A1	F	CACAGGATCAACAGAGGTTGG
	R	GTCCAAGTCGGCATCAGCTA
GAPDH	F	CCAGGTGGTCTCCTCTGACTTCAACA
	R	AGGGTCTCTCTCTTCCTCTTGTGCTC
HIF-1α	F	GCCCCAGATTCAGGATCAGA
	R	TGGGACTATTAGGCTCAGGTGAAC
HIF-2α	F	GCCACCCAGTACCAGGACTACA
	R	CCTCACAGTCATATCTGGTCAGTTCG

2.5. MTT Assay to Test the Cytotoxicity Effects of Doxorubicin on Cells Migrating Different Paper Sheets

The assembled multi-layer migration stack was placed in petri dishes containing DMEM culture medium and incubated at 37 °C for seven days. Then the multi-layer cultures were de-stacked and each sheet was washed with DMEM culture medium. The paper sheets were separately placed into different wells of the culture plate. Cells in each paper layer were treated with 0.5 μM doxorubicin (topoisomerase inhibitor) for 48 h. Relative cell viability was analyzed. In brief, the paper sheets were washed with PBS and placed in a new well. Fresh medium plus 3-(4,5-dimethylthiazol-2-yl)-2,5-diphenyltetrazolium bromide (MTT) were added to each well and incubated at 37 °C for 4 h, after which dimethyl sulfoxide was added and the wells were incubated with shaking for 15 min. Finally, the absorbance at 570 nm was measured in a microplate reader (ELx800TM, Gene, Hong Kong, China) with a reference wavelength of 630 nm. Cell growth inhibition on each paper layer was calculated using non-drug treated cells as reference samples. For example, growth inhibition on L0 was calculated by: [(A570 of L0 without drug treatment − A570 of L0 with drug treatment)/A570 of L0 without drug treatment] × 100%. Growth inhibition on L+1, L-1, L+2, and L-2 were calculated, respectively. Three independent experiments were performed.

2.6. Statistic Analysis

All experiments were performed three times in triplicate. Data are expressed as the mean ± standard deviation. Results were analyzed with the Student's t-test using Origin Statistic software (Origin Lab Corporation, Northampton, MA, USA). p values less than 0.05 were considered statistically significant.

3. Results

3.1. Hydrophobic Parafilm®Patterned Lens Paper for Assembling of Multi-Layer Paper-Based Cell Culture Platform

In this work, hydrophobic Parafilm®film was utilized to fabricate paper-based substrate for cell culture. Parafilm®is a thermoplastic, self-sealing film that offers excellent barrier protection to the contents of tubes, flasks, culture tubes, etc., in daily laboratory usage. The thermal-sensitive film was sandwiched between two lens papers and then fed into a desktop lamination machine which is normally used for the sealing of photography. The temperature of the hot lamination is 110 °C. During the hot lamination process, the melted Parafilm®can penetrate into the lens paper, forming hydrophobic barriers. Figure 2A-a shows the microscopy image of single-layer pristine lens paper. Pores can be observed between fibers. The average pore size was 37 μm ± 26 μm by randomly measuring 50 pores from the microscopy image. Figure 2A-b–d show the Parafilm®-bonded lens paper. A clear boundary, pointed by red arrow, can be observed between pristine lens paper and Parafilm®-patterned area (Figure 2A-b). Five layers of Parafilm®-bonded paper were stacked and fastened by PMMA holder (Figure 2B-a). Food color dye solution was casted on the cell seeding zone. It was observed that solution only wet the hydrophilic zone of each layer ((Figure 2B-c), indicating that hot-lamination assisted Parafilm®patterning can effectively forming hydrophobic area for stopping fluids (Figure 2B-d). The cross section of the Parafilm®-patterned lens paper was shown in Figure 2B-d. The melted Parafilm®can firmly bond two layer of lens paper. The thickness of the hydrophilic area (pristine lens paper) is around 75 μm ± 14 μm, which similar to the thickness of two single pristine lens paper. While, thickness of the Parafilm®-patterned lens paper is around 135 μm ± 11 μm. Matrigel can be safely held within the cell-seeding region (Figure 2B-e). Previously, wax printing [17] or polyvinyl chloride (PVC) sheet [19] were used for generating hydrophobic barrier for paper-based cell culture. Comparing with those reported methods, Parafilm®-assisted fast patterning does not require a wax printer and high temperature-assisted bonding. By using the cost-effective method, lens paper was patterned with hydrophilic areas in different size. Figure 2C shows that a circle zone with a diameter down to 1 mm can be produced (Figure 2C). In this study, to ensure the uniform contact of multi-layer of Parafilm®-patterned lens paper, while provide sufficient area for cell growth, circle zone with a diameter of 6 mm was design and used. Multi-layers of patterned lens paper can be stacked and fastened with a PMMA frame for cell culture.

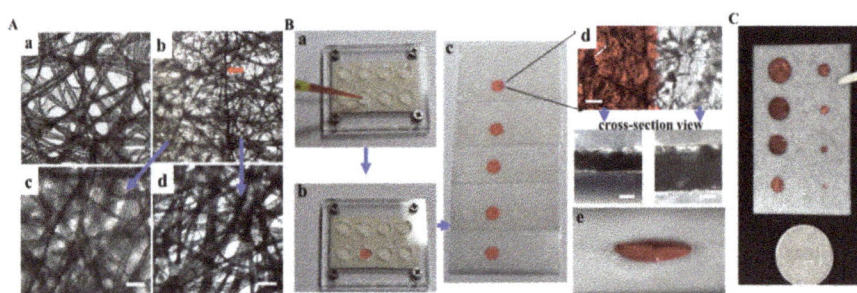

Figure 2. Parafilm®patterned lens paper for assembling of multi-layer paper based cell culture platform. (**A**) Microscopic images of (a) lens paper, (b) Parafilm®impregnated lens paper (red arrow points the edge of Parafilm®), (c) Parafilm®-embedded lens paper, (d) pristine lens paper on Parafilm®-bonded paper, scale bar: 50 μm; (**B**) picture of Parafilm®patterned lens paper: a. assembled multi-layer stack; b. casting solution on cell seeding zone; c. de-stacked multi-layer paper sheet; d. boundary and cross-section of pristine lens paper and Parafilm®-bonded paper, scale bar: 50 μm; e. Matrigel (red) was spotted in hydrophilic region (cell-seeding zone); and (**C**) patterning of hydrophilic region with different size.

3.2. CiGiP Multi-Layer Cultures Assisted Separation and Isolation of Cells According to Their Motility

Cell migration assay was conducted with the stacking CiGiP multi-layer cultures. As illustrated in Figures 1B and 2B, the chip holder used in this study allowed diffusion of the culture medium into the CiGiP structures from both the top and bottom sides, minimizing the nutrition difference between the upper and bottom paper sheets. The cell-seeded paper sheet (L0) was sandwiched in the middle layer of the CiGiP platform, and was the start point of the cell migration. The stacked multi-layer cultures mimic the bulk movement of cells in a tissue-like environment. The multi-layer CiGiP platform was cultured for seven days, and then de-stacking of the paper scaffolds was performed. First, we stained cells on different layers by Calcein-AM/PI kit, which can simultaneously detect live and dead cells. Highly-packed live cells (green) were observed in L0 (Figure 3A). At the same time, dead cells (red) also can be observed after seven days stacking-culture. The cell density sharply decreased in layers above and below L0. The cells traveled into top and bottom layers were visualized by live cell dye Calcein-AM. Since the thickness of the Parafilm®-patterned paper sheet was about 75 μm, the distance the cells travelled within the multi-layer cultures could be quantified. With an increase in traveling distance, the number of cells sharply decreased, indicating heterogeneous motility.

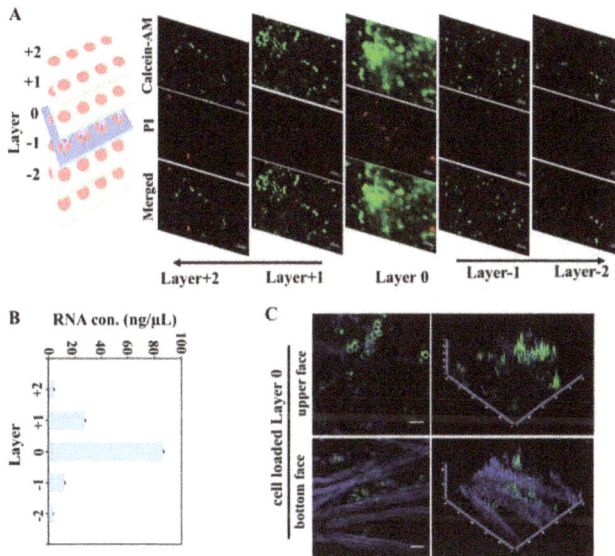

Figure 3. Separation and isolation of cells according to their motility. (**A**) Fluorescent image of cells migrated into different layers of the CiGiP cell migration platform. Prostate cancer DU 145 cells were loaded at the middle layer (Layer 0) of the invasion platform. After seven days culturing, the CiGiP platform was de-stacked. Cells at different layers were stained with Calcein-AM/PI kit. Calcein-AM stains live cell, while PI stains dead cells. Scale bar: 100 μm. (**B**) RNA concentration isolated from cells invaded into different layers of the CiGiP cell invasion platform. (**C**) Confocal microscopy image of cell seeding layer. Eight microliters of cell/Matrigel mixture was spotted into each zone of the seeding layer. The paper sheet was incubated at 37 °C for 12 h. Then the cells on paper sheet was stained with Calcein-AM. The upper face and bottom face of the paper sheet were examined under confocal microscopy (LSM800, Zeiss, Munich, Germany). The 2D images (left panel) were converted into 2.5-D images (right panel) by confocal software Zen. Scale bar: 50 μm.

As shown in Figure 3A and previous studies [12,17], cell staining with dyes facilitated the observation and quantification of live cells in the different layers. The paper sheet was photographed by microscopy or using a scanner. Cell distribution was quantified by image analysis. Due to the 3D structure of the gel embedded paper, it would be difficult to preciously quantify the cells buried in the matrix. To address this concern, we isolated cells from the paper sheet, and measured the levels of total genomic RNA to quantify the cells in each layer. The concentration/amount of RNA in each layer was directly measured from cells isolated from each layer. As shown in Figure 3B, cells in L0 had the highest level of total genomic RNA (86.91 ± 0.18 ng/μL). The RNA amount/concentration sharply decreased in cells from the L+1 and L-1 layers, although the RNA levels in cells from the L+1 layer were higher than those in cells from L-1. We further examined the upper and bottom faces of the cell-seeded sheet (L0) by confocal microscopy (LSM 800, Zeiss, Munich, Germany). As shown in Figure 3C, a higher cell density (green) can be observed from the upper surface of the cell seed layer. The difference between the upper and bottom surfaces of the cell seed layer may due to the gradual gelation of the cell-matrix mixture during penetration through the lens paper, although cell seeding was conducted on an ice-box.

Fluorescent staining assisted live cell imaging and RNA quantification in CiGiP constructs both demonstrate that cells have different motilities. The longer distance the cells travel in the same period of time, the higher their motility capability. Thus the phenotype and molecular characteristics of those cells should be investigated to determine the mechanisms underlying their high motility.

3.3. Elevated ALDH1A1, SOX2, NANOG, and OCT4 Expression in Cells with Higher Motility

By using the CiGiP cell migration platform, subpopulation of cancer cell can be collected from different paper layer. In this study, cell-seeded paper sheet was placed as the middle layer of the paper stack. Culture medium can perfuse from top and bottom side of the hydrophilic region. To exam the effect of stacking caused oxygen and nutrient gradient on cells, the expression of hypoxia-inducible factor 1-alpha (HIF-1α) and hypoxia-inducible factor 2-alpha (HIF-2α) were measured by qPCR. As shown in Figure 4, expression of hypoxia-related HIF-1α and HIF-2α gene is higher at the middle layer than those in outer layer.

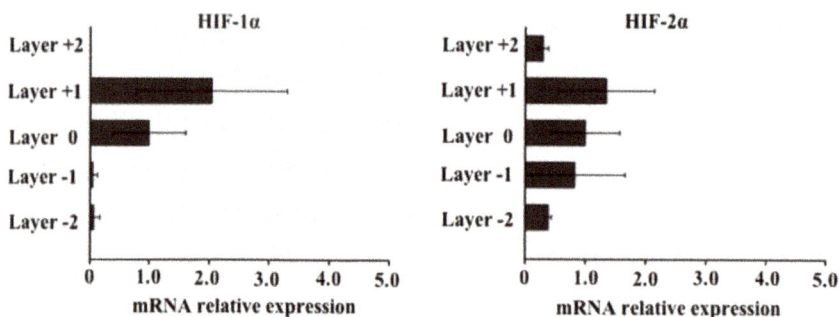

Figure 4. Relative expression hypoxia markers measured by quantitative PCR analysis. Real-time PCR-quantified levels of expression of hypoxia-inducible factor 1-alpha (HIF-1α) and hypoxia-inducible factor 2-alpha (HIF-2α) using 2-ΔΔCt values relative to glyceraldehyde-3-phosphate dehydrogenase (GAPDH) as an internal control. Error bars represent the standard deviation for 24 replicate zones.

To further underline the difference of their motility, expression of several cancer stem cell related biomarker were measured. First, aldehyde dehydrogenase 1 (ALDH1A1), a detoxifying enzyme that oxidizes intracellular aldehydes, was studied. ALDH1A1 is also a modulator of cell proliferation and migration [24]. Upregulation of ALDH1A1 has been observed in primary prostate cancer tissues and metastatic lesions. We performed qPCR to evaluate ALDH1A1 expression on cells collected from different layer of paper. Compared with ALDH1A1 levels at the seeding layer, ALDH1A1 levels in cells at L+2 and L-2 were 60.45 ± 23.92 and 126.72 ± 22.71 fold higher than that of L0, respectively (Figure 5A), indicating a significant ALDH1A1 expression increase in cells with higher motility. In addition, the difference in expression at L+2 and L-2 was most likely due to the fact that the upper face of the cells-seeded sheet (L0) had more cells than the bottom face of this layer (Figure 3C). The cells on the upper face were in direct contact with L1, potentially decreasing the traveling distance and causing variation in motility. In our CiGiP-based cell invasion platform, although only a small fraction of cells travelled through the two layers of paper sheets during the seven days of culture, their ALDH1A1 levels were significantly higher than those with less motility. Growing evidence indicates that ALDH1A1 is a putative CSC marker of several types of solid tumor, including prostate cancer [18,25,26]. According to cancer stem cell (CSC) theory, there is a very small fraction of cancer cells that could initiate cancer and propagate metastasis. The heterogeneous expression pattern of ALDH1A1 from cells with stronger motility suggests the relationship between ALDH1A1, cell motility, and potential stemness. Next, embryonic markers, octamer-binding transcription 4 (Oct-4), SRY (sex determining region Y)-box 2 (SOX2), and NANOG [27,28] were studied. The instrumental roles of those molecules in promoting tumor initiation have been identify. Previous studies also found that the embryonic markers SOX2, NANOG, and OCT4 were elevated in CSCs isolated from human prostate cancer tissue, human prostate tumor models, and some prostate cancer cell lines [29–31]. To investigate the gene expression levels of these putative prostate CSC biomarkers in cells with distinct motility, qPCR measurement was conducted. The results showed that the expression levels of SOX2, NANOG, and OCT4 in cells from the L+2 layer were 162.40 ± 40.28, 57.39 ± 9.63, and 102.98 ± 18.24 fold higher than levels from cells in the L0 layer (Figure 5B–D). The gene expression levels of those putative biomarkers were elevated to 204.15 ± 22.51, 229.95 ± 11.04, and 92.56 ± 24.09 in cells from the L-2 layer. As discussed above, cells that migrate to L-2 have to travel through two layers of Matrigel-impregnated paper sheets, which is probably a longer distance than cells that travel to L+2. Collectively, the elevated ALDH1A1, SOX2, NANOG, and OCT4 expression in cells travelled through different distance would suggest the potential stemness of the small fraction of cells with higher motility.

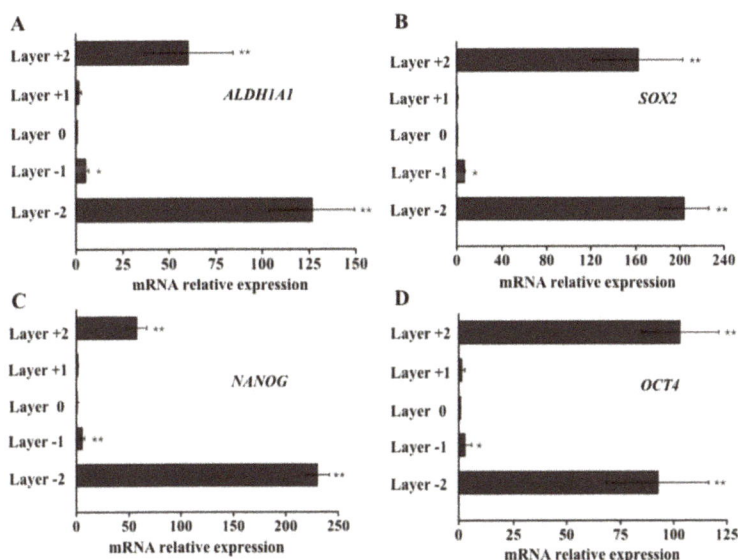

Figure 5. Relative expression of putative CSC-specific markers measured by real-time quantitative PCR analysis. Real-time PCR-quantified levels of expression of ALDH1A1, NANOG, OCT4, and SOX2 transcripts using 2-ΔΔCt values relative to glyceraldehyde-3-phosphate dehydrogenase (GAPDH) as an internal control. Error bars represent the standard deviation for 24 replicate zones, * denotes $p < 0.05$, ** denotes $p < 0.01$.

3.4. Increased Resistance to Doxorubicin of Cells with Higher Motility

According to the CSC theory, CSCs are more resistant to chemotherapy than differentiated cancer cells [32]. Cells were subjected to treatment with the chemotherapy agent doxorubicin, followed by de-stacking of the CiGiP platform to obtain individual paper sheets. The inhibitory effects of doxorubicin on cell proliferation were evaluated. The cell growth inhibition rate at L0 was 42.50% $\pm$ 2.53%. Although only a small fraction of cells successfully invaded to L-2 layer, the cell growth inhibition rate was only 17.85% $\pm$ 1.91%, significantly lower than that at L0 (Figure 6). The decreased cell growth inhibition after doxorubicin treatment indicated drug resistance of cells at L-2, which had stronger motility and elevated expression of putative CSC biomarkers. Thus, the CiGiP invasion platform was successfully used in characterizing and separating cells with high invasion and resistance to chemotherapy reagent from bulk cancer cells. Importantly, the heterogeneous gene expression level in these separated cells may help us gain insights into the mechanisms underlying tumor metastasis and drug resistance.

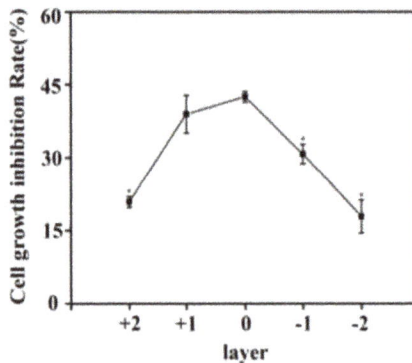

Figure 6. DU145 cells were seeded in layer 0 and the assembled multi-layer migration stack was cultured for seven days. Then each paper sheets were separately placed into culture medium containing 0.5 μM doxorubicin for 48 h. Finally, an MTT assay was conducted with doxorubicin-treated paper sheets. Cell growth inhibition on each paper layer was calculated using non-drug treated cells as reference samples (e.g., growth inhibition on L0 was calculated by: [(A570 of L0 without drug treatment − A570 of L0 with drug treatment)/A570 of L0 without drug treatment] × 100%. Growth inhibition on L+1, L-1, L+2, and L-2 were calculated, respectively. Error bars represent the standard deviation for 24 replicate zones, * denotes $p < 0.01$.

Research studies on cell motility or cell migration are usually conducted using the Boyden chamber transwell or wound healing assay, as these are well-established methods for examining the migration capability of cells, especially after drug challenge [8]. However, in this study, the multi-layer CiGiP cell invasion platform was used to separate and isolate a subpopulation of cells by their motility. The molecular characteristics of these high-motility cells can be further investigated, demonstrating the potential in study the mechanisms underlying heterogeneous cell function. CSCs is one important concepts used to describe a sub-fraction of tumor cells. The CSC hypothesis suggests that tumorigenic stem cells are the source of cancer. The invasion and molecular characteristics of high invasive cells observed in this study suggest a new platform to separate and isolate CSCs. As shown in Scheme 1, a small fraction of cells with higher invasion potential migrated through the paper sheets. This small subpopulation of cells showed CSC characteristics. Thus, the CiGiP cell invasion platform could be used to separate subpopulations of cells according to their motility, and to screen proteins that promote metastasis, thereby facilitating the development of therapeutic strategies for targeting CSCs.

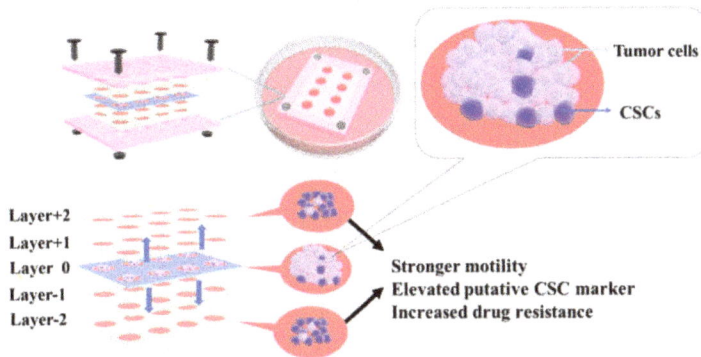

Scheme 1. Schematic drawing of the multi-layer CiGiP cell invasion platform mimicking the in vivo tumor metastasis process. Separation and isolation of subtype of tumors according to their motility can be achieved by de-stacking of the multi-layer invasion platform. The elevated expression of the putative cancer stem cell (CSC) biomarker suggests the relationship between stemness and metastasis.

4. Conclusions

In summary, separating and isolating a subpopulation of cells by their motility was achieved by using the multi-layer CiGiP cell invasion platform. A five-layer CiGiP platform was constructed. The cell-seeding layer was sandwiched in the middle layer. After seven days of culture, only a small fraction of cells invaded through the paper sheets, showing stronger motility than cells that did not invade. The qPCR results showed elevated expression of the putative prostate CSC biomarkers, ALDH1A1, SOX2, NANOG, and OCT4, in cells with higher motility. To the best of our knowledge, this is the first separating and characterizing heterogeneous tumor cells according to their invasion potential. Moreover, the correlation between higher motility and expression of putative prostate CSC biomarkers was demonstrated. There is an urgent need to develop strategies for the prevention of metastasis. The CiGiP cell invasion platform demonstrated in this study could be used to screen proteins that promote metastasis, thereby allowing the development of therapeutic strategies for targeting CSCs.

Author Contributions: Conceptualization, L.Y.; methodology, L.-X.W., Y.Z., and J.-J.F.; formal analysis: L.-X.W., L.Y; writing—original draft preparation: L.-X.W.; writing—review and editing: Z.S.L. and L.Y.; funding acquisition: L.Y.

Funding: This research was funded by National Natural Science Foundation of China (no. 31872753), the Fundamental Research Funds for the Central Universities (XDJK2018B008).

Acknowledgments: This work was supported by Chongqing Engineering Research Center for Micro-Nano Biomedical Materials and Devices.

Conflicts of Interest: The authors declare no conflict of interest.

References

1. Sampieri, K.; Fodde, R. Cancer stem cells and metastasis. *Semin. Cancer Biol.* **2012**, *22*, 187–193. [CrossRef]
2. Friedl, P.; Gilmour, D. Collective cell migration in morphogenesis, regeneration and cancer. *Nat. Rev. Mol. Cell Biol.* **2009**, *10*, 445–457. [CrossRef] [PubMed]
3. Collins, C.; Nelson, W.J. Running with neighbors: Coordinating cell migration and cell-cell adhesion. *Curr. Opin. Cell Biol.* **2015**, *36*, 62–70. [CrossRef]
4. Jung, A.; Brabletz, T.; Kirchner, T. The Migrating Cancer Stem Cells Model—A Conceptual Explanation of Malignant Tumour Progression. In *Cancer Stem Cells*; Springer: Berlin/Heidelberg, Germany, 2007; pp. 109–124.

5. Robinson, B.D.; Sica, G.L.; Liu, Y.F.; Rohan, T.E.; Gertler, F.B.; Condeelis, J.S.; Jones, J.G. Tumor microenvironment of metastasis in human breast carcinoma: A potential prognostic marker linked to hematogenous dissemination. *Clin. Cancer Res.* **2009**, *15*, 2433–2441. [CrossRef] [PubMed]

6. Ehsan, S.M.; Welch-Reardon, K.M.; Waterman, M.L.; Hughes, C.C.; George, S.C. A three-dimensional in vitro model of tumor cell intravasation. *Integr. Biol. (Camb.)* **2014**, *6*, 603–610. [CrossRef] [PubMed]

7. Kalchman, J.; Fujioka, S.; Chung, S.; Kikkawa, Y.; Mitaka, T.; Kamm, R.D.; Tanishita, K.; Sudo, R. A three-dimensional microfluidic tumor cell migration assay to screen the effect of anti-migratory drugs and interstitial flow. *Microfluid. Nanofluid.* **2012**, *14*, 969–981. [CrossRef]

8. Justus, C.R.; Leffler, N.; Ruiz-Echevarria, M.; Yang, L.V. In vitro cell migration and invasion assays. *J. Vis. Exp.* **2014**, *88*, e51046.

9. Chonan, Y.; Taki, S.; Sampetrean, O.; Saya, H.; Sudo, R. Endothelium-induced three-dimensional invasion of heterogeneous glioma initiating cells in a microfluidic coculture platform. *Integr. Biol. (Camb.)* **2017**, *9*, 762–773. [CrossRef]

10. Cui, X.; Guo, W.; Sun, Y.; Sun, B.; Hu, S.; Sun, D.; Lam, R.H.W. A microfluidic device for isolation and characterization of transendothelial migrating cancer cells. *Biomicrofluidics* **2017**, *11*, 014105. [CrossRef]

11. Doyle, A.D.; Petrie, R.J.; Kutys, M.L.; Yamada, K.M. Dimensions in cell migration. *Curr. Opin. Cell Biol.* **2013**, *25*, 642–649. [CrossRef]

12. Derda, R.; Laromaine, A.; Mammoto, A.; Tang, S.K.; Mammoto, T.; Ingber, D.E.; Whitesides, G.M. Paper-supported 3D cell culture for tissue-based bioassays. *Proc. Natl. Acad. Sci. USA* **2009**, *106*, 18457–18462. [CrossRef] [PubMed]

13. Kim, J.B. Three-dimensional tissue culture models in cancer biology. *Semin. Cancer Biol.* **2005**, *15*, 365–377. [CrossRef] [PubMed]

14. Yamada, K.M.; Cukierman, E. Modeling Tissue Morphogenesis and Cancer in 3D. *Cell* **2007**, *130*, 601–610. [CrossRef] [PubMed]

15. Cross, V.L.; Zheng, Y.; Choi, N.W.; Verbridge, S.S.; Sutermaster, B.A.; Bonassar, L.J.; Fischbach, C.; Stroock, A.D. Dense type I collagen matrices that support cellular remodeling and microfabrication for studies of tumor angiogenesis and vasculogenesis in vitro. *Biomaterials* **2010**, *31*, 8596–8607. [CrossRef] [PubMed]

16. Lloyd, C.C.; Boyce, M.W.; Lockett, M.R. Paper-based Invasion Assays for Quantifying Cellular Movement in Three-dimensional Tissue-like Structures. *Curr. Protoc. Chem. Biol.* **2017**, *9*, 75–95. [PubMed]

17. Mosadegh, B.; Lockett, M.R.; Minn, K.T.; Simon, K.A.; Gilbert, K.; Hillier, S.; Newsome, D.; Li, H.; Hall, A.B.; Boucher, D.M.; et al. A paper-based invasion assay: Assessing chemotaxis of cancer cells in gradients of oxygen. *Biomaterials* **2015**, *52*, 262–271. [CrossRef] [PubMed]

18. Yan, W.; Zhang, Q.; Chen, B.; Liang, G.-T.; Li, W.-X.; Zhou, X.-M.; Liu, D.-Y. Study on Microenvironment Acidification by Microfluidic Chip with Multilayer-paper Supported Breast Cancer Tissue. *Chin. J. Anal. Chem.* **2013**, *41*, 822–827. [CrossRef]

19. Simon, K.A.; Mosadegh, B.; Minn, K.T.; Lockett, M.R.; Mohammady, M.R.; Boucher, D.M.; Hall, A.B.; Hillier, S.M.; Udagawa, T.; Eustace, B.K.; et al. Metabolic response of lung cancer cells to radiation in a paper-based 3D cell culture system. *Biomaterials* **2016**, *95*, 47–59. [CrossRef]

20. Park, H.J.; Yu, S.J.; Yang, K.; Jin, Y.; Cho, A.N.; Kim, J.; Lee, B.; Yang, H.S.; Im, S.G.; Cho, S.W. Paper-based bioactive scaffolds for stem cell-mediated bone tissue engineering. *Biomaterials* **2014**, *35*, 9811–9823. [CrossRef]

21. Sapp, M.C.; Fares, H.J.; Estrada, A.C.; Grande-Allen, K.J. Multilayer three-dimensional filter paper constructs for the culture and analysis of aortic valvular interstitial cells. *Acta Biomater.* **2015**, *13*, 199–206. [CrossRef]

22. Justice, B.A.; Badr, N.A.; Felder, R.A. 3D cell culture opens new dimensions in cell-based assays. *Drug Discov. Today* **2009**, *14*, 102–107. [CrossRef] [PubMed]

23. Ng, K.; Gao, B.; Yong, K.W.; Li, Y.; Shi, M.; Zhao, X.; Li, Z.; Zhang, X.; Pingguan-Murphy, B.; Yang, H.; et al. Paper-based cell culture platform and its emerging biomedical applications. *Mater. Today* **2017**, *20*, 32–44. [CrossRef]

24. Ma, I.; Allan, A.L. The Role of Human Aldehyde Dehydrogenase in Normal and Cancer Stem Cells. *Stem Cell Rev.* **2011**, *7*, 292–306. [CrossRef] [PubMed]

25. Li, T.; Su, Y.; Mei, Y.; Leng, Q.; Leng, B.; Liu, Z.; Stass, S.A.; Jiang, F. ALDH1A1 is a marker for malignant prostate stem cells and predictor of prostate cancer patients' outcome. *Lab. Investig.* **2010**, *90*, 234–244. [CrossRef] [PubMed]

26. Tomita, H.; Tanaka, K.; Tanaka, T.; Hara, A. Aldehyde dehydrogenase 1A1 in stem cells and cancer. *Oncotarget* **2016**, *7*, 11018–11032. [CrossRef] [PubMed]

27. Rodda, D.J.; Chew, J.L.; Lim, L.H.; Loh, Y.H.; Wang, B.; Ng, H.H.; Robson, P. Transcriptional regulation of nanog by OCT4 and SOX2. *J. Biol. Chem.* **2005**, *280*, 24731–24737. [CrossRef]

28. Tay, Y.; Zhang, J.; Thomson, A.M.; Lim, B.; Rigoutsos, I. MicroRNAs to *Nanog*, *Oct4* and *Sox2* coding regions modulate embryonic stem cell differentiation. *Nature* **2008**, *455*, 1124–1128. [CrossRef]

29. Jeter, C.R.; Liu, B.; Liu, X.; Chen, X.; Liu, C.; Calhoun-Davis, T.; Repass, J.; Zaehres, H.; Shen, J.J.; Tang, D.G. NANOG promotes cancer stem cell characteristics and prostate cancer resistance to androgen deprivation. *Oncogene* **2011**, *30*, 3833–3845. [CrossRef]

30. Linn, D.E.; Yang, X.; Sun, F.; Xie, Y.; Chen, H.; Jiang, R.; Chen, H.; Chumsri, S.; Burger, A.M.; Qiu, Y. A Role for OCT4 in Tumor Initiation of Drug-Resistant Prostate Cancer Cells. *Genes Cancer* **2010**, *1*, 908–916. [CrossRef]

31. Jia, X.; Li, X.; Xu, Y.; Zhang, S.; Mou, W.; Liu, Y.; Liu, Y.; Lv, D.; Liu, C.H.; Tan, X.; et al. SOX2 promotes tumorigenesis and increases the anti-apoptotic property of human prostate cancer cel. *J. Mol. Cell Biol.* **2011**, *3*, 230–238. [CrossRef]

32. Postovit, L.M.; Costa, F.F.; Bischof, J.M.; Seftor, E.A.; Wen, B.; Seftor, R.E.; Feinberg, A.P.; Soares, M.B.; Hendrix, M.J. The commonality of plasticity underlying multipotent tumor cells and embryonic stem cells. *J. Cell. Biochem.* **2007**, *101*, 908–917. [CrossRef] [PubMed]

micromachines

MDPI

Article

Microfluidic Analyzer Enabling Quantitative Measurements of Specific Intracellular Proteins at the Single-Cell Level

Lixing Liu [1,2,†], Beiyuan Fan [1,2,†], Diancan Wang [3,†], Xiufeng Li [1,2], Yeqing Song [3], Ting Zhang [1,2], Deyong Chen [1,2], Yixiang Wang [3,*], Junbo Wang [1,2,*] and Jian Chen [1,2,*]

1   State Key Laboratory of Transducer Technology, Institute of Electronics, Chinese Academy of Sciences, Beijing 100190, China; liulixing16@mails.ucas.ac.cn (L.L.); fanbeiyuan@ucas.ac.cn (B.F.); lixiufeng13@mails.ucas.ac.cn (X.L.); 1141230132@ncepu.edu.cn (T.Z.); dychen@mail.ie.ac.cn (D.C.)
2   University of Chinese Academy of Sciences, Beijing 100190, China
3   Peking University School of Stomatology, Beijing 10081, China; bjdxwdc@gmail.com (D.W.); duo.k.tong@gmail.com (Y.S.)
*   Correspondence: kqwangyx@bjmu.edu.cn (Y.W.); jbwang@mail.ie.ac.cn (J.W.); chenjian@mail.ie.ac.cn (J.C.); Tel.: +86-10-82195537 (Y.W.); Tel.: +86-10-58887191 (J.W.); Tel.: +86-10-58887256 (J.C.)
†   These authors contributed equally to this work.

Received: 12 October 2018; Accepted: 8 November 2018; Published: 12 November 2018

Abstract: This paper presents a microfluidic instrument capable of quantifying single-cell specific intracellular proteins, which are composed of three functioning modules and two software platforms. Under the control of a LabVIEW platform, a pressure module flushed cells stained with fluorescent antibodies through a microfluidic module with fluorescent intensities quantified by a fluorescent module and translated into the numbers of specific intracellular proteins at the single-cell level using a MATLAB platform. Detection ranges and resolutions of the analyzer were characterized as 896.78–6.78 × 10^5 and 334.60 nM for Alexa 488, 314.60–2.11 × 10^5 and 153.98 nM for FITC, and 77.03–5.24 × 10^4 and 37.17 nM for FITC-labelled anti-beta-actin antibodies. As a demonstration, the numbers of single-cell beta-actins of two paired oral tumor cell types and two oral patient samples were quantified as: 1.12 ± 0.77 × 10^6/cell (salivary adenoid cystic carcinoma parental cell line (SACC-83), n_{cell} = 13,689) vs. 0.90 ± 0.58 × 10^5/cell (salivary adenoid cystic carcinoma lung metastasis cell line (SACC-LM), n_{cell} = 15,341); 0.89 ± 0.69 × 10^6/cell (oral carcinoma cell line (CAL 27), n_{cell} = 7357) vs. 0.93 ± 0.69 × 10^6/cell (oral carcinoma lymphatic metastasis cell line (CAL 27-LN2), n_{cell} = 6276); and 0.86 ± 0.52 × 10^6/cell (patient I) vs. 0.85 ± 0.58 × 10^6/cell (patient II). These results (1) validated the developed analyzer with a throughput of 10 cells/s and a processing capability of ~10,000 cells for each cell type, and (2) revealed that as an internal control in cell analysis, the expressions of beta-actins remained stable in oral tumors with different malignant levels.

Keywords: instrumentation; microfluidic flow cytometry; intracellular proteins; absolute quantification

---

1. Introduction

Single-cell protein expressions provide key insights in studying cellular heterogeneities such as tumour heterogeneities and immune response variations [1–3]. Currently, flow cytometry is the golden instrument for single-cell protein analysis where cells stained with fluorescence-labelled antibodies rapidly travel through a capillary while the fluorescent intensities are measured [4,5]. Using calibration beads with the numbers of surface proteins under well controls, quantitative flow cytometry enables the absolute quantification of single-cell surface proteins while the copy numbers of intracellular proteins at the single-cell level remain elusive due to the lack of calibration approaches [6–9].

Microfluidics is an approach to processing fluids based on microfabricated channels (1–100 μm) [10,11], and due to their dimensional comparisons with biological cells, microfluidic instruments have been developed for single-cell protein analysis [12,13]. More specifically, barcoding microchips with a commercial brand of "Isoplexis" were developed where individual cells are confined within microchambers, and the absolute quantification of specific intracellular proteins is realized by cell lysis and the captures of target cellular proteins by preprinted antibodies on the bottom surfaces of the microchips [14–17]. In a second microfluidic approach, with a commercial brand of "single-cell westerns", key steps of settling single cells into microwells, lysis in situ, gel electrophoresis, photoinitiated blotting to immobilize proteins, and antibody probing are included, enabling the quantitative analysis of cellular proteins [18–20]. However, in comparison to flow cytometry, these microfluidic instruments still suffer from limited throughputs since they cannot characterize single cells in a continuous fluid flow.

In order to address this issue, previously we developed a constriction channel-based microfluidic platform to quantify specific intracellular proteins of single cells in a high-throughput manner [21,22]. In this modified flow cytometry, cells stained with fluorescence-labelled antibodies are forced to deform through a constriction channel (a microfabricated channel with a cross-sectional area smaller than a cell), with raw fluorescent profiles collected. Meanwhile, solutions of fluorescent antibodies are flushed through this constriction channel to produce calibration curves, enabling the translation of raw fluorescent signals into copy numbers of specific intracellular proteins.

In this study, the instrumentation of the aforementioned approach was demonstrated, where the functionalities of individual modules were realized and the key parameters of the assembled instrument (e.g., detection resolutions, ranges, and throughputs) were characterized. In addition, the developed instrument was used to quantify numbers of single-cell beta-actins from two paired oral tumor cell types and two oral patient samples as a validation of the developed instrument. Furthermore, based on the results of this instrument, the expressions of beta-actins at the single-cell level remained stable within oral tumor cells, validating the use of beta-actins as internal controls in cell analysis. The structure of this paper is as follows: In Section 2, working mechanisms of the developed instrument are described. In Section 3, module functionalities, instrument operation, and data processing are described in detail. In Section 4, absolute quantification of beta-actins at the single-cell level are demonstrated. Conclusion and future work are included in Section 5.

2. Schematics and Working Mechanisms

Figure 1A,B shows schematics and a prototype of the microfluidic flow cytometry enabling the measurement of copy numbers of specific intracellular proteins at the single-cell level. The developed instrument had five key components, including a microfluidic module composed of a constriction channel with a microfabricated chrome window; a fluorescent module composed of an inverted microscope, a light source of a light-emitting diode (LED), a photomultiplier tube (PMT), and a data acquisition card (DAQ); a pressure module composed of a pressure controller; a LabVIEW platform (National Instruments, Austin, TX, USA) for instrument operation; and a MATLAB platform (The MathWorks, Inc., Natick, MA, USA) for data processing.

The flow chart of the developed flow cytometry is shown in Figure 1C. In operations, under the control of the LabVIEW platform for instrument operation, the pressure module flushed cells stained with fluorescence-labelled antibodies through the constriction channel of the microfluidic module. While the cell traveled through the fluorescent detection region defined by the constriction channel and the patterned chrome window, raw fluorescent signals were collected by the fluorescent module. Since the cross-sectional area of the constriction microchannel was smaller than cells, cells fully filled the constriction channel during their squeezing processes, and thus (1) cells traveled through the fluorescent detection region one by one, enabling single-cell protein quantification, and (2) solutions with fluorescence-labelled antibodies were flushed into the constriction channel directly, based on the pressure module, to form calibration curves.

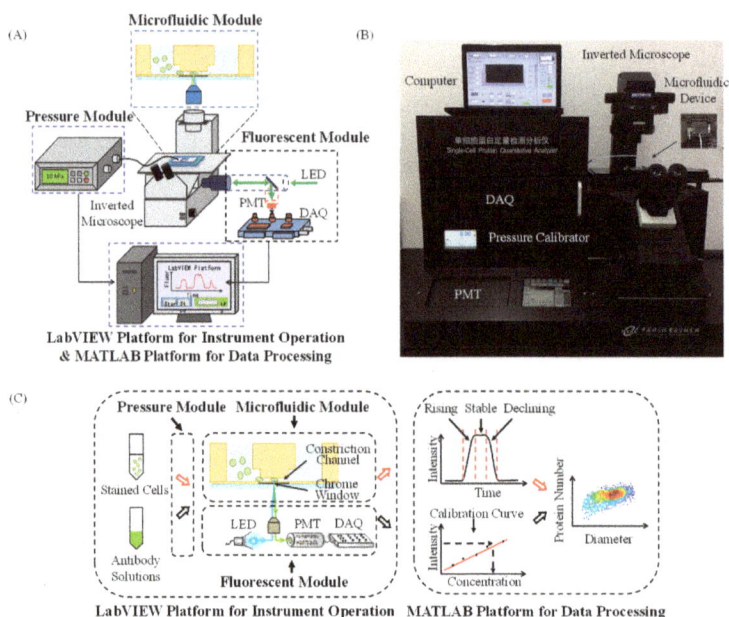

Figure 1. (**A**) Schematics and (**B**) a prototype of the fluorescent microfluidic flow cytometry enabling the measurement of copy numbers of specific intracellular proteins at the single-cell level. The developed instrument had five key components: A microfluidic module, a fluorescent module, a pressure module, a LabVIEW platform for instrument operation, and a MATLAB platform for data processing. (**C**) Working flow chart of the developed microfluidic instrument. Under the control of the LabVIEW platform for instrument operation, the pressure module flushed cells stained with fluorescence-labelled antibodies through the constriction channel of the microfluidic module while fluorescent intensities were quantified by the fluorescent module and then were further translated to the copy number of specific intracellular proteins at the single-cell level, leveraging the MATLAB platform for data processing.

Using the MATLAB platform for data processing, the fluorescent profile of a traveling cell through the fluorescent detection region of the microfluidic module could be divided into a rising domain, a stable domain, and a declining domain (see Figure 1C). More specifically, in the rising domain, there was a gradual increase in the fluorescent intensity, indicating that the deformed cell with fluorescence-labelled antibodies gradually filled the fluorescent detection region defined by the constriction channel and the patterned chrome window. As the deformed cell further moved in the constriction microchannel, there was a duration with the stable fluorescent intensity, which indicated the full occupation of the fluorescent detection region. As for the declining domain, it corresponded to the gradual decrease in fluorescent intensities in the leaving of the deformed cell. Based on this analysis, the diameters of cells and peaking values of fluorescent signals could be obtained through interpreting fluorescent profiles. Furthermore, leveraging the calibration curve, these raw parameters could be translated into the copy numbers of a targeted protein at the single-cell level. For detailed steps of pulse processing, please refer to previous publications [21,22].

3. Module Functionality, Operation, and Data Processing

3.1. Microfluidic Module

The microfluidic module consisted of a constriction channel with a cross-section area of 8 μm × 8 μm and a chrome window of 2.5 μm. The proposed device was fabricated based on conventional microfabrication techniques in which the patterned polydimethylsiloxane (PDMS, Dow Corning Corp., Midland, MI, USA) layer, including constriction channels, was replicated from a double-layer SU-8 (MicroChem Corp., Westborough, MA, USA) mould and bonded with a quartz slide that was first patterned with a chrome layer and then coated with a thin layer of unpatterned PDMS (see Figure 2A). For detailed information, please refer to the Supplementary Materials of this paper and previous publications [21,22]. The fabricated microfluidic devices are shown in Figure 2B, where multiple constriction channels could be fabricated in one single device.

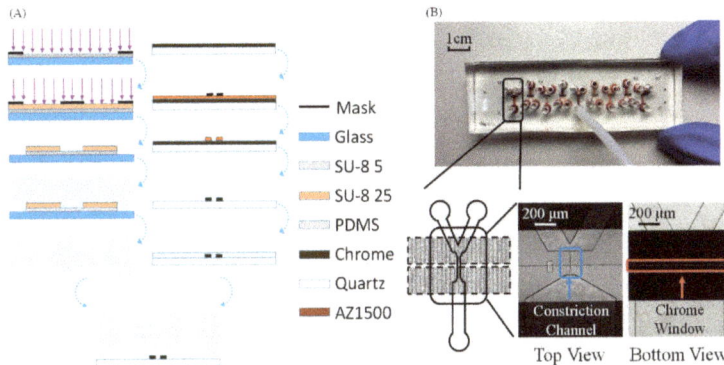

Figure 2. (**A**) Fabrication process and (**B**) a prototype device of the microfluidic module composed of a polydimethylsiloxane (PDMS)-based constriction channel with a patterned chrome window on quartz. The fabrication was based on conventional lithography including key steps of SU-8 exposure, PDMS molding, chrome patterning on quartz, and the bonding of the constriction channel layer and the chrome layer.

3.2. Instrument Operation

Figure 3A and Video I show the LabVIEW platform for instrument operation, which regulated the fluorescent and pressure modules. There were two key parameters in controlling the fluorescent module, the "gain voltage" of PMT (H10722-01, Hamamatsu, Japan) and the "sampling rate" of DAQ (PCI-6221, National Instruments, Austin, TX, USA) (see top-left of Figure 3A). An increase in the gain voltage of PMT could increase the amplification ratio of weak fluorescent signals, which at the same time increased background noises. After a balance between fluorescent amplifications and basal noise levels, in this study a gain voltage of 0.8 V was used, producing an amplification ratio of 3×10^5 and a basal noise level of 5 mV. In addition, durations of travelling cells were several milliseconds, and thus a sampling rate of 100 kHz was used, enabling the collection of at least 100 points for individual fluorescent pulses. Note that the physical components of the fluorescent module also included the light source of LED (M470L3-C1, Thorlabs, Newton, NJ, USA) and an inverted microscope (IX 83, Olympus, Tokyo, Japan), which did not need controls from the LabVIEW-based operation platform.

In regulating the pressure module, the output pressure value generated by the pressure calibre (Druck PACE-5000, Burlington, VT, USA) was defined in the LabVIEW platform, with the real-time pressure values measured and displayed (see bottom-left of Figure 3A). In this instrument, high-value negative pressures were usually preferred for high-throughput analysis, and thus buttons for directly generating pressure values (e.g., −10 kPa, −20 kPa, and −30 kPa) were included in this interface. In addition, facing channel blockages, high-value positive pressures were also needed to blow away blocking particles, and thus buttons for directly generating pressure values (e.g., 10 kPa, 20 kPa, and 30 kPa) were also included in this interface.

(A)

(B)

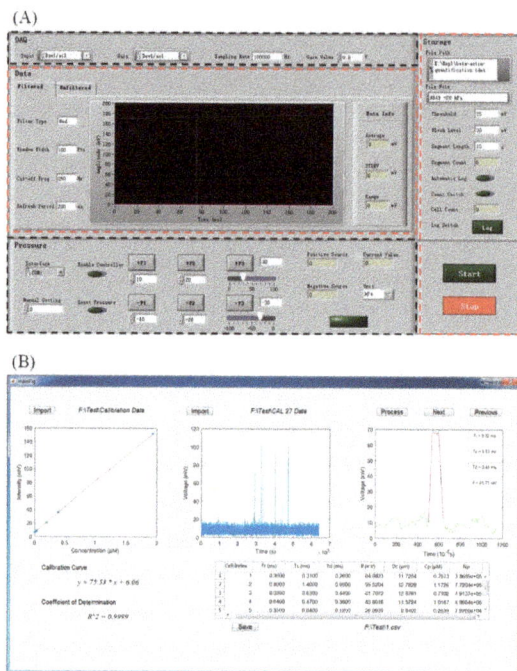

Figure 3. (**A**) The interface of the LabVIEW Platform for instrument operation mainly included the top-left and bottom-left areas for the regulations of the fluorescent and the pressure modules, respectively. The middle area displayed the sampled voltages collected by the data acquisition card (DAQ) in a real-time manner, and the right area displayed the parameters for the storage of the collected signals. (**B**) The interface of the MATLAB-based data processing mainly included the import of the calibration curve (left), the import of the raw voltage data indicating fluorescent intensities (middle), and the curve-fitting of individual fluorescent pulses (right). Based on these steps, the numbers of protein copies for individual cells were obtained and displayed at the bottom of this interface.

Under the control of the LabVIEW platform for instrument operation, fluorescent and pressure modules were well coordinated to collect fluorescent signals of stained individual cells in a semiautomatic manner (see Video I Instrument Operation). In this video clip, each pulse represented a traveling cell, and the time durations for individual fluorescent pulses were between 1 and 10 ms. Thus, the throughput of this instrument was estimated at 10 cells/s when the time gaps among incoming cells were taken into consideration. Note that the throughput of this system was still two orders lower than the throughput of conventional flow cytometry since in this study, in order to improve signal-noise ratios of fluorescent intensities for travelling cells, a median filtration was conducted to process raw voltage signals collected by DAQ where fluorescent pulses with time durations lower than 0.5 ms were treated as noises and thus removed, limiting the further improvements in detection throughputs of this system.

3.3. Data Processing

Figure 3B and Video II show the MATLAB platform for semiautomatic data processing, which translated raw fluorescent signals into copy numbers of specific intracellular proteins. As shown in Video II Data Processing, in the first step, the calibration curve illustrating the relationship between concentrations of specific antibody and fluorescent intensities was loaded into the MATLAB platform. Then, raw fluorescent signals were loaded into the MATLAB platform and pulses were processed in a time sequence. Each pulse was first divided into a rising domain, a stable domain, and a declining domain, and then was processed to cell sizes and fluorescent intensities, which were further translated into copy numbers of specific intracellular proteins based on the calibration curve. Leveraging the MATLAB platform for data processing, it took ~10 min to process ~10,000 fluorescent pulses, realizing the quantitative measurements of specific intracellular proteins at the single-cell level. For detailed information on the processing of fluorescent pulses, please refer to previous publications [21,22].

4. Demonstration

Based on the calibration curves of three types of fluorescence-labelled macromolecules, the detection ranges and resolutions of the proposed instrument were first characterized. More specifically, the detection ranges and resolutions of Alexa 488, FITC, and FITC-labelled anti-beta-actin antibodies were quantified as 896.78–6.78 × 10^5 and 334.60 nM, 314.60–2.11 × 10^5 and 153.98 nM, and 77.03–5.24 × 10^4 and 37.17 nM, respectively. These results indicated that in comparison to Alexa 488, the same numbers of FITC molecules could produce higher fluorescent intensities. As to the comparison of FITC and FITC-labelled anti-beta-actin antibodies, since there were multiple FITC molecules for each antibody, the detection resolution for FITC-labelled anti-beta-actin antibodies was much lower than for FITC.

As a functional demonstration, the developed microfluidic instrument was used to characterize the numbers of beta-actins at the single-cell level. Note that as an internal control in cell analysis, the expressions of beta-actins were assumed to be stable within individual cells, which was questionable due to recently reported quantitative results at the single-cell level [23].

In order to address this issue, in this study two paired oral tumor cell types were characterized, where single-cell beta-actin numbers were obtained as 1.12 ± 0.77 × 10^6/cell (salivary adenoid cystic carcinoma parental cell line (SACC-83), n_{cell} = 13,689) versus 0.90 ± 0.58 × 10^6/cell (salivary adenoid cystic carcinoma lung metastasis cell line (SACC-LM), n_{cell} = 15,341), and 0.89 ± 0.69 × 10^6/cell (oral carcinoma cell line (CAL 27), n_{cell} = 7357) versus 0.93 ± 0.69 × 10^6/cell (oral carcinoma lymphatic metastasis cell line (CAL 27-LN2), n_{cell} = 6276) (see Figure 4 and Supplementary Materials for cell culture and preparation).

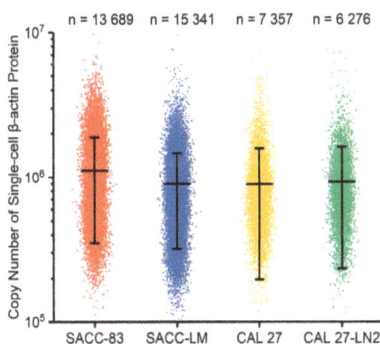

Figure 4. Scatter plots of the copy numbers of single-cell β-actin proteins for salivary adenoid cystic carcinoma parental cell line (SACC-83) (n_{cell} = 13,689), salivary adenoid cystic carcinoma lung metastasis cell line (SACC-LM) (n_{cell} = 15,341), oral carcinoma cell line (CAL 27) (n_{cell} = 7357), and oral carcinoma lymphatic metastasis cell line (CAL 27-LN2) (n_{cell} = 6276). These results indicated that the developed instrument was capable of collecting beta-actins from ~10,000 single cells.

These experimental results validated the developed microfluidic platform, which could collect the numbers of beta-actins from ~10,000 cells in total for each cell type. Meanwhile, the values of quantified single-cell beta-actins were around 1 × 10^6/cell for these four cell types, whereas no significant differences were located for oral tumor cells with different malignant levels. Thus, these results indicated that the expression levels of beta-actins at the single-cell level remained stable for oral tumor cell types.

Furthermore, the developed microfluidic instrument was used to characterize two patient samples of oral tumors, where single-cell beta-actin copy numbers were quantified as $0.86 \pm 0.52 \times 10^6$/cell ($n_{cell}$ = 365) and $0.85 \pm 0.58 \times 10^6$/cell ($n_{cell}$ = 177) (see Figure 5 and Supplementary Materials for cell preparation and data processing). These results further validated the functionality of the developed instrument, which was capable of processing real clinical samples. Furthermore, the quantified numbers of beta-actins of these two oral tumor samples were comparable with the results of oral tumor cell lines of CAL 27, confirming again that the expression levels of beta-actins at the single-cell level remained stable within oral tumor cells. Note that in processing clinical samples, the total numbers of characterized cells were two orders lower than the counterparts of cultured cell lines. In comparison to experiments with cell lines, clinical samples were noticed to block the constriction channels more often, limiting the processing throughput. Since this blockage may have resulted from cell clusters in clinical samples, in future studies size-based filtrations are recommended to remove cell aggregations before the clinical samples are flushed into this microfluidic platform.

In comparison to previous papers [21,22], where the pressure sources and fluorescent detections were controlled manually, the LabVIEW-based platform developed in this study enabled the system to function in a continuous manner and collect data from ~10,000 cells for each cell type with much lower experimental durations. In addition, the semiautomatic platform functioned in a robust manner and processed the real clinical samples, which was not reported in previous publications. Furthermore, in comparison to previous papers [21,22] where the fluorescent pulses were processed manually, the MATLAB platform for data processing developed in this study enabled the semiautomatic processing of fluorescent pulses, which significantly decreased the time durations of data processing.

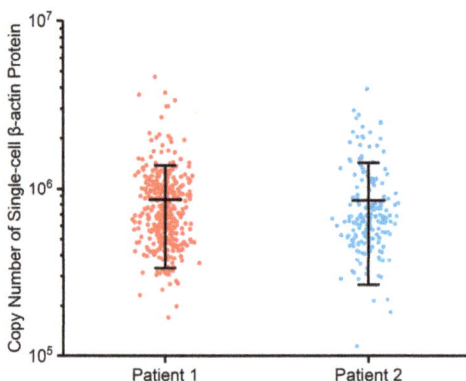

Figure 5. Scatter plots of the copy numbers of single-cell β-actin proteins from two oral tumor patients. These results indicated that the developed instrument could be used to absolutely quantify specific intracellular proteins of patient samples at the single-cell level.

5. Conclusions and Future Work

In this study, the instrumentation of a microfluidic analyzer was demonstrated, which enabled the measurement of single-cell numbers of specific intracellular antibodies at a throughput of roughly 10 cells/s. The functionalities of individual modules were validated and integrated to form the prototype instrument, where the numbers of beta-actins of two paired oral tumor cell types and two samples of oral tumor patients were collected and compared. Future developments of the instrument are aimed at the absolute quantification of multiple types of intracellular proteins simultaneously (e.g., beta-actins, alpha-tubulins, and beta-tubulins) at a higher throughput (i.e., 1000 cells/s).

Supplementary Materials: The following are available online at http://www.mdpi.com/2072-666X/9/11/588/s1.

Author Contributions: Conceptualization, L.L., D.W., Y.W., and J.C.; Data curation, L.L., X.L., and T.Z.; Formal analysis, L.L. and T.Z.; Funding acquisition, D.W., D.C., Y.W., J.W., and J.C.; Investigation, L.L., B.F., and T.Z.; Methodology, L.L., B.F., and J.C.; Project administration, D.C. and J.W.; Resources, D.W., Y.S., and Y.W.; Software, X.L.; Supervision, D.W., D.C., Y.W., J.W., and J.C.; Validation, L.L. and D.C.; Visualization, L.L. and B.F.; Writing–original draft, L.L., Y.W., and J.C.; Writing–review and editing, J.W. and J.C.

Funding: This research was funded by the National Natural Science Foundation of China (grant number 61431019, 61671430) and the Chinese Academy of Sciences Key Project Targeting Cutting-Edge Scientific Problems (grant number QYZDB-SSW-JSC011).

Acknowledgments: The authors would like to acknowledge the Instrument Development Program, the Youth Innovation Promotion Association and Interdisciplinary Innovation Team of the Chinese Academy of Sciences, and Instrument Development of the Beijing Municipal Science and Technology Commission.

Conflicts of Interest: The authors declare no conflict of interest.

References

1. Altschuler, S.J.; Wu, L.F. Cellular Heterogeneity: Do Differences Make a Difference? *Cell* **2010**, *141*, 559–563. [CrossRef] [PubMed]
2. Wu, M.; Singh, A.K. Single-cell protein analysis. *Curr. Opin. Biotechnol.* **2012**, *23*, 83–88. [CrossRef] [PubMed]
3. Su, Y.; Shi, Q.; Wei, W. Single cell proteomics in biomedicine: High-dimensional data acquisition, visualization, and analysis. *Proteomics* **2017**, *17*, 1600267. [CrossRef] [PubMed]
4. De Rosa, S.C.; Herzenberg, L.A.; Herzenberg, L.A.; Roederer, M. 11-color, 13-parameter flow cytometry: Identification of human naive T cells by phenotype, function, and T-cell receptor diversity. *Nat. Med.* **2001**, *7*, 245–248. [CrossRef] [PubMed]
5. Perfetto, S.P.; Chattopadhyay, P.K.; Roederer, M. Seventeen-colour flow cytometry: Unravelling the immune system. *Nat. Rev. Immunol.* **2004**, *4*, 648–655. [CrossRef] [PubMed]
6. Zenger, V.E.; Vogt, R.; Mandy, F.; Schwartz, A.; Marti, G.E. Quantitative flow cytometry: Inter-laboratory variation. *Cytometry* **1998**, *33*, 138–145. [CrossRef]
7. Schwartz, A.; Fernández-Repollet, E. Quantitative flow cytometry. *Clin. Lab. Med.* **2001**, *21*, 743–761. [PubMed]
8. Marti, G.E.; Zenger, V.E.; Vogt, R.; Gaigalas, A. Quantitative flow cytometry: History, practice, theory, consensus, inter-laboratory variation and present status. *Cytotherapy* **2002**, *4*, 97–98. [CrossRef] [PubMed]
9. Maher, K.J.; Fletcher, M.A. Quantitative flow cytometry in the clinical laboratory. *Clin. Appl. Immunol. Rev.* **2005**, *5*, 353–372. [CrossRef]
10. Whitesides, G.M. The origins and the future of microfluidics. *Nature* **2006**, *442*, 368–373. [CrossRef] [PubMed]
11. Wootton, R.C.R.; deMello, A.J. Exploiting elephants in the room. *Nature* **2010**, *464*, 839–840. [CrossRef] [PubMed]
12. Wei, W.; Shin, Y.S.; Ma, C.; Wang, J.; Elitas, M.; Fan, R.; Heath, J.R. Microchip platforms for multiplex single-cell functional proteomics with applications to immunology and cancer research. *Genome Med.* **2013**, *5*, 75–87. [CrossRef] [PubMed]
13. Yu, J.; Zhou, J.; Sutherland, A.; Wei, W.; Shin, Y.S.; Xue, M.; Heath, J.R. Microfluidics-Based Single-Cell Functional Proteomics for Fundamental and Applied Biomedical Applications. *Annu. Rev. Anal. Chem.* **2014**, *7*, 275–295. [CrossRef] [PubMed]
14. Fan, R.; Vermesh, O.; Srivastava, A.; Yen, B.K.H.; Qin, L.; Ahmad, H.; Kwong, G.A.; Liu, C.; Gould, J.; Hood, L.; et al. Integrated barcode chips for rapid, multiplexed analysis of proteins in microliter quantities of blood. *Nat. Biotechnol.* **2008**, *26*, 1373–1378. [CrossRef] [PubMed]
15. Ma, C.; Fan, R.; Ahmad, H.; Shi, Q.; Comin-Anduix, B.; Chodon, T.; Koya, R.C.; Liu, C.; Kwong, G.A.; Radu, C.G.; et al. A clinical microchip for evaluation of single immune cells reveals high functional heterogeneity in phenotypically similar T cells. *Nat. Med.* **2011**, *17*, 738–743. [CrossRef] [PubMed]
16. Shi, Q.; Qin, L.; Wei, W.; Geng, F.; Fan, R.; Shin, Y.S.; Guo, D.; Hood, L.; Mischel, P.S.; Heath, J.R. Single-cell proteomic chip for profiling intracellular signaling pathways in single tumor cells. *Proc. Natl. Acad. Sci. USA* **2012**, *109*, 419–424. [CrossRef] [PubMed]

17. Lu, Y.; Xue, Q.; Eisele, M.R.; Sulistijo, E.S.; Brower, K.; Han, L.; Amir, E.D.; Pe'er, D.; Miller-Jensen, K.; Fan, R. Highly multiplexed profiling of single-cell effector functions reveals deep functional heterogeneity in response to pathogenic ligands. *Proc. Natl. Acad. Sci. USA* **2015**, *112*, E607–E615. [CrossRef] [PubMed]

18. Hughes, A.J.; Spelke, D.P.; Xu, Z.; Kang, C.-C.; Schaffer, D.V.; Herr, A.E. Single-cell western blotting. *Nat. Methods* **2014**, *11*, 749–755. [CrossRef] [PubMed]

19. Kang, C.-C.; Lin, J.-M.G.; Xu, Z.; Kumar, S.; Herr, A.E. Single-Cell Western Blotting after Whole-Cell Imaging to Assess Cancer Chemotherapeutic Response. *Anal. Chem.* **2014**, *86*, 10429–10436. [CrossRef] [PubMed]

20. Lin, J.-M.G.; Kang, C.-C.; Zhou, Y.; Huang, H.Y.; Herr, A.E.; Kumar, S. Linking invasive motility to protein expression in single tumor cells. *Lab Chip* **2018**, *18*, 371–384. [CrossRef] [PubMed]

21. Li, X.; Fan, B.; Cao, S.; Chen, D.; Zhao, X.; Men, D.; Yue, W.; Wang, J.; Chen, J. A microfluidic flow cytometer enabling absolute quantification of single-cell intracellular proteins. *Lab Chip* **2017**, *17*, 3129–3137. [CrossRef] [PubMed]

22. Li, X.; Fan, B.; Liu, L.; Chen, D.; Cao, S.; Men, D.; Wang, J.; Chen, J. A Microfluidic Fluorescent Flow Cytometry Capable of Quantifying Cell Sizes and Numbers of Specific Cytosolic Proteins. *Sci. Rep.* **2018**, *8*, 14229. [CrossRef] [PubMed]

23. Fan, B.; Li, X.; Liu, L.; Chen, D.; Cao, S.; Men, D.; Wang, J.; Chen, J. Absolute Copy Numbers of β-Actin Proteins Collected from 10,000 Single Cells. *Micromachines* **2018**, *9*, 254. [CrossRef]

MDPI

St. Alban-Anlage 66

4052 Basel

Switzerland

Tel. +41 61 683 77 34

Fax +41 61 302 89 18

www.mdpi.com

Micromachines Editorial Office

E-mail: micromachines@mdpi.com

www.mdpi.com/journal/micromachines